Progress on Nutrient Composition, Meat Standardization and Grading, Processing and Safety in Different Types of Meat Sources

Progress on Nutrient Composition, Meat Standardization and Grading, Processing and Safety in Different Types of Meat Sources

Editor

Nelson Huerta-Leidenz

MDPI • Basel • Beijing • Wuhan • Barcelona • Belgrade • Manchester • Tokyo • Cluj • Tianjin

Editor
Nelson Huerta-Leidenz
Department of Animal and Food
Sciences. Texas Tech Universit
USA

Editorial Office
MDPI
St. Alban-Anlage 66
4052 Basel, Switzerland

This is a reprint of articles from the Special Issue published online in the open access journal *Foods* (ISSN 2304-8158) (available at: https://www.mdpi.com/journal/foods/special_issues/grading_nutrient_processing_safety_meat).

For citation purposes, cite each article independently as indicated on the article page online and as indicated below:

LastName, A.A.; LastName, B.B.; LastName, C.C. Article Title. *Journal Name* **Year**, *Volume Number*, Page Range.

ISBN 978-3-0365-2542-6 (Hbk)
ISBN 978-3-0365-2543-3 (PDF)

Cover image courtesy of Mr. Martin Ceballos.

Contents

About the Editor

Nelson Huerta-Leidenz Emeritus Professor of Universidad del Zulia (LUZ), Venezuela, and Researcher, International Center for Food Industry Excellence, Department of Animal and Food Sciences, Texas Tech University, USA. He earnt his degrees from LUZ (Veterinary Medicine) and Texas A&M University (M.S. and Ph.D.). From 1978 to 1997, he was a full-time professor at Agronomy Faculty of LUZ. His research interests include applied growth, meat yield and quality, and meat grading of large ruminants. He has generated 106 refereed journal articles, edited 4 books and authored or coauthored 11 book chapters, in addition to other contributions. He is a member of publishing committees of different scientific journals and has served as a referee of several international journals. He was a President of the Mexican Meat Science and Technology Association (2014–2016) and is an Honorary Member of the Venezuelan Beef Council (CONVECAR). Since 2010, he has served as an invited board member of the International Stockmen's Educational Foundation. He has received recognition from the American Meat Science Association (2010 International Lectureship Award) and Mexican Meat Council (2015 Academic Merit Award).

Editorial

Progress on Nutrient Composition, Meat Standardization, Grading, Processing, and Safety for Different Types of Meat Sources

Nelson Huerta-Leidenz

Department of Animal and Food Sciences, Texas Tech University, P.O. Box 42141, Lubbock, TX 79409-2141, USA; nelson.huerta@ttu.edu; Tel.: +1-956-250-4337

1. Introduction

Beef contains a plethora of healthy nutrients and it is the highest valued livestock-based food product. However, other meats (such as pork and poultry) and co-products of the meat industry can also be nutrient-dense with advantageous sensorial and technological qualities. The current standardization and grading schemes can assist in describing meats for their eating quality and/or fabrication yield, but more innovative, objective technologies are much needed to improve the segregation of the heterogeneous supply of carcasses and cuts into more homogeneous groups as regards quality and/or yield. In terms of food safety, while there have been significant improvements in reducing foodborne illnesses in the meat industry, the morbidity and mortality attributed to *Salmonella*, *E. coli* O157: H7, *Listeria monocytogenes*, *Campylobacter jejuni*, and other pathogens remain an issue with serious socioeconomic impacts. There is, therefore, a need to evaluate antimicrobials and technologies to assist in mitigating these recurrent problems in public health. This Special Issue of *Foods* was designed to cover scientific and technological advances in selected topics of global importance for the progress of the livestock and meat industries. Therefore, in this Special Issue, we have included contributions that encompass key current research on nutrient composition, instrumental meat grading, and food safety. This collection of scientific articles and reviews is fundamentally a profile of a much broader perspective of animal and food sciences applied in trans-cultural settings.

Citation: Huerta-Leidenz, N. Progress on Nutrient Composition, Meat Standardization, Grading, Processing, and Safety for Different Types of Meat Sources. *Foods* **2021**, *10*, 2128. https://doi.org/10.3390/foods10092128

Received: 6 September 2021
Accepted: 7 September 2021
Published: 9 September 2021

2. A Summary of the Research in this Special Issue

The summary and comments about the papers published in this Special Issue deal mostly with findings directly related to the themes of progress on nutrient composition, meat standardization, grading, and safety for different types of meat sources. While included in some of the research, findings related to sensory quality and other aspects outside of the Special Issue themes are not addressed in this preface in order to focus specifically on the targeted topics. However, we cordially invite the readers to discover the entire body of knowledge compiled in these articles, which extends beyond the defined parameters of this Special Issue.

2.1. Progress on Nutrient Composition

Two state-of-the-art extensive reviews are published on this topic, one [1] informing on the genetic and nutritional strategies available to enhance the nutritional quality of red meat (168 reviewed papers), and the other [2] attempting to characterize the quality and nutrient composition of meat produced in the tropics (147 reviewed papers). The original articles that dealt with the nutrient composition of meats and co-products, separately or concurrently with aspects of eating quality, were the results of research carried out in various countries, in different species and environments, including Chile [3] (the effect of supplementing pigs with brown seaweed on quality traits and nutrient composition of pork), Canada [4] (the impact of feedlot diets containing various levels of barley on the nutritional quality of *Bos taurus* beef), Spain and USA [5] (a comparison of the veal produced by bullock calves in the Pyrenees [PGI-Certified *Ternera de Navarra* (*CTNA*)] and US Angus certified beef), Colombia [6] (characterization of the lipid profile of visceral fats by-products of chicken), and Venezuela [7] (multivariate relationships between nutrient composition and carcass characteristics of *Bos indicus*-type cattle raised on pasture in a tropical environment).

Juárez et al. [1] update us on the available genetic and nutritional strategies to enhance the nutritional quality of red meat (beef, lamb, and pork). This subject is particularly pertinent today, given the numerous studies suggesting that red meat consumption may have negative effects on human health and the environment. This review identifies information gaps on the evaluation of genetic parameters related to meat composition, as well as multiple bioethical challenges linked to new trends in genetic engineering, while also demonstrating that much progress has been achieved regarding the dietary manipulation of the nutrient content of meat. The authors [1] note that most studies used approaches that independently assess the impact of genetics or nutritional strategies, but few explored the interactions between these two factors.

The most notable findings suggesting potential benefits in genetic manipulation [1] were: (a) differences between breeds or heritability values reported for concentrations of certain vitamins (E and B) and minerals (copper, total iron (myoglobin), and selenium) in beef; there are no reports on the heritability of vitamin A content in muscle, (b) genetic manipulation of zinc content in lambs and pigs, uncertain in cattle, (c) highly variable heritability values are reported for total protein content and individual amino acids, (d) there is the potential to manipulate individual amino acid concentrations by genetic and dietary means, (e) well-known differences between and within breeds in terms of total variability of intramuscular fat (IMF), with very obvious cases of genetic groups with greater levels of marbling [1], and (f) genetics not appearing as a primary factor in the accumulation and proportions of *trans* fatty acids in beef.

Regarding the potential for dietary manipulation of micronutrients, the main findings of the review by Juárez et al. [1] were: (a) pasture feeding increases the levels of vitamins or their precursors (tocopherol, b-carotene, thiamine, riboflavin, lutein, retinol, a-tocopherol, and g-tocopherol) in beef or suckling veal under grass-feeding systems, (b) there is a favorable response to the specific supplementation of vitamins (higher concentration of vitamin E in cattle and of B9 and B12 in pigs) or minerals (selenium and iodine in cattle) but not of zinc, and it is pointed out that it is easier to reduce the iron content in beef than to increase it, (c) in pigs, feed additives derived from algae have an impact on the nutrient potential of B vitamin concentrations, (d) monogastric diets are usually complemented with B vitamins, and although additional increases have a small impact on the muscle concentration of B9 and B12 vitamins, vitamin B2 levels do not seem to be affected by higher supplementation, (e) gastrointestinal–pancreatic control of zinc absorption hinders its manipulation through the diet, and (f) most studies have reported little or no effect on zinc muscle concentrations after zinc supplementation, particularly in pigs.

Jerez et al. [3], when evaluating the addition of two amounts (2 or 4%) of brown seaweed to the regular diet of finishing pigs, found that total lipids and microminerals such as Cu, Zn, and Mn decreased in the muscle of pigs fed the higher percentage of seaweed (4%), despite a small but significant increase in total ash. However, the fatty acid composition of pork was not influenced by the inclusion of the brown seaweed additive at any level. Inexplicably, these authors [3] found a higher cholesterol content in the meat of pigs fed the diet with the higher content (4%) of brown seaweed, suggesting that there are some components of algae that impact cholesterol content.

The modification of the fatty acid profile of beef by dietary means still faces the difficulty of a buffering effect of the extensive ruminal biohydrogenation which transforms unsaturated fatty acids into saturated fatty acids (SFA) [1]. However, pasture feeding and feeding management have assisted in increasing beneficial biohydrogenation intermediates.

For example, supplementing ingredients rich in polyunsaturated fatty acids (PUFA), such as flaxseed co-extruded with peas, before feeding hay, instead of feeding hay and a mixture of supplements, leads to a substantial increase in vaccenic acid and conjugated linoleic acid (CLA), and these differences are related to changes in the microbial population in the rumen. According to Juárez et al. [1], those ingredients or additives in the cattle diet that modify the rumen microbiome may have a greater impact than a direct fatty acid supplementation.

Concentrate-based diets have been associated with a decrease in PUFA/SFA ratios, but, as the finishing period progresses, there is a higher conversion of SFA to monounsaturated fatty acids (MUFA), and the relative rates are influenced by breed [1]. The study conducted by Barragan-Hernandez et al. [4] in this Issue of *Foods* examined the effect of grain type in the diet (corn vs. barley vs. a barley/corn mix) on the sensory attributes, volatile compounds, and beef flavor profile of steers. The authors also examined the normalized partial profile (% of total) of fatty acids presumably responsible for variations in flavor. They [4] reported that the corn-based diet elicited a significantly lower proportion of n-3 fatty acids, a higher proportion of stearic acid (considered to have a neutral cholesterolemic effect), and a higher value of the n-6/n-3 ratio compared to the other two treatments (respective means of 8.32, 6.22, and 7.26 for the corn-, barley-, and barley/corn mix-based diets). Although significant, differences in the n-6/n-3 ratio could be considered low in magnitude and possibly irrelevant from the point of view of human health; authors [4] did not discuss this. In fact, although some authors recommend that the average value of the n-6/n-3 ratio should not be greater than 5.1, the recommendation of the WHO is that the value of this index should not exceed 10 [3], which indicates that the n-6/n-3 values for the different grain diets tested by Barragan-Hernandez et al. [4] are within the safe range.

According to Juárez et al. [1], the dietary effects observed in IMF tend to have a greater impact on the most abundant fat deposits, which is important to try to improve the lipid profile and achieve certain marketing (particularly, health) claims. In fact, these authors [1] suggest an alternative to ground beef, based on feeding a small group of animals with diets characterized by a high concentration of certain beneficial fatty acids or selecting carcasses naturally presenting this feature and then mixing the fat from those animals or carcasses with lean meat from the regular population. For their part, Peña-Saldarriaga et al. [5] with a similar proposal, pointed out the potential of using fat that can be removed with high yields as a co-product of chicken. According to their analyses, these underutilized fats contain—without significant variation due to environmental factors—a lower concentration of myristic acid (an undesirable saturate) and a greater proportion of essential polyunsaturated fatty acids (PUFA, approximately 40% of the total UFA) when compared with pork fat or beef tallow (the latter having less than 20%). The authors [5] suggest using these fats in the formulation of meat products such as sausages, replacing other sources of animal fat (for example, bird skin) commonly used in the poultry industry.

Another review and one research article deal with the composition of nutrients in beef produced in the tropics. When trying to characterize this type of meat through an extensive review of the literature, Rubio et al. [2] pointed out that, in general, (a) there is little variation in macronutrients (almost all protein content values are in the range of 20% to 24%, with some differences due to castration, genetic influences, or to the finishing on pastures vs. grain), (b) the proximate component that showed the largest variation was the IMF content that varied from 1.0% to 8.9%, but most of the literature indicates a lean beef, with <3.6% of IMF, (c) low marbling scores (IMF), which are typically observed in cattle influenced by *B. indicus*, are attributed to the reduced volume of adipocytes, and not to the quantity of cells, (d) there are few studies on the mineral content of cattle raised in tropical environments, and of note is that the feeding system (pasture with or without supplementation) had little to do with the beef mineral content, but an age effect was observed when comparing grazing cattle at 17, 19, and 24 months of age, (e) the impact of climatic conditions on the edible tissues of tropical cattle should be considered, in particular, pastures suffer a seasonal effect on their quantities and qualities, and fluctuations in feed quality can cause mineral imbalances throughout the year, and (f) there are indications that the IMF of steers with genetic predominance of *B. indicus* contains more myristic, palmitic, linoleic, and linolenic acids, but less stearic acid than their counterparts with predominance of *B. taurus*.

According to Rubio et al. [2], the complexity of the research on the impact of genetics on the fatty acid composition of tropical cattle meat is greater than previous studies suggest. In one of the few studies reported, 14 of the 43 individual fatty acids and fatty acid indices in the IMF were affected by an interaction between the genetic pool and the cattle finishing system. Of the 29 fatty acids and fatty acid indices for which the interaction was not significant, 11 were influenced by the genetic group, and 25 by the finishing system. In general, with the exceptions of cholesterol, 18: 1 *trans*-6, -7, -8, 18: 1 *trans*-12, and the 22: 5, n-3/18: 3, n-3 ratios, all fatty acids and individual indices were affected by at least one of the factors considered or by their interactions. Differences between genetic groups were lower with pasture finishing, but under grain finishing, *B. indicus* showed higher amounts of SFA and stearic acid and lower concentrations of fatty acids synthesized from linoleic and gamma-linolenic acids. In general, animals finished on grass produced meat with lower levels of IMF, *trans* fatty acids, and SFAs and higher content of CLA and long-chain PUFA (20: 5, n-3 and 22: 5, n-3). Rich diets in forage favor the growth of the fibrolytic microorganisms responsible for CLA production, and forage-fed livestock has higher concentrations of linoleic, stearic, arachidonic acids (20: 4, n-6), eicosapentaenoic (20: 5, n-3) and docosapentaenoic (22: 5, n-3) acids in the meat than animals fed with concentrates [2]. However, IMF contents are often low in grass-fed beef (<2 g/100 g of fresh muscle), and this meat cannot therefore be considered a significant source of CLA. Again, it should be noted that climatic variations in tropical regions can greatly affect the quality of the grass and hence its nutritional contributions. A review in this Special Issue [2] cautions against genetic manipulation based on a selection of zebu cattle (Nellore) with lower body fat to a given weight, because such a genetic approach would decrease the proportion of MUFA (oleic acid) in the subcutaneous fat depot, with concomitant increases in saturated fatty acids, such as stearic and other, less healthy, saturated fatty acids such as myristic and palmitic acids.

The potential of carcass traits for assisting in the prediction of beef chemical components was evaluated by Arenas de Moreno et al. [7]. In this study, the authors performed an analysis of hierarchical conglomerates and canonical correlations to explore multivariate relationships between selected traits of the beef carcass derived from cattle fed on tropical pastures and chemical components (proximate, minerals, and lipids) in *longissimus lumborum* muscle (LL). The statistical approach is demonstrated as a powerful tool to study the relationship between a selected set of carcass traits and the proximate, lipid, and mineral components, particularly when there is a certain degree of interaction between the three groups of chemical variables. The association of carcass traits and minerals was poor. However, the analyses pointed out an important relationship of backfat thickness and marbling scores with the content of total lipids and fatty acids in the LL. In their conclusions, the authors argue in favor of backfat thickness, rather than marbling, as the most feasible potential predictor for performing future regression analyses attempting to explain the variation in lipid composition of this type of livestock.

2.2. Progress on Meat Standardization and Grading

Product consistency and differentiation are proved tactics for succeeding in meat marketing and trade. The certification of beef carcasses serves for a series of marketing programs. Several certification programs are based on a set of specifications (that may include a breed) to make marketing statements on certain characteristics, especially quality. In the USA, these specifications go beyond the requirements required for the grades offered by the official grading system and are the basis for the different branding programs endowed by the USDA. Certified Angus Beef (CAB) is the most recognized meat branding program in the USA. On the other hand, Protected Geographical indications (PGI) commonly used in the EU are certification programs based on original, identifiable characteristics of a product derived from a specific location in order to protect its quality and reputation. These two distinct marketing strategies have the same purpose: product differentiation. Beriain et al. [7] compared the veal produced by bullock calves in the Pyrenees [PGI-Certified *Ternera de Navarra* (*CTNA*)] to US-CAB. Physicochemical and sensory traits were assessed in Spain (Navarra) and USA. The authors found noticeable contrasts (i.e., marbling, IMF, and other proximate components) which are explained not only by the animals' distinct genetic make-up but also by their dissimilar age, sex, and management. The authors highlighted that the taste panels in the two countries agreed that the *CAB* striploins outperformed the *CTNA* samples in juiciness, tenderness, and flavor, notwithstanding the similarities between CAB and CTNA in total and soluble collagen contents.

Segura et al. [8] determined the potential of computer vision systems (CVS), namely, the whole-side carcass camera (HCC), to the rib-eye camera (CCC), and the dual-energy X-ray absorptiometry (DXA) technology, to predict the composition of wholesale cuts and carcasses of mature cows. This comparative study [8] was carried out in Canada where the classification system segregates mature carcasses (for example, cows) as Canada D, with a series of designations (Canada D1 to D4) for the types of carcass that do not count on a method for predicting the cut-out yield before fabrication. The technologies used in the study [8] provided estimation values of the total amount of tissues and a general description of the composition of the entire carcass and its primal cuts without requiring the destructive procedure of dissection. The DXA technology could be considered the gold standard for estimating carcass composition. The primary estimates of DXA, on average, had higher R^2 values for fat (0.95), lean (0.97) and bone (0.82) than those of CVS, and even exceeded the prediction equations using all variables retrieved by the cameras (HCC and CCC). However, to date, DXA has been limited by practical restrictions in industry implementation (horizontal table scan, room temperature operation, and scan speed in minutes instead of seconds). Instead, CVS technologies (HCC and CCC cameras) are being widely implemented in the USA and Canada. According to their findings, the authors determined the feasibility of using HCC to predict the composition of carcass and wholesale cuts, and the combination of the two CVS technologies led to significant improvements in the predictions, in particular, for the lean/fat ratios, suggesting that the dual CVS approach is an alternative for improving the accuracy of predicting the composition of carcasses and primal cuts of cull cows. The assessment of different types of instrumental grading would allow not only finding out the best technology for differentiating these products in the marketplace but also identifying new opportunities for the future development of automation in the meat industry, an emerging need in the pandemic era.

2.3. *Progress in Food Safety for Different Types of Meat Sources*

Six articles are presented on food safety. One article evaluates four antimicrobials on refrigerated pork loins [9], another assesses the antimicrobial application mode (immersion vs. spray) to reduce *Campylobacter jejuni* in chicken wings [10], four deal with physical and/or chemical interventions, such as the use of UV-C solely or in conjunction with antimicrobials [11,12], refrigeration technologies (dry chilling vs. spray chilling) combined with hot water washing by bio-mapping of indicator organisms on beef striploins during storage [13], and in-plant validation of a novel aqueous ozone generation technology compared to lactic acid solutions for suppressing the growth of natural microbiota, i.e., *E. coli* O157:H7 and *Salmonella* surrogates, on beef carcasses and trimmings [14].

Antimicrobial sprays evaluated by Vargas et al. [9] on pork loins subjected to four refrigerated storage periods (1, 14, 28, and 42 days) included: cold water (control), 1,3-dibromo-5,5-dimenthylhydantoin at 225 ppm (Bovibrom-225), the same active principle as Bovibrom-225 but at 500 ppm (Bovibrom-500), chlorine dioxide at 3 ppm (Fit Fresh), and Rhamnolipid at 750 ppm (Natural Washing Solution). Initial counts did not differ between treatments, while as for after-treatment interventions, the treatment with Natural Washing Solution did not effectively reduce the counts of APC-mesophilics, APC-psychrotrophs, and coliforms ($p < 0.01$). The antimicrobials Bovibrom-500 ppm, Fit Fresh, and Natural Washing Solution were the best in maintaining reduced microbial counts when compared to control treatments of pork loins after 14 days of storage under refrigerated conditions at 0–4 °C.

Gonzalez et al. [10] inoculated surfaces of fresh, skin-on, chicken wings with 3.9 log colony-forming units [CFU]/mL) of a mixture of six *Campylobacter jejuni* strains of poultry origin. The inoculated wings were left untreated, to serve as controls, or they were treated by immersion or spray application of water, a blend of sulfuric acid and sodium sulfate (pH 1.2; SSS), formic acid (1.5%; FA), peroxyacetic acid (550 ppm; PAA), PAA (550 ppm) acidified with SSS (pH 1.2; SSS-aPAA), or PAA (550 ppm) acidified with formic acid (1.5%; FA-aPAA). All five chemical interventions were efficacious ($p < 0.05$) in reducing *C. jejuni* populations on chicken wings, with larger immediate reductions by immersion than by spraying. Acidification of PAA (550 ppm) with SSS or FA did not enhance the immediate (0 h) bactericidal effects of non-acidified PAA. However, the combination of the acidified PAA treatments and the subsequent chilled storage conditions (4 °C, 24 h) likely prevented the recovery of sub-lethally injured bacterial cells. As a result, chicken wings treated with SSS-aPAA or FA-aPAA and stored at 4 °C for 24 h showed the lowest pathogen levels.

Calle et al. [11] evaluated the use of UV-C LED light for the destruction of *Salmonella* present on chicken breast and food contact surfaces. The antimicrobial properties of UV light have been explored elsewhere, mainly for applications in liquids, contact surfaces, and packaging materials, where its effectiveness has been demonstrated. However, the most common application involves the use of mercury lamps. The growing interest in ultraviolet (UV) light was driven by its FDA approval in 1997 for surface decontamination of foods. According to the literature [11,12], several facts of UV irradiation use in food safety are reported: (a) pathogens absorb UV light, and thymine-dimers molecular lesions in the DNA are formed via photochemical reactions, ultimately leading to cell death, (b) UV light is currently used to control pathogens in water and for the decontamination of food contact surfaces and food packaging materials, (c) UV light-emitting diodes (LED) are increasingly being used as substitutes for mercury lamps, conventional sources of UV-light, for their smaller size and lesser generation of heat, (d) the emission spectrum of UV-LED can be tuned to emit UV light of specific wavelengths between 250 and 280 nm, which are the most effective at driving the photochemical reactions leading to the formation of thymine dimers, (e) UV-LED devices are more robust, durable, and safe compared to mercury lamps because they do not contain glass tubes that may break and contaminate workstations with mercury, (f) UV-C band irradiation stands out for its low cost, with no potentially hazardous chemical residues, and low carbon footprint [12].

In the USA, Calle et al. [11] have shown that UV-LED could be used to disinfect skinless chicken breast (CB) as well as food contact surfaces such as stainless steel (SS) and high-density polyethylene (HD) inoculated with *Salmonella enterica*. The greatest reductions were obtained after 180 s of exposure on HD (5.2 Log CFU/cm^2), followed by 60 s on SS (3.5 Log CFU/cm^2), and 900 s on CB (3.0 Log CFU/cm^2). The best reductions were obtained when UV-C LED was applied on SS. For example, 60 s of exposure yielded 3.48, 2.05, and 1.77 Log CFU/cm^2 on SS, CB, and HD, respectively. The porosity of surfaces appears to play a role in the effectiveness of the UV-C LED light, since bacterial cells appear to be shielded by hollow surfaces, as observed in electron micrographs.

The most typical chemical interventions to reduce *Salmonella* and other pathogens in poultry and red meat products involve the application of treatments at different steps of processing, which include the use of organic acids, inorganic compounds, chlorine-based treatments, and phosphate-based products, among other compounds [11]. However, consumers seem to have adverse opinions about the use of such chemicals in foods [11], whereas lactic acid (LA) application at a maximum concentration of 5% (m/v) is generally accepted because it does not present risks to consumer health [12]. It is known that *Listeria monocytogenes* can survive and grow in vacuum-packaged meat cuts stored at temperatures between 0 and 4 °C. In Uruguay, Brugnini et al. [12] studied the combined effect of

UV-C and the application of lactic acid on the inactivation of *Listeria monocytogenes* and lactic acid bacteria (LAB) in vacuum-packaged beef. Surface response analysis indicated that a maximum log reduction for *L. monocytogenes* (1.55 ± 0.41 log CFU/g) and LAB (1.55 ± 1.15 log CFU/g) with minimal impact on meat color was achieved with 2.6% LA and 330 mJ/cm2 UV-C. This strategy could be useful to ensure beef safety and to help extend the shelf life of vacuum-packaged beef to safely reach distant markets. These two studies [11,12] further support the use of UV as a "no-touch" technology in the food industry to effectively sanitize high-touch surfaces where there may be a higher risk of meat contamination from pathogens. UV disinfecting technologies have been used for a number of years and they could be more effective with improved features in the future, given their constant innovation.

Casas et al. [13] evaluated the impact of spray and dry chilling combined with hot water treatments on the levels of microbial indicators during refrigerated storage of beef striploins at an Australian beef processing plant. A total of 200 carcasses were evaluated. Samples were taken before and after (washed samples) carcass hot wash and 24 h after subjecting the carcasses to spray vs. dry chilling. The hot water carcass wash consisted of spraying water at 85 ± 2 °C onto the surface of the carcasses. The spray chilling method consisted of continuously spraying water at 0–2 °C at 15 min intervals, during 18–24 h storage. Dry chilling consisted of 18–24 h storage in a refrigerated room at 0 °C with constant airflow, while the sprayers were turned off. The excised striploins were cut into four sections that were individually vacuum-packaged to be sampled after 0, 45, 70, and 135 days of storage and distribution under refrigeration. Aerobic plate counts (APC), enterobacteria, *Escherichia coli*, coliforms, and psychrotroph (PSY) counts were evaluated for each sample. Under the conditions evaluated in this study, the hot water carcass intervention was not found to significantly reduce APC and PSY counts compared to the no-wash treatments. Despite significantly reducing a small number of bacterial species on the surface of the carcass, washing may also redistribute the bacteria throughout the whole carcass surface and can contribute to further microbial attachment, growth, and development during prolonged storage. The authors [13] concluded that the optimal shelf life of striploins can be achieved using dry chilling air systems, which will guarantee the required 130 days of shelf life for the export of fresh, never frozen beef from Australia to the EU. The use of spray chilling schemes increases the available water for the growth of bacteria, resulting in higher growth rates of bacteria during long-term refrigerated storage and, therefore, in a reduced shelf life.

Casas et al. [14] assessed the antimicrobial efficacy of an aqueous ozone solution (BioSafe) and lactic acid solutions on the natural microbiota and *E. coli* O157: H7 and *Salmonella* surrogates in beef carcasses and trimmings at a commercial meat processing plant. The lactic acid operating parameters applied in the plant for this study included treatment with a 2–5% lactic acid solution sprayed at a temperature of 43–55 °C, with a spray pressure of 15 psi. The operational parameters of the ozone intervention included generators that use air oxygen molecules (O_2) passed through a crown field, which divides them into individual oxygen atoms (O). These individual O atoms combine with an O_2 molecule to form an ozone molecule (O_3). After the intervention and immediate reaction with organic matter, O_3 becomes oxygen again, without leaving byproducts or harmful waste, according to the description of the manufacturer and the patented technology developed. Ozone and lactic acid interventions significantly reduced ($p < 0.003$) bacterial counts in carcasses and trimmings. Furthermore, lactic acid further reduced APC and coliforms in trim samples as compared to the ozone intervention ($p < 0.009$). Ozone significantly reduced ($p < 0.001$) the concentration of *Salmonella* surrogates. According to the plant's historical data, a reduction ($p < 0.001$) of presumptive *E. coli* O157: H7 in trimmings was recorded after a full year of implementing the ozone intervention. These results are very promising, since the use of ozone in combination with organic acids would allow a more efficacious, safe approach for the decontamination of beef carcasses and products. According to the authors [14], this new technology for ozone generation and its application as an antimicrobial can become an alternative that may also act synergistically with existing interventions, minimizing the risk of *Salmonella* and *E. coli* O157: H7.

Funding: This editorial article received no external funding.

Acknowledgments: My deep appreciation to all the authors and institutions for their support in bringing together this Special Issue. I am also grateful to Nuria Prieto, Alexandra Calle, Caterina Rufo, Manuel Juarez, and Oscar Lopez Campos for their assistance in proofreading the excerpts of their own articles. Finally, I would like to recognize the valuable help that I received from Tania Ngapo for editing the last version of the manuscript.

Conflicts of Interest: The author declares no conflict of interest.

References

1. Juárez, M.; Lam, S.; Bohrer, B.M.; Dugan, M.E.R.; Vahmani, P.; Aalhus, J.; Juárez, A.; López-Campos, O.; Prieto, N.; Segura, J. Enhancing the Nutritional Value of Red Meat through Genetic and Feeding Strategies. *Foods* **2021**, *10*, 872. [CrossRef] [PubMed]
2. Rubio-Lozano, M.S.; Ngapo, T.M.; Huerta-Leidenz, N. Tropical Beef: Is There an Axiomatic Basis to Define the Concept? *Foods* **2021**, *10*, 1025. [CrossRef] [PubMed]
3. Jerez-Timaure, N.; Sánchez-Hidalgo, M.; Pulido, R.; Mendoza, J. Effect of Dietary Brown Seaweed (*Macrocystis pyrifera)* Additive on Meat Quality and Nutrient Composition of Fattening Pigs. *Foods* **2021**, *10*, 1720. [CrossRef] [PubMed]
4. Barragán-Hernández, W.; Dugan, M.E.R.; Aalhus, J.L.; Penner, G.; Vahmani, P.; López-Campos, Ó.; Juárez, M.; Segura, J.; Mahecha-Ledesma, L.; Prieto, N. Effect of Feeding Barley, Corn, and a Barley/Corn Blend on Beef Composition and End-Product Palatability. *Foods* **2021**, *10*, 977. [CrossRef] [PubMed]
5. Peña-Saldarriaga, L.M.; Fernández-López, J.; Pérez-Alvarez, J.A. Quality of Chicken Fat by-Products: Lipid Profile and Colour Properties. *Foods* **2020**, *9*, 1046. [CrossRef] [PubMed]
6. Arenas de Moreno, L.; Jerez-Timaure, N.; Huerta-Leidenz, N.; Giuffrida-Mendoza, M.; Mendoza- Vera, E.; Uzcátegui-Bracho, S. Multivariate Relationships among Carcass Traits and Proximate Composition, Lipid Profile, and Mineral Content of *Longissimus lumborum* of Grass-Fed Male Cattle Produced under Tropical Conditions. *Foods* **2021**, *10*, 1364. [CrossRef] [PubMed]
7. Beriain, M.J.; Murillo-Arbizu, M.T.; Insausti, K.; Ibañez, F.C.; Cord, C.L.; Carr, T.R. Physicochemical and Sensory Assessments in Spain and United States of PGI-Certified *Ternera de Navarra* vs. Certified Angus Beef. *Foods* **2021**, *10*, 1474. [CrossRef] [PubMed]
8. Segura, J.; Aalhus, J.L.; Prieto, N.; Larsen, I.L.; Juárez, M.; Lãpez-Campos, Ó. Carcass and Primal Composition Predictions Using Camera Vision Systems (CVS) and Dual-Energy X-ray Absorptiometry (DXA) Technologies on Mature Cows. *Foods* **2021**, *10*, 1118. [CrossRef]
9. Vargas, D.A.; Miller, M.F.; Woerner, D.R.; Echeverry, A. Microbial Growth Study on Pork Loins as Influenced by the Application of Different Antimicrobials. *Foods* **2021**, *10*, 968. [CrossRef] [PubMed]
10. Gonzalez, S.V.; Geornaras, I.; Nair, M.N.; Belk, K.E. Evaluation of Immersion and Spray Applications of Antimicrobial Treatments for Reduction of *Campylobacter jejuni* on Chicken Wings. *Foods* **2021**, *10*, 903. [CrossRef]
11. Calle, A.; Fernandez, M.; Montoya, B.; Schmidt, M.; Thompson, J. UV-C LED Irradiation Reduces *Salmonella* on Chicken and Food Contact Surfaces. *Foods* **2021**, *10*, 1459. [CrossRef] [PubMed]
12. Brugnini, G.; Rodríguez, S.; Rodríguez, J.; Rufo, C. Effect of UV-C Irradiation and Lactic Acid Application on the Inactivation of *Listeria monocytogenes* and Lactic Acid Bacteria in Vacuum-Packaged Beef. *Foods* **2021**, *10*, 1217. [CrossRef] [PubMed]
13. Casas, D.E.; Manishimwe, R.; Forgey, S.J.; Hanlon, K.E.; Miller, M.F.; Brashears, M.M.; Sanchez-Plata, M.X. Biomapping of Microbial Indicators on Beef Subprimals Subjected to Spray or Dry Chilling over Prolonged Refrigerated Storage. *Foods* **2021**, *10*, 1403. [CrossRef] [PubMed]
14. Casas, D.E.; Vargas, D.A.; Randazzo, E.; Lynn, D.; Echeverry, A.; Brashears, M.M.; Sanchez-Plata, M.X.; Miller, M.F. In-Plant Validation of Novel On-Site Ozone Generation Technology (Bio-Safe) Compared to Lactic Acid Beef Carcasses and Trim Using Natural Microbiota and Salmonella and *E. coli* O157:H7 Surrogate Enumeration. *Foods* **2021**, *10*, 1002. [CrossRef] [PubMed]

MDPI

Review

Tropical Beef: Is There an Axiomatic Basis to Define the Concept?

Maria Salud Rubio Lozano [1], Tania M. Ngapo [2] and Nelson Huerta-Leidenz [3,*]

1 Meat Science Laboratory, Centro de Enseñanza Práctica e Investigación en Producción y Salud Animal, Facultad de Medicina Veterinaria y Zootecnia, Universidad Nacional Autónoma de México, Cruz Blanca 486, San Miguel Topilejo, Mexico D.F. 14500, Mexico; msalud65@gmail.com

2 Saint-Hyacinthe Research and Development Centre, Agriculture and Agri-Food Canada, 3600 Boulevard Casavant Ouest, Saint Hyacinthe, QC J2S 8E3, Canada; tania.ngapo@canada.ca

3 Department of Animal & Food Sciences, Texas Tech University, Lubbock, TX 79409-2141, USA

* Correspondence: nelson.huerta@ttu.edu

Abstract: Cattle production in tropical regions has been estimated to account for just over half of cattle worldwide, yet it has not been demonstrated that sufficient similarities in the cattle exist to describe tropical cattle and, even less so, to characterize the meat from these animals. The aim of this review is to investigate the quality and nutrient composition of meat from cattle raised in the Tropics to determine if there is an axiomatic basis that would allow the definition of a concept of "tropical beef". Tropical beef is the meat obtained from cattle raised in tropical environments, the population of which remains largely uncharacterized. Production systems in the Tropics are highly diverse but converge on the use of indigenous and *Bos indicus* breeds or *Bos indicus*-influenced crossbreeds under pasture feeding regimes. While some systems allow cattle to be slaughtered at ≤2 years of age, most often animals are ≥3 years. These production systems generally produce lean, low-yielding carcasses and tough (>46 N), lean (≤3.6% intramuscular fat) meat with a macronutrient composition otherwise similar to beef from animals raised elsewhere (72–74% moisture and 20–24% protein). Fatty acid profiles depend on the breed and production systems, while mineral content is influenced by the environment. Although lean and tough, tropical beef is highly acceptable to the consumers it serves, is culturally and traditionally relevant and, in many countries, contributes to food security. Consolidating the findings from animal and meat science studies in the Tropics has allowed the demonstration of an axiomatic basis defining "tropical beef" as a concept.

Keywords: tropical; beef; meat quality; nutrient; composition

Citation: Rubio Lozano, M.S.; Ngapo, T.M.; Huerta-Leidenz, N. Tropical Beef: Is There an Axiomatic Basis to Define the Concept? *Foods* **2021**, *10*, 1025. https://doi.org/10.3390/foods10051025

Academic Editor: Andrea Garmyn

Received: 12 April 2021
Accepted: 29 April 2021
Published: 9 May 2021

Publisher's Note: MDPI stays neutral with regard to jurisdictional claims in published maps and institutional affiliations.

1. Introduction

It has been estimated that cattle production in tropical regions accounts for just over half of the cattle worldwide, equating to greater than 805 M head [1]. For such a significant source of beef, the volume of meat-related scientific literature actually undertaken in the Tropics is relatively modest, with most works focused primarily on animal production. Some of this animal science literature uses the terms "tropical beef" or "tropical cattle" to describe cattle raised in and/or adapted to tropical environments [2–7]. However, given that there is much geographic, cultural and economic variation in these environments, it is not surprising that this research varies widely in all aspects of animal production. Regardless, the undefined global terms of "tropical beef" and "tropical cattle" are often cited as descriptors in distinct studies. Yet, while there are commonalities among studies, it has not been demonstrated that sufficient similarities in the cattle exist to describe tropical cattle and, even less so, to characterize the meat from these animals.

The Tropics are the region of Earth surrounding the equator delimited at ±23.5 degrees in latitude by the Tropic of Cancer to the north and the Tropic of Capricorn to the south. The region constitutes 36% of the Earth's landmass and includes more than 130 countries from Africa, America, Asia and Oceania, either wholly or partially [8,9]. According to the Köppen classification, there are three categories of tropical climates based on rainfall dynamics and an average annual temperature always above 18 °C: (a) wet equatorial climate (rainforest), (b) tropical monsoon and trade–wind littoral climate (monsoon), and (c) tropical wet–dry climate (savannah) [10,11].

By 2050, global meat consumption is expected to increase by 30% and at least 70% of the increase in beef production required to meet the growing demand is expected to come from the tropical and subtropical regions of the world [12]. Unless there are major changes in production systems, environmental conditions will always determine the types of livestock that can be used in the harsh tropical regions, even though these types may not necessarily meet the growing demand for meat and milk [13]. However, it is not only the climate that dictates animal production in many of these countries. In 2020, the United Nations estimated that some 43% of the world's population, almost 3.8 billion people, live in the Tropics [9]. Of these, about 99% live in a nation considered to be "developing", which includes 85% of the poorest people in the world. People living in the Tropics are also far more culturally diverse than in the rest of the world, exemplified by the fact that these regions account for more than 80% of all living languages [14].

These climatic, cultural and economic conditions have driven production systems in the Tropics to concord in the use of breeds (generally, *Bos indicus* and *Bos indicus* crosses), management systems (extensive and semi-intensive) and feed (pastures and finishing grain) and, consequently, produce carcasses of similar quality [15,16]. However, although similar, each region represents an important source of variation to provide meat that is acceptable to the consumer and is culturally and traditionally relevant [17]. These animals are often dual-(milk and meat) or multi-purpose (milk, beef, draught, fuel and fertilizer) and have important functions ranging from the provision of food and income to socioeconomic, cultural and ecological roles of farming communities [18,19]. Tropical cattle production systems make an important contribution to household food security and income for smallholder beef production. However, the majority of the tropical cattle populations remain largely uncharacterized, and the meat quality of these populations is even less explored.

The aim of this review is to investigate the quality and composition of meat from cattle raised in the Tropics to determine if there is an axiomatic basis that would allow the definition of the concept of "tropical beef", where beef refers to the meat and not the animal. Literature cited in the present review was gathered through a range of databases, including, but not exclusively, Scopus, Food Science and Technology abstracts, Agricola, Biological abstracts, CAB abstracts and OVID medline, as well as extended library and online searches for texts on cattle raised in the Tropics. Keywords used included, but were not limited to, variations of Tropics, beef, cattle, breeds, *Bos indicus*, zebu, sanga, Criollo, meat, quality, nutrition, nutrient, composition, carcass, fatty acids, intramuscular fat, minerals, eating quality, tenderness, feedlot, pasture and production systems. The references from the articles obtained by this method were used to identify additional relevant material.

2. Cattle Production in the Tropics

It is inherent that, in order to define a concept of "tropical beef", characterization of the cattle from which the meat is derived is first required. Reviewing similarities and differences between cattle production systems in the Tropics allows a description of tropical cattle and provides context in defining the resulting beef, given that almost all aspects of animal production impact meat quality and nutrient composition to some degree. The importance of a holistic approach to understanding tropical beef quality and composition is exemplified in the description of the strong growth of Brazilian cattle production described by Ferraz and de Felício [20] as being based on a triploid of Nellore-cerrado-*Brachiaria* grass (that is, breed-environment and production system-feed) since 1970. A large body of scientific work reports on genetics and production of cattle in tropical environments. It is beyond the scope of this study to review these aspects, but rather, the focus of this section is to provide context in order to define and understand the characteristics of the meat obtained from "tropical cattle".

2.1. Breeds

The cattle breeds of preference for production in tropical climates are generally *Bos indicus* or *Bos indicus* crossbreeds. Indeed, Meat and Livestock Australia [21] actually describes *Bos indicus* as tropical breed cattle genetically adapted to survive and produce under adverse conditions, including heat and poor-quality pastures. Unique evolutionary traits of *Bos indicus* breeds, also known as zebu, to tropical climates are well-documented and include resistance to some ecto- and endoparasites and endemic diseases, heat and drought tolerance and other harsh environmental conditions, such as limited water, poor pasture and high humidity [16,22–30]. Although adapted to the local environments, *Bos indicus* cattle are often poor milk and meat producers [31]. Furthermore, poor production performance traits, carcass conformation, and meat marbling content and eating quality are also generally associated with these breeds [16,20,24,32–35]. Consequently, crosses with *Bos taurus* breeds are much studied, given that crossbreeding represents a proven strategy to improve the adaptation almost immediately. Indeed, heterosis has been demonstrated to influence cattle body temperature maintenance, reproduction, survival and, to a lesser extent, temperament in subtropical or stressful environmental conditions, such as toxic fescue [36–38]. However, while crossbreeding might improve the carcass, meat and sensory quality traits, generally, the higher the proportion of *Bos taurus*, the lesser the adaptability to the tropical environment [33,39].

The most commonly used *Bos indicus* breeds in research appear to be Nellore and Brahman, likely a result of the use of these breeds in large-scale commercial meat chains. Indeed, studies on Nellore are predominantly from the research undertaken in Brazil, the country with the world's largest commercial herd, of which the Brazilian Zebu Breeders Association claim that 80% has influence from zebu cattle, and the breed with the largest number of animals is the Nellore [20]. However, there are other breeds of significant number in Brazil, including Guzerat, Gyr, Indubrasil and Tabapua, and interest has also been shown in *Bos taurus* breeds adapted to tropical environments, such as Brazilian Caracu, as well as the introduction of breeds like Senepol and Bonsmara and composite programs, such as Montana Tropical [20]. Crossbreeds and composites are prevalent throughout cattle production, and research reports on tropical herd improvement by crossbreeding date back over a century. Some crosses, such as the Senepol and Bonsmara, have even been developed to recognition as breeds in their own right, including, for example, Brangus, Santa Gertrudis and Charbray [21]. In many countries, artificial selection and management interventions have resulted in marked productivity improvements and by extension, economic performance for commercial cattle breeds [40,41]. However, this is not universal, and for example, in Africa, the focus of selection has predominantly been on survival, in often unpredictable, harsh and changing environmental conditions, and not consistently for productivity gains [42].

In addition to the predominantly *Bos indicus* indigenous breeds and crosses found in tropical countries, in Latin America, there is a group of *Bos taurus* cattle referred to as Criollo. Criollo have the ability to adapt to harsh arid landscapes with minimal human intervention [43]. Some Criollo cattle have been developed into unique breeds, such as the tropically adapted Romosinuano of Colombia, while others are responsible for the genetics that led to the Texas Longhorn. While not common, some small, isolated populations of Criollo in Mexico have not been crossbred at all [43]. These cattle are not part of Mexico's commercial market, due, in large part, to being light-muscled and having non-uniform conformation [43,44]. As for indigenous cattle, many of the characteristics and traits attributed to Criollo have yet to be verified scientifically.

While there is an overwhelming amount of research reporting on genetic selection and production of tropical breeds, these encompass but a few of the many breeds that are found in tropical climates, most of which are little described in the scientific literature, if at all. For example, a survey conducted as part of a large effort to systematically collate information aimed at assessing the status of the cattle genetic resources of sub-Saharan Africa describes 145 cattle breeds/strains little reported elsewhere [27,45]. The large number of indigenous cattle breeds would suggest that there is significant genetic diversity of cattle in many parts of the world, yet many cattle breeds face extinction [42]. However, artificial selection and management have often been achieved at the cost of reduced genetic diversity and, in some cases, fertility [40,41]. For example, to compensate for the relatively low production potential of indigenous cattle, crossbreeding with exotic breeds is commonly practiced in Africa, with minimal within breed selection for the indigenous breeds [42]. The end result is a continual erosion and loss of cattle diversity, including for adaptive traits. To give an indication of the scale of this loss of diversity, in 1999, it was reported that 32% of indigenous African cattle breeds were in danger of extinction [45].

To summarize, there are a vast number of breeds of cattle grown in tropical regions, of which the majority appear to be *Bos indicus* or *Bos indicus*-influenced. There are also some tropically adapted *Bos taurus* breeds and types, but these appear of little interest to large-scale commercial operations. A large body of scientific literature reports on the breeding and genetics of cattle in the Tropics, but until recently, the focus of this research has been primarily on improving production traits for financial gain. However, in parallel to the rapid evolution of genetic research tools, realities of climate change and ever-increasing erosion of diversity, so too has the focus of research evolved and nowadays encompasses carcass conformation, meat and sensory quality attributes, production traits that reduce the environmental footprint of production [44] and breed classification and description [27,45].

2.2. Production Systems

Beef cattle farming systems and supply chains vary according to geographical regions, availability of resources, infrastructure, urbanization and markets [46]. It is not surprising, therefore, that, in tropical countries, production systems run the gamut from large commercial operations specifically for meat production to farmers with but a few mixed-purpose cattle. As for breeding and the genetics of cattle in tropical climates, there is a vast amount of publications on animal production. Brief descriptions taken from select reviews serve to illustrate the diversity of the production systems in tropical climates.

In a review of Brazilian cattle production, Ferraz and de Felício [20] described that, at 305 M head of cattle, Brazil is second only to India (325 M head) in total cattle herd size and has the largest commercial cattle herd in the world. Cattle are raised on 1.8 million farms, ranging from small beef farms of less than 500 head per household per year to commercial operations with over 4000 head per year. While extensive production systems are the norm, an estimated 10% of Brazil's meat production in 2019 was finished in feedlots as a means to limit the weight loss common in the dry season. To minimize the impact of the marked decrease in tropical forage quality and availability in the dry season, three production systems are employed [20]. The first is a complete pasture-based system in which controlled mating is used to start the calving season in November/December to February; calves are weaned May–June and kept on dryland pasture until the next rainy season in October. The animals lose weight during the dry season, and about half are slaughtered at 24–30 months, the balance at 36–42 months. In the second, finishers buy two-year-old steers and finish them in better pastures for one year. And, in the third system, calves are supplemented during pre-weaning to produce heavier weaned calves that go directly to one of three finishing schemes: (a) at 8 months and 240 kg live weight, animals (generally crossbreeds) are transferred to feedlots for 120 days and slaughtered at 420 kg, (b) weaned animals (pure and crossbreeds) are sent to pasture for a growing phase from 18–24 months, then transferred to feedlots, and (c) animals (mostly *Bos indicus* and some crossbred steers) are kept on pasture and slaughtered at 30–42 months and 450–500 kg. The average slaughter age for cattle in Brazil is 4 years [20].

In a more recent review of another Latin American country, Parra-Bracamonte, Lopez-Villalobos, Morris and Vázquez-Armijo [47] described cattle production systems in Mexico, which has around 31.7 M cattle. The five most-important beef production states in Mexico (Veracruz, Jalisco, San Luis Potosí, Tabasco and Chiapas) are in tropical and subtropical regions. Mexican beef originates from cow–calf production systems, which provide cattle for feeder or feedlot systems and for live export. Cow–calf operations in Mexico consist of purebred, multiplier and dual-purpose systems [47], the latter two systems being found in tropical regions. Nearly 90% of farms within these systems have more than 20 ha. Multiplier cow–calf herds are the most numerous and located in all agro-climatic regions. These are extensive pasture-based farms in which the main product is weaned calves. Dual-purpose cattle farming systems producing milk and meat comprise almost 9% of the cattle in Mexico [47]. These types of farms are located mostly in tropical regions. Meat produced for the domestic market is generally finished in feedlots, and all geographical regions have feedlot systems, but the levels of management and days of fattening vary with the region. In the tropical regions, longer periods at pasture and fewer days of fattening in feedlots are usual (for example, 70–90 days), compared to feedlots in the northern temperate, arid and semi-arid regions (for example, 130–150 days) [48].

Unlike Brazil and Mexico, Australia has only a very small proportion of wet tropics, and most of its beef production comes from a dry tropical environment, characterized by distinct wet and dry seasons [49]. Northern Australian grazing lands, including tropical regions, collectively support about 14 million head or 60% of the national beef herd [50]. The production systems are similar to those of Brazil, where millions of hectares are used for grazing, with few feedlots. Individual properties range from less than 1000 ha and fewer than 1000 head to over 1.5 M ha with more than 40,000 cattle. Traditionally, reducing the stocking rates to maximize the head performance on native tropical pastures has been the option of preference, slaughtering animals at 550–600 kg live weight at about 4 years of age [51]. Nowadays, feedlot or supplementary feeding strategies may be used to finish animals at a younger age and to improve the carcass and fat cover [20].

Indonesia provides a complete contrast to the above countries. In 2017, Agus and Widi [52] reported that the cattle population totaled about 16.6 M head. Of these, 90% are held by smallholder farming systems, with about 6.5 M farmers living in rural areas. The remaining 10% are from more commercial farmers (<1% of all farmers) and large beef cattle companies. Smallholder farmers are those who keep between two and four head of cattle and use stall feeding in Java where the land is scarce to 50 head or more extensively grazed in other areas. The definition of small holder is a stark contrast to those in Australia and Brazil with 500–1000 head of cattle. While most other reviews have focused on the commercial production of cattle, Agus and Widi [52] noted the importance of livestock for smallholder livelihoods around the world. For poor households in Indonesia, as in many other tropical countries, the non-income benefits of keeping livestock are particularly important. These farmers keep cattle to produce meat for the urban market, to support cropping with manure, to provide draught power and as assets. These sentiments are also reiterated by Mwai, Hanotte, Kwon and Cho [42], who describe that, across the African continent, cattle remain major sociocultural assets, play important social–cultural roles in many African societies (such as, marriage and initiation), represent an important source of animal protein (dairy and beef), provide draught power and supply fertilizer through manure, which is also used as fuel by some communities. In Indonesia, both stall feeding and extensive systems use low-quality feed, mainly from crop residues as well as agricultural byproducts and other nonconventional feedstuffs, such as oil palm leaves, cassava foliage, cotton seed meal, seaweed and food wastes. In terms of feedlot operations, about 75% of cattle imported from Australia are destined for feedlot in Indonesia.

These selected reviews provide an overview of the enormous diversity of cattle production in tropical countries. Yet, there are similarities. In general, production is extensive and often on forage of relatively poor nutritional quality. Age at slaughter varies with animals achieving slaughter weight at ages greater than in temperate or sub-tropical climates, often around 4 years unless semi-intensive or intensive production systems are used. Nutrient and feed supplementation or introduction of legumes or specialized crops into pastures is recommended in some regions and particularly during dry seasons. Possibly as important as the introduction of *Bos taurus* genetics to improve carcass conformation, increase fat deposition and decrease age at slaughter, is the introduction of feedlots to tropical beef production. Although incipient, increasingly, cattle are finished in feedlots, particularly as a means to meet the demands of export markets. Alternative finishing options include the transfer to farms of higher quality forage and/or supplementation. However, all of these interventions are costly, and for many but large commercial operations, the cost may be prohibitive.

3. Carcass and Meat Quality

The most reported of carcass characteristics in research on tropical beef are the slaughter weight and dressing percentage, often included as an extension of cattle production studies. It is therefore not surprising that these characteristics are at the core of the scientific literature of beef carcass and meat quality research in tropical Africa, of which reports are relatively scarce. A study in Ghana found carcass weights of zebu cattle (156 kg average) heavier than sanga (93 kg), which were, in turn, heavier than West African Shorthorn (73 kg), these being from slaughter weights of 309, 201 and 162 kg, respectively [53]. In a review of Shorthorn cattle production in West and Central Africa, carcass weights ranged from 80 to 200 kg [54], and it was concluded that, owing to their small size, the performance of Shorthorn cattle was generally low. However, the dressing percentages (ranging from 42% to 55%) were similar to those of other breeds within and outside the region. In Uganda, it was also found that, while the indigenous genotypes are well-adapted to the tropical production environment, slow growth rates and smaller mature body weights limit their potential for meat production [55]. Here, beef production is described as evolving from traditional pastoral practices to sedentary semi-intensive systems on private ranches. Reflecting what is actually happening across the region, a study was undertaken comparing three locally available genotypes (pure Boran, Ankole x Holstein Freisian cross and a composite genotype) and finished either in pasture or in feedlots (60, 90 or 120 days) that use locally available agro-industrial byproducts. Bulls were 12 to 20 months old at slaughter and the average final live weights ranged from 198 to 238 kg on pasture compared to 221 to 279 kg in feedlots. Similarly, hot carcass weights were also higher for those animals fattened in feedlots (115–153 kg vs. 99–114 kg). Slaughter characteristics did not vary with genotype. In a Cameroonian study, breed also had only a limited effect on the carcass characteristics of cattle harvested in a local slaughterhouse [56]. In this study, 1953 carcasses from three local zebu breeds, Gudali, White Fulani and Red Mbororo, were evaluated, and body condition score, carcass weight and carcass conformation were highest in castrated males, while heifers had the highest fatness levels and bulls, the lowest. It was concluded that the month of year greatly influences the carcass weight, which increased from March to September and decreased from September to March. In an earlier study, an average loss in body weight of 13.3 kg/month was reported from December to April due to the poor quality of forage coinciding with the dry season (November to March) [57]. While breeds showed limited differences in carcass traits in these African studies, an impact on meat toughness was observed in a Cameroonian study [56]. Gudali meat was tougher (unaged shear force of 112 N) than White Fulani (72 N) and Red Mbororo (78 N). In a Beninese study, the tenderness of meat from Borgou, Lagunaire and Zebu Fulani cattle did not significantly differ, but did decrease from 91–122 N at slaughter to 37–66 N after 8 days aging [58]. In both of these studies, bulls were raised in pasture and selected at a local slaughterhouse at 3–5 years of age [56,58].

While only but a few studies from tropical Africa are reported, the challenges of achieving profitable slaughter weights in beef cattle production in pasture is a common research theme in studies of tropical beef production. When striving to meet markets where consumer demand for tenderness is a priority, meat from young animals is a prerequisite, exacerbating the need for increased live weight gains in tropical cattle. Indeed, Poppi, Quigley, da Silva and McLennan [49] illustrated that the target market determines the growth path, so that, for example, a targeted high slaughter weight (undefined) can be achieved at 3.5–4.5 years of age of the animal from extensive range land pastures in Australia with minimal inputs, providing a profitable production system in Northern Australia to meet the ground meat market in North America. However, this growth path cannot attain more profitable markets with exigent meat quality demands. This challenge has driven large-scale research programs in the region over the last 40 years.

Since the 1980s, in Australia, crosses of Brahman with other breeds have been investigated to improve the production and carcass characteristics in tropical regions. Ball and Johnson [59] found Brahman crossbred cattle to have higher saleable beef yield (1–3%) over Hereford cattle under tropical conditions. In a series of experiments, Wythes, Shorthose, Dodt and Dickinson [60] observed that slaughter and carcass weights, backfat thickness and shear force values of unaged meat were generally similar between steers of *Bos taurus* and crosses of *Bos indicus* × *Bos taurus*. In a follow-up study, chronological age and dentition had no significant impact on shear force values (76–99 N) of *M. longissimus dorsi* from cows and steers of Brahman and Brahman x Shorthorn or Hereford crossbreeds [61]. It was concluded that overall toughness of meat from cattle slaughtered in Northern Australia was of much greater concern than the minor differences between genotypes. Newman, Burrow, Shepherd and Bindon [62] noted that purebred Brahman had the highest peak shear force measurements (52–59 N, aging not specified) when compared to progeny of Brahman females mated to sires of eight different breeds (Brahman, Santa Gertrudis, Belmont Red, Angus, Hereford, Shorthorn, Charolais or Limousin). Furthermore, the average peak force values in Brahman cattle were considered above acceptable values for tenderness (no acceptability threshold was provided). In this study, it was also reported that European and British sire breeds produced consistently heavier carcasses than those from the progeny of tropically adapted breeds or Brahman sires [63]. When domestic market carcass weight (220 kg) was targeted, very small differences were found between sire breeds for carcass yield traits. However, when export market carcass weights (280 kg and 340 kg) were achieved, crossbreeds of Brahman and Charolais or Limousin produced leaner carcasses and greater yield percentages than other crossbreeds. In addition, differences in intramuscular fat (IMF) percentage among sire breeds were not observed at a 220 kg carcass weight (1.65%), but at 280 and 340 kg, increases in the IMF (2.28% and 2.85%, respectively) were consistent with increasing age [62]. It was concluded that the common practice of incurring fixed costs of slaughtering animals at lighter weights for the Australian domestic market to ensure a tender product is a fallacy and that considerable cost savings might accrue to processors and retailers who slaughter animals at heavier weights without any detrimental effects on meat tenderness.

In the same study, pasture and feedlot-finished steers and heifers were compared [62,63]. While much of the tropical beef research is focused on the use of crossbreeding to improve production and meat quality traits, over the last couple of decades, research on the use of feedlots, particularly for the finishing phase, has also come to the forefront. These workers found that animals finished at pasture were considerably older (739–805 days) and leaner (8.0–13.6 mm fat at P8 and 1.58–1.74% IMF) than those finished in feedlots (626–672 days, 11.5–15.8 mm fat at P8 and 2.09–2.30% IMF) and had larger eye muscle areas, higher retail beef yield percentages and the greatest weight of retail primals [62,63]. The meat from pasture-finished animals was also consistently tougher than that from feedlot-finished heifers (55 vs. 47 N, aging not specified).

In stark contrast to these earlier publications is a study reporting that Senepol × Brahman steers produced a more tender meat than purebred Brahman steers (shear force values after 14 days of aging of 34 N and 39 N, respectively) [64]. In addition, other than hump height, most of the carcass measures were similar for the two genotypes, and it was suggested that this crossbreed demonstrated a viable method to improve the meat quality of cattle produced in Northern Australia. These animals were raised in pasture and finished in feedlot. It was noted that all the meat from the purebred Brahman was relatively tender when compared to values that have been found for other Brahmans. The good tenderness results found for both genotypes in this study were considered likely due to the young slaughter age achieved (average estimated age of 21.5 month and hot carcass weight of 238 kg), and it was concluded that Brahman cattle with good meat quality can be produced by production systems that give good growth rates and minimize the age at slaughter. However, it was noted that this may not be possible on many extensive properties in northern Australia where growth rates are low, and it was cautioned that changing the growth path of Brahmans for slaughter at a younger age would not overcome grading penalties incurred as a consequence of a higher hump.

As for most of the carcass and meat quality research in tropical environments, in the Brazilian tropics, the use of crossbreeds is the primary focus of much of the published research. Norman and de Felicio [65] observed that, although some differences in the carcass composition of Nellore, Guzerat, Charolais and Canchim bulls could be attributed to breed effect, most were caused by the varying nutritional status of the animals pre-slaughter. Furthermore, lower hindquarter (45–46 kg vs. 47–48 kg) and higher forequarter (39 vs. 36–37 kg) weights were observed in the *Bos indicus* animals, attributed, at least in part, to earlier sexual maturity. Maggioni, Marques, Rotta, Perotto, Ducatti, Visentainer and do Prado [66] found that greater daily weight gains of bulls of crossbreeds (1/2 Nellore × 1/2 European or 1/4 Nellore × 3/4 European bulls) resulted in better carcass conformation (good vs. regular), thicker subcutaneous fat (3.38 vs. 1.92 mm) and a higher marbling score (light vs. trace) than those of purebred Nellore. Pflanzer and de Felicio [67] found that if slaughtering Nellore steers at a young age, the animals need to be fattened in order to achieve an acceptable marbling level. Bressan, Rodrigues, Rossato, Ramos and da Gama [68] found that meat from feedlot-finished animals was more tender than that from pasture-finished animals (55 vs. 59 N after 10 days of aging). However, these workers also found that *Bos taurus* cattle had lower shear force than *Bos indicus* (54 vs. 60 N), without reporting the actual breeds, other than to note that they were commercial bulls.

In a study of crosses of another extensively used breed in Brazil, Guzerat (Guzerat × Holstein, Guzerat × Nellore and 1/2 Simmental + 1/4 Guzerat + 1/4 Nellore), the three-cross had heavier cold carcass weights and greater rib-eye areas than the other crosses [69]. The crosses with Nellore were also tougher than that with Holstein (50.9 and 50.1 vs. 43.1 N shear force, respectively) [69]. Interestingly, one study compared a *Bos indicus* × *Bos indicus* crossbreed (Brahman × Nellore) with Angus × Nellore and purebred Nellore [70]. The carcass weights of both crossbreeds were heavier than those of purebred Nellore, and the proportion of carcasses grading Choice or Prime was greater in Angus × Nellore cattle than in the Brahman × Nellore or purebred Nellore cattle (26%, 12% and 16%, respectively). Steaks from Angus × Nellore calves were more tender than Nellore steaks, with the Brahman × Nellore steaks being intermediate (33, 42 and 39 N, respectively, after 14 days of aging). Significant variation among Nellore sires was observed for slaughter weight, dressing percentage, carcass weight, *longissimus* muscle area and marbling score, but not for backfat or shear force. The percentage of carcasses of Nellore cattle grading Choice or Prime ranged from 0% to 61.5%, and it was concluded that, while *Bos indicus* cattle have inferior carcass and meat quality relative to Angus × Nellore crossbreeds under tropical conditions, there is substantial variation within the Nellore breed for these traits, and several sires had a proportion of their progeny comparable in meat tenderness to those of Angus sires.

It should be noted that, in addition to shear force, a range of meat quality traits have been measured in studies of tropical beef, including ultimate pH, meat color, cooking and thawing losses, water-holding capacity, myofibrillar fragmentation index and sarcomere length, with few differences observed. Shear force is the exception, with a general consensus that *Bos indicus* breeds produce tough meat in tropical environments and tougher meat than *Bos taurus* breeds, often well-exceeding the minimum shear force of "very tough meat" (for example, 46 N [71], although it has also been reported as low as 38 N [72]), as is evident when compared in a tabulated form (Table 1). It is also apparent in the literature that many of the shear force measures are made without prior aging of the meat. If not aged, even a normally tender cut of beef can be expected to be tough. However, shear force measures at 1 to 2 days postmortem in many of these tropical countries reflect the local market in which beef aging is oftentimes rarely undertaken, such as in Venezuela [72], Costa Rica [73] and Mexico [74].

Table 1. Research on the impact of production factors, quality grade and aging time (treatments under study in bold) on the Warner Bratzler shear force (or equivalent) in meat from beef raised in tropical environments [1].

Country	Breed or Purchased Meat	*n*	Sex	Age (Months)	Feed	Other	Shear Force (N)	Days p.m. [2]	Muscle [3]	Ref
No Comparison										
Brazil	Nellore	1329	Bulls	21–24	Feedlot-finished		59	2	LT	[75]
							58	7		
							62	9		
Mexico [4]	Purchased steaks	20					41		LD	[76]
Puerto Rico	Non-specified "typical" breeds	105	Male, female	≤30, ≥36	Pasture		46	1	LL	[77]
							43		St	
							49		Sm	
Breed										
Brazil	**Nellore**	306	Steers, heifers	22–24	Feedlot-finished		42 [a]	15	LD	[70]
	Nellore × Angus						33 [b]			
	Brahman × Nellore						39 [ab]			
Brazil	***Bos taurus***	160	Bulls	26–40	Pasture or concentrate-finished		74	1	LT	[68]
	Bos indicus						88	1		
	Bos taurus						54	10		
	Bos indicus						60	10		
Brazil	**Guzerat × Holstein**	36	Bulls	26	Feedlot-finished		43 [a]	1	LL	[69]
	Guzerat × Nellore						51 [b]			
	$\frac{1}{2}$Simmental × $\frac{1}{4}$Guzerat × $\frac{1}{4}$Nellore						50 [b]			
Australia	**Brahman × Shorthorn**	170	Bulls	36–84	Pasture		75	1	LD	[60]
	Shorthorn						75			
Australia	**Brahman × Shorthorn**	240	Cows	60–120	Pasture		89	1	LD	[60]
	Shorthorn						96			
Australia	**Brahman**	50	Steers	About 22	Feedlot-finished		39 [a]	15	LD	[64]
	Brahman × Senepol						34 [b]			
Mexico	**Brahman**	20	Bulls	18–24	Feedlot	β-adrenergic agonist used	54 [a]	2		[74]
	Charolais						49 [b]			
Mexico	**Zebu**	90	Bulls		Concentrate + forage finishing		80 [a]	2	LD	[78]
	Zebu × Holstein					β-adrenergic agonist used, acetate + estradiol implants	62 [b]			
	Zebu × American Brown Swiss						63 [b]			
	Zebu × European Brown Swiss						63 [b]			
	Holstein						61 [b]			
	European Brown Swiss						68 [b]			
Benin	**Zebu Fulani**	25	Bulls	60	Pasture		102	1	LT	[58]
	Lagunaire						91	1		
	Borgou						122	1		
	Zebu Fulani						60	3		
	Lagunaire						51	3		
	Borgou						90	3		
	Zebu Fulani						38	9		
	Lagunaire						37	9		
	Borgou						66	9		

Table 1. *Cont.*

Country	Breed or Purchased Meat	*n*	Sex	Age (Months)	Feed	Other	Shear Force (N)	Days p.m. [2]	Muscle [3]	Ref
Cameroon	**Goudali**	60	Bulls	36–60	Pasture		112^{b}	8	LT	[56]
	White Fulani						72^{b}			
	Red Mbororo						78^{b}			
Venezuela	**Brahman,**	71	Bulls		Pasture-finished, Supplement-finished		40^{ab}	2	LD	[79]
	Brahman × Angus,						36^{c}			
	Brahman × Gelbvieh						38^{bc}			
	Brahman × Limousin,						36^{bc}			
	Brahman × Romosinuano						35^{c}			
	Brahman × $\frac{3}{4}$ ***Bos taurus***						43^{a}			
Sex										
Brazil	Nellore, Nellore × Angus, Brahman × Nellore	306	**Steers**	22–24	Feedlot-finished		39	15	LD	[70]
			Heifers				36			
Brazil	Angus × (Limousin × Nellore), Angus × (Simmental × Nellore)	225	**Bulls**	12.5, 18	Pasture-finished, feedlot-finished		84^{a}	1	LD	[80]
			Heifers				94^{b}			
Costa Rica	$\frac{3}{4}$Brahman × $\frac{1}{4}$Charolais	47	**Steers-3**	61–66	Pasture		86^{e}	2, 7, 14, 28	LL	[73]
			Steers-7				93^{f}		LL	
			Steers-12				91^{ef}		LL	
			Bulls				100^{g}		LL	
			Steers-3				64^{bc}		GM	
			Steers-7				66^{bcd}		GM	
			Steers-12				70^{cd}		GM	
			Bulls				73^{d}		GM	
			Steers-3				61^{b}		St	
			Steers-7				63^{bc}		St	
			Steers-12				63^{bc}		St	
			Bulls				61^{b}		St	
			Steers-3				40^{a}		PM	
			Steers-7				39^{a}		PM	
			Steers-12				38^{a}		PM	
			Bulls				39^{a}		PM	
Age or Dentition (Permanent Incisors)										
Australia [6]		198	Cows	**2 teeth**	Pasture		99	1	LD	[61]
				4 teeth			97			
				6 teeth			82			
				8 teeth			96			
Australia [6]		168	Steers	**4 teeth**	Pasture		82	1	LD	[61]
				6 teeth			77			
				8 teeth			76			
Brazil	Nellore	60	Steers	**20–24**			68	15	LD	[81]
				30–36			68			
				42–48			57			
Puerto Rico	Non-specified "typical" breeds	105	Male, female	**≤30,**	Pasture		30^{a}	1	LL, St, Sm	[77]
				≥36			46^{b}			

Table 1. *Cont.*

Country	Breed or Purchased Meat	*n*	Sex	Age (Months)	Feed	Other	Shear Force (N)	Days p.m. [2]	Muscle [3]	Ref
Feed										
Brazil	*Bos indicus, Bos Taurus*	160	Bulls	26–40	**Pasture-finished**		85	1	LT	[68]
					Concentrate-finished		77	1		
					Pasture-finished		59	10		
					Concentrate-finished		55	10		
Brazil	Nellore	30	Bulls	22	**No glycerin in dry feed**		47	1	LD	[82]
					7.5% glycerin		46			
					15% glycerin		37			
					22.5% glycerin		44			
					30% glycerin		40			
Brazil	Nellore	60	Bulls	22	**No glycerin in dry feed**		30	1	LD	[83]
					Glycerin + corn		32			
					Glycerin + soybean hulls		28			
Australia	Brahman, Brahman × (Brahman x Santa Gertrudis, sanga × Belmont Red, Angus, Hereford, Shorthorn, Charolais, or Limousin)	349	Heifers, Steers	22–24	**Pasture-finished (heifers)**		55		LD	[63]
					Feedlot-finished (heifers)		48			
					Feedlot-finished (steers)		47			
Venezuela	Brahman, Brahman × (Gelbvieh, Romosinuano, Limousin, Angus or $\frac{3}{4}$ *Bos Taurus*)	71	Bulls		**Pasture-finished**		58^{a}	2	LD	[79]
					Supplement-finished		67^{b}			
Other: Fat Class (F), Ageing Time of the Meat										
Brazil	Nellore	60	Steers	22–48		**F: Slight**	70^{a}	15	LD	[81]
						F: Average	59^{b}			
Costa Rica	$\frac{3}{4}$Brahman × $\frac{1}{4}$Charolais	47	Steers castrated at 3, 7 or 12 months, bulls	61–66	Pasture		102^{i}	**2**	LL	[73]
							96^{h}	**7**	LL	
							96^{h}	**14**	LL	
							76^{f}	**28**	LL	
							83^{g}	**2**	GM	
							68^{e}	**7**	GM	
							64^{de}	**14**	GM	
							57^{c}	**28**	GM	
							66^{e}	**2**	St	
							61^{cd}	**7**	St	
							62^{cd}	**14**	St	
							60^{cd}	**28**	St	
							44^{b}	**2**	PM	
							38^{ab}	**7**	PM	
							38^{ab}	**14**	PM	
							36^{a}	**28**	PM	

[1] Shear force data in a given study with differing superscripts (down a column) are significantly different ($p \leq 0.05$). [2] Number of days postmortem (p.m.) at which the shear force measure was undertaken. [3] *Longissimus dorsi* (LD), *Longissimus thoracis* (LT), *Longissimus lumborum* (LL), *Semimembranosus* (Sm), *Semitendinosus* (St), *Gluteus medius* (GM) and *Psoas major* (PM). [4] Veracruz data only. [5] Castrated at 3 months (Steers-3), 7 months (Steers-7) or 12 months (Steers-12). [6] Northern Queensland data only.

In Venezuela and Mexico, it is also reported that the occurrence of steers in the cattle population is atypical, as castration is rarely practiced [84,85]. In beef production, it is generally accepted that bulls provided adequate nutrition grow faster and more efficiently and produce carcasses with less fat than steers [86,87]. A higher proportion of cuts derived from the forequarter and a retained percent yield of total retail lean product at different weight ranges (from 163 to 365 kg carcass weight) has also been shown in bulls [88]. However, meat from steers is often preferred by consumers over meat from bulls because of the improved sensory traits, particularly tenderness [89,90]. In a Costa Rican study, it was reported that late castration (>12 months of age) had been reintroduced as a production tool to potentially increase the fatness and meat quality of subprimals of steers while taking advantage of the growth rates and efficiency of bulls [73]. However, few differences were observed in carcass and sub-primal yield traits of 3/4 Brahman ×1/4 Charolais bulls and steers raised on pasture and slaughtered at about 400kg live weight. *Longissimus lumborum* steaks from steers were more tender than those from bulls, whether castrated at 3, 7 or 12 months of age (100 vs. 86, 93 and 91 N, respectively), but *gluteus medius* was only significantly more tender from steers castrated at 3 months of age (64 vs. 73 N in bulls), and *semitendinosus* (about 62 N) and *psoas major* (about 39 N) were not different at any castration age. It was also observed that all but *psoas major* were significantly tougher at 2 days than 7, 14 and 28 days of aging, and for all four muscles, there were no significant differences between shear force values at 7 and 14 days of aging. Tenderness of the *longissimus lumborum* (76 N) and *gluteus medius* (57 N) was significantly improved at 28 days, but was still very tough in all except the *psoas major* (36 N; 60 N for *semitendinosus*).

In Mexico, young feedlot-finished bulls of six genotypes (zebu, European Brown Swiss, Holstein, zebu × European Brown Swiss, zebu × American Brown Swiss and zebu × Hereford) showed few significant differences in carcass and meat characteristics [78]. Of note was the higher shear force of the zebu (80 N) than all other genotypes (61–68 N). Another Mexican study similarly found that meat from feedlot-raised *Bos indicus* cattle was tougher than that from *Bos indicus* × *Bos taurus* crosses (73 vs. 55 and 59 N, respectively, after 14 days of aging), although all the samples were tough.

In a study of grading criteria of 23,484 beef carcasses in a commercial abattoir in a tropical region of Mexico, a beef carcass classification norm was implemented using five evaluation criteria applied in sequence: (1) maturity (age), (2) conformation (muscularity), (3) lean color, (4) fat color and (5) distribution of the subcutaneous fat cover [91]. The carcasses were classified as 13.4% Select, 45.8% Standard, 27.4% Commercial, 10.6% Out of Grade and 2.7% Veal, with no carcasses attaining the highest quality, Supreme grade. Based on maturity, 79.2% of the carcasses met the specifications for Supreme, but when the next criterion, conformation, was evaluated, only 0.5% of the carcasses graded Supreme. Using commercially purchased steaks, it was also found that beef from the central and southern regions of Mexico (regions where tropical production is prevalent) had greater shear force values than those from the northern (non-tropical) regions (46–47 vs. 36 N, respectively) [92]. Interestingly, while consumers also found beef from the north more tender than that from the other two regions, the overall desirability ratings were not significantly different. In addition, it has been observed that beef produced in the north of Mexico, which is largely based on feedlots, yields carcasses with a whiter fat than from the central and southern regions, where production relies more on pastures [84]. This finding corroborates other studies comparing feedlot and pasture-fed cattle [93,94], and with the emergence of feedlots, it is curious that the fat color is rarely, if ever, reported in the research on tropical beef.

In Venezuela, an analysis of carcass data from 590 bulls, steers and heifers showed that the dressing percentage of Zebu-type cattle outperformed the dairy/dual-purpose-type (64% vs. 54%), noting a wide range of values for the slaughter weights (285–657 kg), carcass weights (146–444 kg) and dressing percentages (47–71%) [95]. In the Venezuelan llanos, Brahman crossbreeds (× Romosinuano, Limousin, Angus, Gelbvieh or 3/4 *Bos taurus*) finished on pasture with supplementation achieved market weight (500 kg) with a desirable conformation at an earlier physiological age (shortened by 43 days) than those finished on pasture without supplementation [79]. The supplemented animals resulted in heavier (287 vs. 279 kg carcass weight) and fattier (1.26 vs. 0.88 cm backfat thickness) carcasses, but no differences were found for the low-yielding *longissimus* muscle area (79 cm^2). Unexpectedly, supplementation produced meat with higher shear force values than pasture-only finishing (67 vs. 58 N), although all were very tough.

All of these studies illustrate that, while the limited carcass and meat quality research reports from tropical countries appear somewhat scattered and often use few animals, there are recurrent findings. Interventions are much-studied, and success is apparent in the use of crossbreeds, young bulls and feedlot finishing. The results from other studies, such as those of late castration and pasture finishing with supplementation, are not as promising, but the research is very limited to date. Regardless of the intervention, of which most are costly, tropical production systems generally result in low slaughter weights, lean carcasses and tough meat. Even with aging, crossbreeds, electrical stimulation and feedlot finishing, in general, tropical beef is very tough. In Brazil, Ferraz and de Felício [20] suggest a tender meat is achieved after aging during transport to export markets, but evidence is lacking in the scientific literature. Furthermore, this focus on the export market is indicative of much of the research on beef quality in tropical environments.

Research on tropical beef meat quality has been reported from a very limited number of countries—generally, those for whom export markets are of interest and research funding, resources and infrastructure follow as the sector strives to meet importers' quality criteria. There is, therefore, a bias in the type of research undertaken targeting the quality criteria of non-tropical countries. Meat toughness may be a limiting factor for these more valuable export markets, but one can question the implementation of costly interventions for domestic markets. In many tropical countries, not only is the meat inherently tough, be it a consequence of tradition, food hygiene, lack of resources and infrastructure or for some other reason, meat is not aged. However, methods of food preparation often negate the necessity for a tender meat, and this is reflected in consumer perceptions. For example, it is reported that beef is rarely aged in Mexico [96], yet, in a survey of 488 Mexican consumers, 89% stated that the beef they buy is almost always or always of good quality, while only 1% reported that it is almost never or never good quality [97]. When asked how they prepare beef, the most popular methods were roasting, stewing and boiling (42%, 44% and 37%, respectively, noting that consumers could answer as many responses to this question as were appropriate). Only 26% said they fry beef and 6% grill. In the same survey, 59% of consumers preferred beef steaks with no marbling.

4. Nutrient Composition

Given the significance of marbling in export criteria as a meat quality indicator and the role of fat in the human diet, it is not surprising that there are a number of studies reporting the IMF content, generally with moisture and protein analyses, and fatty acid composition of beef from cattle raised in tropical environments. There are also a few studies of the mineral content, but no reports of amino acids or vitamins were apparent.

4.1. Macronutrients

Almost all the reports on meat from beef raised in the Tropics that include proximate analyses are from Latin America (Table 2). Compiled in tabulated form, it is readily apparent that there is generally little variation in the macronutrients of cattle grown in tropical regions. With the exception of three findings, the moisture ranges from 71% to 76%, and three quarters of the reported data are between 72% and 74%. The three excluded findings appear to be outliers, reporting very low moisture values [96,98]. Two of these data points are from a study comparing the use of anabolic steroid implants where low moisture contents (about 60%) and concomitant high protein contents (about 36%) were reported [96]. The third is from a study comparing meat from cattle raised on pasture with supplementation to feedlots, and the moisture contents of the latter were reported as 67.3%, likely, at least in part, a consequence of the high IMF content [98]. Indeed, in all of the studies reporting differences in moisture content, these data correspond with opposing differences in IMF content.

Table 2. Research on the impact of production factors, quality grade or muscle (treatments under study in bold) on the moisture, fat and protein in meat from beef raised in tropical environments.

Country	Breed or Purchased Meat	*n*	Sex	Age (Months)	Feed	Other	Moisture (%)	Protein (%)	IMF (%)	Ref
No Comparison										
Venezuela	Brahman × (Brahman, Black Angus, Red Angus, Romosinuano or Charolais)	17	Bulls, steers	24	Pasture		75.3	20.9		[99]
Venezuela	Zebu crossbreeds, dairy crossbreeds	145	Bulls, steers, heifers	30–48	Pasture		73.9			[100, 101]
Venezuela	Brahman, Zebu × Brown Swiss, *Bos taurus* (Angus, Limousin, Gelbvieh, Criollo Romosinuano) × Brahman	20	Bulls, steers	30–60	Pasture		73.0	22.3	2.6	[102]
Venezuela	Purchased meat	20					74.2	22.4	3.6	[103]
Mexico [2]	Purchased steaks	20					73.5	19.4	1.9	[76]
Mexico [3]	Purchased steaks	20					72.2	22.3	3.6	[92]
Indonesia	Simmental × Ongole						72.4	21.8	3.5	[104]
Breed										
Brazil	**Nellore**	45	Bulls	24	Feedlot-finished		72.2	25.1 ^{a}	1.7	[66]
	$\frac{1}{2}$Nellore × $\frac{1}{2}$European						73.2	23.8 ^{b}	2.0	
	$\frac{1}{4}$Nellore × $\frac{3}{4}$European						73.5	23.7 ^{b}	1.8	
Brazil	***Bos taurus***	160	Bulls	26–40	Pasture- or concentrate-finished		73.3	19.7	5.0 ^{a}	[68]
	Bos indicus						72.7	19.8	5.7 ^{b}	
Brazil	**Nellore**	18	Steers	25	Pasture-finished		74.1 ^{a}	23.4 ^{a}	2.7 ^{a}	[105]
	Simmental × Nellore						73.9 ^{ab}	23.0 ^{ab}	3.1 ^{a}	
	Santa Gertrudis × Nellore						73.3 ^{b}	22.7 ^{b}	3.6 ^{b}	
Cameroon	**Goudali**	50	Bulls	20–41	Pasture		76.6	20.1	0.60	[106]
	Italian Simmental × Goudali						76.0	20.5	0.76	
Cameroon	**Goudali**	60	Bulls	36–60	Pasture		74.9	22.1 ^{a}	1.1	[56]
	White Fulani						75.5	21.5 ^{b}	1.4	
	Red Mbororo						75.8	21.6 ^{b}	0.9	
Mexico	**Zebu**	90	Bulls		Concentrate + forage-finished	β-adrenergic agonist used; trenbolone acetate + estradiol implants	74.1	20.3	1.7	[78]
	Zebu × Holstein						74.3	20.8	1.7	
	Zebu × American Brown Swiss						74.4	20.6	1.7	
	Zebu × European Brown Swiss						74.0	20.3	2.0	
	Holstein						75.2	19.8	1.6	
	European Brown Swiss						75.4	20.8	1.3	
Mexico	**Brahman**	20	Bulls	18–24	Feedlot	β-adrenergic agonist used	73.7 ^{a}		2.9 ^{a}	[74]
	Charolais						75.1 ^{b}		2.4 ^{b}	
Benin	**Zebu Fulani**	25	Bulls	60	Pasture			21.7	1.25	[58]
	Lagunaire							19.5	0.44	
	Borgou							20.7	0.61	
Sex										
Brazil	Nellore crosses	30	**Steers Imm** [4]	36	Pasture + supplement-finished		75.7	21.1	1.6 ^{a}	[107]
			Steers Surg				74.7	20.9	2.2 ^{b}	
			Bulls				76.2	19.9	1.2 ^{c}	

Table 2. *Cont.*

Country	Breed or Purchased Meat	*n*	Sex	Age (Months)	Feed	Other	Moisture (%)	Protein (%)	IMF (%)	Ref
Brazil	Zebu × Aberdeen Angus	27	**Steers**	27	Pasture + supplement-finished		73.1 ^{a}	23.8 ^{a}	2.0 ^{a}	[108]
			Bulls				76.2 ^{b}	22.9 ^{b}	1.0 ^{b}	
Venezuela	Brahman	34	**Steers**	19, 24	Pasture				2.1 ^{a}	[109]
			Bulls						1.8 ^{b}	
Age										
Brazil	Nellore	60	Steers	**20–24**			72.3 ^{a}		4.2 ^{a}	[67]
				30–36			71.9 ^{ab}		5.0 ^{ab}	
				42–48			71.0 ^{b}		5.7 ^{b}	
India	Kangayam	12		**12–18**			76.1 ^{a}	20.7 ^{a}	2.1 ^{a}	[110]
				>36			74.0 ^{b}	22.0 ^{b}	2.9 ^{b}	
Puerto Rico	Non-specified "typical" breeds	105	Male, female	**≤30**	Pasture		74.6	20.1	1.9 ^{a}	[77]
				≥36			73.8	20.6	2.7 ^{b}	
Venezuela	Brahman	34	Bulls, steers	**19**	Pasture				1.4	[109]
				24					2.0	
Feed										
Brazil	*Bos indicus, Bos Taurus*	160	Bulls	26–40	**Pasture-finished**		73.8 ^{a}	21.4 ^{a}	3.0 ^{a}	[68]
					Concentrate-finished		72.2 ^{b}	18.2 ^{b}	7.7 ^{b}	
Brazil	Nellore	60	Bulls	22	**No glycerin in dry feed**				2.5	[83]
					Glycerin + corn				3.0	
					Glycerin + soybean hulls				2.9	
Brazil	Nellore	30	Bulls	22	**No glycerin in dry feed**		76.3	21.8	2.1	[82]
					7.5% glycerin		75.1	22.1	2.5	
					15% glycerin		76.0	21.2	2.3	
					22.5% glycerin		75.	22.4	2.3	
					30% glycerin		75.8	21.8	1.9	
Venezuela	Brahman, Angus, Romosinuano, Senepol, Simmental, commercial zebu crosses	89	Bulls, steers		**Pasture**	Implants (Ralgro, Revalor)	73.9 ^{a}	21.4		[96, 111]
					Pasture + supplement		74.2 ^{b}	21.7		
Venezuela	Criollo Limonero	23	Steers	36	**Pasture**	-	71.9	22.4	2.9	[112, 113]
					Pasture + concentrate		71.5	22.9	3.1	
					Pasture + legume		72.2	22.3	3.1	
Mexico	"Multi-racial"	80	Steers	22–38	**Pasture + supplement**		71.6 ^{a}	21.3	5.6 ^{a}	[98]
					Feedlot		67.3 ^{b}	22.7	8.9 ^{b}	
Mexico	$\frac{3}{4}$ Zebu, $\frac{3}{4}$ *Bos taurus* (Holstein crosses)	52	Steers		**Pasture-finished**		71.3	20.8	2.3	[114]
					Feedlot-finished		73.8	22.2	2.2	
Other: Fat Class (F), Carcass Grade (C), Muscle (M), Implants (I)										
Brazil	Nellore	60	Steers			**F: Slight**	72.3 ^{a}		4.2 ^{a}	[67]
						F: Medium	71.1 ^{b}		5.7 ^{b}	
Venezuela	Angus, $\frac{3}{4}$ Brahman (n = 18); purchased meat (n = 40)	58	Steers		Pasture + supplement-finished	**C[5]: BF A**	74.4	20.5	3.5	[115, 116]
						C: BF AA	74.3	20.4	4.0	
						C: LD A	74.7	21.9	2.0	
						C: LD AA	74.0	21.5	3.0	

Table 2. *Cont.*

Country	Breed or Purchased Meat	*n*	Sex	Age (Months)	Feed	Other	Moisture (%)	Protein (%)	IMF (%)	Ref
Venezuela	Brahman, Angus, Romosinuano, Senepol, Simmental, commercial zebu crosses	77	Bulls		Pasture ± supplement	**I [6]: Ral-Ral**	60.0	35.5	3.6 [a]	[96]
						I: Rev-Ral	59.4	35.9		
Venezuela	Brahman, Angus, Romosinuano, Senepol, Simmental, commercial zebu crosses	89	Bulls, steers		Pasture ± supplement	**I [7]: Ral-Ral**			1.3 [a]	[111]
						I: Rev-Ral			1.4 [b]	
Breed × Production System Interaction										
Brazil	**Nellore**	134	Bulls	23.5, 27.5	**Feedlot-finished**				2.7 [a]	[117]
	Simmental × Nellore				**Feedlot-finished**				2.1 [b]	
	Nellore				**Pasture-finished**				1.3 [c]	
	Simmental × Nellore				**Pasture-finished**				1.6 [c]	

[1] Moisture, protein or fat content means in a given study with differing superscripts (down a column) are significantly different ($p \leq 0.05$). [2] Veracruz data only. [3] Villahermosa data only. [4] Chemical (Imm) or surgical (Surg) castration. [5] *Biceps femoris* (BF) or *longissimus dorsi* (LD) and Venezuelan carcass grade AA or A. [6] Ralgro (Ral) implants consisting of zeranol, an anabolic agent; Revalor (Rev) implants consisting of the anabolic steroid trenbolone acetate and the estrogen hormone, estradiol.

While almost all of the protein content values are in the range from 20% to 24%, some differences were observed between bulls and steers (22.9% vs. 23.8%, respectively [108]), breeds of Nellore and Nellore × European crosses (25.1% vs. about 23.8% respectively) [66] and pasture and grain finishing (21.4% vs. 18.4%, respectively [68]). However, these differences in the protein contents were relatively small, and the majority of the reports found no differences in the studies that covered a range of factors, including age at slaughter, fat class, carcass grade, sex, breed, feeding system and muscles.

The IMF content showed the greatest variation of the proximate analyses, ranging from 1.0% to 8.9%. However, the majority of the studies indicates a lean meat, with more than three-quarters of the data reporting an IMF of ≤3.6%. These findings are in accord with the low marbling and lean carcasses reported in tropical beef [62,63,66,67,84,92,94]. Given the importance of marbling score in the global marketplace, studies have been undertaken to better understand the low marbling scores in *Bos indicus*-influenced cattle compared to *Bos taurus* cattle [118–120]. In a review of these studies, it was noted that no strong relationship between the capacity to synthesize fatty acids *de novo* and the marbling score or adipocyte volume was reported, and it was concluded that the low marbling scores typically observed in *Bos indicus*-influenced cattle are mainly attributed to their smaller intramuscular adipocyte volume compared with *B. taurus* breeds [16]. Given that it is the volume of adipocytes, and not the quantity, that is of most consequence in explaining the difference in the genotypes, this reduced volume would also explain low IMF contents of tropical beef.

While generally low in cattle in tropical environments, there are reported differences of IMF contents with production characteristics. In three studies comparing slaughter ages, IMF content was observed higher in older animals. Cattle described as typical for Puerto Rico that were at least 3 years of age at slaughter had higher IMF than those of up to 2.5 years at slaughter (1.9% vs. 2.7%) [77]. In southern India, Kangayam bulls > 3 years of age had higher IMF contents than those 12–18 months old (2.89% vs. 2.09%) [110]. And, in Brazil, a progression of increasing fat contents with the age of Nellore steers was observed when grouped as 20–24 months (4.2%), 30–36 months (5.0%) and 42–48 months (5.7%) [67].

Differences in IMF content with sex are also reported in two Brazilian studies. One study found that pasture-finished Nellore × Aberdeen Angus steers had higher IMF content than bulls (1.96% and 0.95%) [108], while another reported that whole bulls had the lowest IMF content (1.23%), surgically castrated steers the highest (2.17%) and chemically castrated steers were intermediate and significantly different from both (1.61%) [107]. Increased accumulation of IMF through fat deposition induced by castration is primarily a result of an altered hormonal balance [121].

In terms of breed, the IMF content findings are inconsistent (Table 2). Lower IMF content was reported in Brahman than Charolais bulls (2.4% vs. 2.9%) [74], and in 25-month-old Nellore steers (2.65%) compared to Santa Gertrudis × Nellore crosses (3.64%), while the IMF content of Simmental × Nellore steers was intermediate and not different from either [105]. Two studies found no differences in IMF contents, both reporting low values (1.3–2.0% IMF) [66,78]. In one of these studies, the breeds were not reported [66], while in the other bulls of zebu, Holstein and European Brown Swiss purebreds and zebu crosses with Holstein, American Brown Swiss or European Brown Swiss were used. The lack of effect of genotype in the latter study was explained a result of the use of the combination of an anabolic implant and a β-agonist in the diet, which can significantly reduce the accumulation of fat in the muscle [122]. Lastly, one study reported a higher IMF content in *Bos indicus* than *Bos taurus cattle* (5.7% vs. 5.0%), but the breeds of cattle were not given. These IMF values are notably high compared to the other studies [68].

Finally, there are two reports on the impact of the feeding system on IMF content. Grain-finished bulls (*Bos indicus* and *Bos taurus*) produced meat with higher IMF contents than those finished on pasture (7.7% vs. 3.0%) [68]. Meat from steers (defined as "multiracial") raised in feedlot had higher IMF content than those raised on pasture with supplementation (8.9% vs. 5.58%), both values being relatively high [98]. These findings are in accordance with the generally accepted conclusion that diets with a high content of concentrate cause rapid growth of cattle, and this is associated with a greater deposition of IMF [123].

4.2. Fatty Acids and Cholesterol

Beef IMF, regardless of where the cattle is raised, is comprised of over 20 individual fatty acids, of which six contribute more than 90% of the total fatty acid (TFA) content, myristic (C14:0), palmitic (C16:0), palmitoleic (C16:1), stearic (C18:0), oleic (18:1 cis-9) and linoleic (C18:2) acids [20]. Fatty acid profiles can vary with factors such as breed, feed and sex of the animal.

Given the decades of research on crossbreeding as a means to improve production traits and meat quality, including through increased fatness and marbling of otherwise lean *Bos indicus* cattle, it is not surprising that many of the studies on fatty acids focus on the impact of breed. In a Cameroonian study, while some differences were found, overall, genotype had a limited effect on the fatty acid profile of crossbred Simmental × Gudali and purebred Gudali bulls [106]. All of the animals were raised on pasture and were slaughtered at 20–41 months of age. These workers suggested that similar diets explained the lack of differences in the fatty acid profiles between breeds. Indeed, the levels of α-linolenic acid (C18:3n-3) and linoleic acid (C18:2n-6), both of dietary origin, were similar between the genotypes. High levels of PUFA (17.8% of TFA) were explained [106], not only by the effect of the relatively high proportion of phospholipids in muscle expected in the very lean animals but, also, by the high content of PUFA, particularly PUFA n-3, that characterizes fresh forage from pasture [124].

In another Cameroonian study using three local *Bos indicus* breeds, it was also found that breed had a limited effect on the fatty acid profile of meat from Gudali and Red Mbororo bulls raised on pasture and slaughtered at 3–5 years of age [125]. The only difference, albeit small, was a lower concentration of stearic acid (C18:0) in the Gudali bulls (18.0% vs. 19.8% of TFA). Gudali also had a lower concentration of stearic acid compared to White Fulani (20.7% of TFA). It was suggested that these differences could be a consequence of differences in gastrointestinal tract and rumen volume among breeds, which can influence the ruminal microbial ecosystem [68]. Ruminal biohydrogenation of dietary fat was concluded to have occurred to a lower extent in Gudali compared to in White Fulani cattle, given that, in addition to the aforementioned lower proportion of stearic acid in the IMF of Gudali, a higher proportion of α-linolenic acid (C18:3n-3; 2.34% vs. 1.61% of TFA) was found. Furthermore, docosapentaenoic acid (C22:5n-3), which is derived from α-linolenic acid, was also higher in Gudali than in White Fulani beef (1.62% vs. 1.17% of TFA). As already mentioned, α-linolenic acid is exclusively of dietary origin, and in addition, stearic acid is an end product of the biohydrogenation of dietary unsaturated fatty acids (UFA) [125].

Other differences in the fatty acid profiles of the Gudali and White Fulani were evident, while the Red Mbororo was generally not significantly different from either. Gudali bulls had higher tridecanoic acid (C13:0) (0.25% vs. 0.11% of TFA) and lower pentadecanoic acid (C15:0) (0.29% vs. 0.37% of TFA) relative to White Fulani. The pentadecanoic acid findings were explained by genetic differences between the breeds related to *de novo* C15:0 syntheses from propionate in adipose tissue [126]. Total SFA was lower and PUFA and n-3 PUFA were higher in Gudali compared to White Fulani. The SFA and MUFA were positively correlated with IMF and the PUFA was negatively correlated, suggested a consequence of the decrease in the phospholipids/neutral lipids ratio that arises from an increase in the IMF [127]. Reported PUFA/SFA ratios of 0.29 [125] and 0.33–0.36 [106] are lower than the minimum PUFA/SFA ratio of 0.45 recommended for human health [128]. The inability to achieve the recommended PUFA/SFA ratio is well-documented in both *Bos taurus* [129–131] and *Bos indicus* [68,107] cattle, a consequence of the extensive biohydrogenation of the dietary UFA by rumen microorganisms.

In another African study, further differences in the fatty acid profiles of two *Bos taurus* breeds (Borgou and Lagunaire), and Zebu Fulani were observed [58]. These cattle were raised on pasture and slaughtered at about 5 years of age. Zebu Fulani had higher contents of myristic (C14:0; 2.08% vs. 0.37% and 1.42% of TFA, respectively) and palmitic acids (C16:0; 22.4% vs. 14.5% and 19.3% of TFA, respectively) than Lagunaire and Borgou breeds. The content of α-linolenic acid varied from 2.46% to 3.81% of TFA but did not differ with breed. Zebu Fulani had the highest proportion of SFA (49.7% vs. 35.6% and 43.0% of TFA, respectively) and lowest total n-6 fatty acids (10.2% vs. 20.3% and 16.4% of TFA, respectively) when compared to Lagunaire and Borgou bulls. The ratio of PUFA/SFA fatty acids varied from 0.04 to 0.57 and was higher in Borgou than in Lagunaire.

In a Brazilian study, it was found that Nellore × Santa Gertrudis steers had higher SFA (508 vs. about 474 g/kg total fatty acid) and conjugated linoleic acid (CLA) isomer C18:2 cis-9, trans-11 (9.9 vs. 8.4 and 9.2 mg/g fat, respectively) and lower PUFA (46 vs. about 70 g/kg of TFA) than Nellore or Simmental × Nellore [105]. No differences in n-6 fatty acids were observed, but Simmental × Nellore cattle had lower n-3 than Nellore or Santa Gertrudis. In a Mexican study, it was found that fatty acid profiles of IMF from crossbreeds of 3/4 zebu or 3/4 European (based on Holstein crosses) differed, regardless of pasture or feedlot finishing [114]. The IMF from steers with *Bos indicus* dominance contained more myristic (31 vs. 26 mg/g fat), palmitic (255 vs. 238 mg/g fat), linoleic (64 vs. 38 mg/g fat) and linolenic acids (13 vs. 6 mg/g fat), but less stearic acid (191 vs. 226 mg/g fat) than steers with *Bos taurus* dominance.

While these few studies illustrate the complexity of research on the impact of breed on the fatty acid composition of beef, recent work demonstrates that it is even more complex than these studies would suggest. Indeed, a Brazilian study of pasture- and grain-finished purebred Nellore and crossbred Simmental × Nellore bulls found that, among the 43 individual fatty acids and indices of fatty acids in the IMF, 14 were affected by an interaction between the genetic group and the finishing system [117]. For the major groups of fatty acids, the interaction of the genetic group and the finishing system influenced the totals of the SFA and PUFA, while, for MUFA, only the effect of genetic group was significant. Of the 29 fatty acids and indices of fatty acids where the interaction was not significant, 11 were influenced by the genetic group and 25 by the finishing system. Overall, with only the exceptions of cholesterol, 18:1 trans-6,-7,-8, 18:1 trans-12 and the ratio 22:5n-3/18:3n-3, all the individual fatty acids and indices were affected by at least one of the factors considered or their interactions. These findings serve to illustrate that the importance of studying animals in their production environment cannot be overstated. The most prominent feature was the impact of the finishing system on the IMF and the fatty acid profile, but differences among the genetic groups were important. Differences among the genetic groups were minor with pasture finishing, but in grain finishing, the *Bos indicus* showed higher amounts of SFA and stearic acid and lower concentrations of fatty acids synthesized from linoleic and α-linolenic acids. Generally, animals finished on pasture produced meat with lower IMF, trans fatty acids and SFA contents and higher CLA and long-chain PUFA (20:5 n-3 and 22:5 n-3).

A number of studies have evaluated the impact of feed on fatty acid composition, and in particular, the manipulation of animal feed has been used as a method to improve the nutritional quality of meat. In general, pasture systems (not necessarily tropical) lead to an increase in PUFA in bovine meat compared to grain-based diets [132,133]. Diets rich in forage favor the growth of fibrolytic microorganisms responsible for the rumen production of CLA [134], and cattle fed forage have higher concentrations of linoleic, stearic, arachidonic (20:4n-6), eicosapentaenoic (20:5n-3) and docopentanaenoic (22:5n-3) acids in meat than those fed concentrates [135]. However, it must be kept in mind that pastures in climatically different environments can vary enormously. Furthermore, climatic variations in tropical regions can greatly impact pasture. For example, a Venezuelan study concluded that variations in linoleic acid and CLA contents in IMF could be explained by slaughter in two different seasons [136]. Brahman crosses were slaughtered at 17 and 19 months of age in April and June 1998, which coincided with the "El Niño" climatic effects, characterized by a prolonged dry season. A 24-month-old group was slaughtered between November and December 1998 under less severe environmental conditions. During drought periods, the fiber contents in plants in pastures increase. In addition, a higher quantity of soluble fiber creates an environment that promotes a greater production or a decreased utilization of CLA by rumen bacteria [137], explaining, at least in part, the higher CLA content in meat from the younger animals (1.76 mg/g IMF at 17 months, 1.98 mg/g IMF at 19 months and 1.20 mg/g IMF at 24 months). It should be noted that these workers also concluded that, considering the sparingly low lipid concentrations (<2 g/100 g of fresh muscle), none of the meat could be considered a significant source of CLA. And, while one might think that feedlots would reduce variation among climatic regions given the global trade of grain, factors other than those related to climate may be significant. For example, it is reported that due to limitations of grain processing, the utilization of starch in feedlots in Brazil is not optimal, and levels fed are much lower than those used in North American feedlots [138]. Furthermore, there is a significant incorporation of forage and byproducts in the feeds of feedlot-raised beef.

In Mexico, a study found that steers of 3/4 zebu and 3/4 European cattle in a feedlot system consumed greater amounts of fatty acids compared to those on pasture (157 vs. 116 g/day SFA, 154 vs. 77 g/day MUFA and 189 vs. 135 g/day PUFA), but the latter consumed more alpha-linolenic acid [114]. Pasture-fed animals deposited lesser proportions of myristic and palmitic acids (32.4 and 255.4 mg/g fat, respectively) in the IMF than feedlot animals (21.8 and 236.5 mg/g fat, respectively), even though the consumption of these two fatty acids was similar with production system. No differences in the n-6/n-3 ratio (about 7.2) or total content of CLA were observed (14.4–16.8 mg/g fat). It was suggested that CLA is more likely located in the subcutaneous fat than in the IMF.

In another Mexican study, the concentration of the CLA isomer, C18:2 *cis*-9, *trans*-11, in beef from grazing cattle was slightly more than double that in meat from cattle-raised in a feedlot system [98]. In this experiment, steers defined as "multiracial" were raised on pasture to a final live weight of 320 kg or in feedlot to 480 kg. In the pasture-fed beef, concentrations of pentadecanoic (C15:0), heptadecanoic (C17:0) and stearic acids and CLA isomers (C18:2, n-6, C18:3, n-3, C18:2 *cis*-9 and *trans*-11) were higher, while beef from the feedlot system had higher concentrations of myristic, myristoleic (C14:1), oleic (C18:1 *cis*-9), elaidic (C18:1, n-9) and *cis*-10 heptadecanoic acids (C17:1*cis*-10).

In a Colombian study, meat from zebu cattle raised in four production systems (two silvopastoral systems, improved pasture and a traditional grazing system) was compared [139]. Myristic and palmitic acids were higher in meat from the traditional (3.65 and 32.6 g/100 g of TFA, respectively) than the improved pasture system (2.82 and 28.9 g/100 g of TFA, respectively), while linolenic acid was lower (0.96 vs. 2.30 g/100 g of TFA). The results from the silvopastoral systems differed, with one system showing a similar fatty acid profile to the traditional pasture system.

Aside from breed and feed, the potential of castration as a means to improve the meat quality in tropical beef is a research area that has piqued some interest. Higher levels of stearic acid were found in the IMF of meat from castrated Nellore cattle (around 21% of TFA) than from intact bulls (18.5% of TFA) in a Brazilian study comparing two methods of castration [107]. The linoleic acid concentration was lower in surgically castrated cattle than chemically castrated or intact bulls (3.5% vs. 4.5% and 5.1% of TFA, respectively). Regardless of the type of castration, the content of linolenic (about 0.9% vs. 1.8% of TFA), eicosapentaenoic (C20:5n-3; about 0.13% vs. 0.19% of TFA) and docosapentaenoic acids (C22:5n-3; 0.7–1.1% vs. 1.6% of TFA) was higher in IMF of meat from bulls compared to steers. The IMF from the surgically castrated animals contained less PUFA than that from chemically castrated cattle (8.2% vs. 9.7% of TFA), which was less than that from bulls (12.0% of TFA). This was explained by the fact that bulls exhibit a greater musculature development than steers [109], and PUFA is a major component of phospholipids in the cellular membranes of muscle tissues [140]. Steers also had lower SFA contents (49.0%

and 52.2% of TFA for chemical and surgical castration, respectively) than bulls (47.3% of TFA), resulting in a smaller PUFA/SFA ratio of surgically castrated cattle than the bulls (0.16 vs. 0.25) and that from chemically castrated cattle being no different from either (0.20). The PUFA/SFA ratio increased as the fat content decreased, given that, at low levels of IMF, the contribution made by phospholipids is proportionately greater, and these are more unsaturated than the triacylglycerols, which themselves increased in proportion as the total lipid increased [141]. The n-6/n-3 ratios were higher in castrated animals (about 3.0) than intact bulls (2.1), but all were low.

Similar trends were observed in a Brazilian study of crossbred zebu x Aberdeen Angus on supplemented pasture [108]. The myristic (1.3% vs. 1.7% of TFA), linoleic (6.6% vs. 4.2% of TFA) and linolenic acids (1.2% vs. 0.6% of TFA) were all higher in bulls than steers. In addition, PUFA were higher in bulls (14.1% vs. 8.0% of TFA), resulting in a higher PUFA/SFA ratio than in steers (0.29 vs. 0.19). And, the n-6/n-3 ratios were lower in bulls than steers (2.4 vs. 3.0). Also observed were lower palmitic acid (23.5% vs. 25.0% of TFA) and MUFA (36% vs. 39% of TFA) in the IMF of bulls than steers.

The cholesterol content of IMF has also been measured in some studies. In a Venezuelan study, feed supplementation in pasture finishing had no impact on the cholesterol content (around 29 mg/100 g muscle tissue) of meat from Criollo Limonero steers slaughtered at 36 months of age after finishing on one of three systems: pasture, pasture supplemented with concentrate or pasture supplemented with legumes [112]. Another Venezuelan study found that cholesterol in meat from Brahman-influenced bulls and steers raised on pasture increased from 54.5 mg/100 g tissue at 19 months of age to 69.04 mg/100 g at 24 months of age [109]. A Brazilian study reported a lower cholesterol content in Nellore (46.6 mg/100 g muscle) and Nellore × Simmental steers (46.9 mg/100 g muscle) than in Nellore × Santa Gertrudis steers (48.3 mg/100 g muscle) [105].

Finally, it must be noted that there are a number of studies aimed at identifying the means to genetically select traits in tropically adapted cattle, including a couple on fatty acid composition. Some workers have suggested that the demonstrated existence of genetic variation in the IMF from feedlot-finished Nellore steers allows the possibility to increase the proportion of healthy and favorable beef fatty acids through selection [142]. Others have found that the selection for decreased subcutaneous fat resulted in decreased proportions of oleic acid with concomitant increases in stearic, myristic and palmitic acids [143]. It was suggested that selection for decreased fat at a given weight will result in a decrease in the proportion of MUFA in the subcutaneous fat in the carcass, with a corresponding increase in the proportions of less healthy SFA.

4.3. Minerals

In a recent review on minerals in meat, particularly from animals raised in tropical regions, Ribeiro, Mourato and Almeida [144] described how concentrations are influenced by a vast array of factors, including species, sex, genotype, production stage, region, climate, tissue characteristics and animal management practices, namely rearing systems and nutrition. However, there are few studies on mineral contents of meat from beef raised in tropical environments.

Investigating the use of trace elements as fingerprints for traceability of Brazilian beef, mineral content was analyzed in meat from cattle raised in five different environments, of which three were tropical [145]. While relatively few differences were observed, these differences were sufficient to conclude that chemical traceability of Brazilian beef according to the biome of origin was feasible. Differences in mineral contents of the muscle tissues included the largest mass fractions of bromine and selenium in the Amazônia biome (tropical rainforest) and the lowest mass fraction of zinc in the Pantanal biome (tropical savanna).

In Venezuela, no differences in mineral contents of meat from Criollo Limonero steers fattened on three different feed systems were observed [112,113], nor was any difference found in mineral content with carcass grade of bulls, steers and heifers obtained from a commercial slaughterhouse in Venezuela [146]. In another study on carcass grade of

Venezuelan beef, subprimal cuts were obtained from graded carcasses [115,116]. Again, mineral content did not vary with the carcass grade, but did show lower concentrations of Ca, Fe and Zn and higher P and K than meat imported from the US.

Also in Venezuela, no impact of sex on mineral content was observed in meat from zebu-influenced steers and bulls slaughtered at 17, 19 and 24 months [147]. However, an age effect was found for all of the minerals studied. As animal age increased from 17 to 24 months, concentrations of Na (59.5 vs. 71.2 mg/100 g fresh tissue), K (321 vs. 400 mg/100 g fresh tissue), Ca (3.54 vs. 8.17 mg/100 g fresh tissue), Mg (21.8 vs. 23.6 mg/100 g fresh tissue) and Zn (3.66 vs. 3.83 mg/100 g fresh tissue) increased, while those of P (203 vs. 195 mg/100 g fresh tissue), Mn (0.02 vs. 0.01 mg/100 g fresh tissue), Fe (2.34 vs. 1.75 mg/100 g fresh tissue) and Cu (0.16 vs. 0.09 mg/100 g fresh tissue) decreased. When the values at 19 months of age were included, it was noted that variations in mineral contents did not always show an apparent trend. While statistical differences were not identified, Na, P, Mn, Fe and Cu contents were similar at 17 and 19 months of age, and K and Zn were intermediary between 17 and 24 months, but Ca and Mg were the highest at 19 months. It was suggested that an effect of feed (produced by a drought period) could have impacted the findings. The amount and quality of grasses were reduced for the animals slaughtered at 17 and 19 months of age, while the slaughter of the 24-month-old animals occurred under less severe environmental conditions, a phenomenon that could have produced changes in the intramuscular accumulation of some minerals.

Indeed, Ribeiro, Mourato and Almeida [144] describe how the impact of climatic conditions on the soil and pasture is a determinant factor for the mineral concentrations of edible tissues. In tropical countries, pastures suffer a seasonal effect on their quantities and qualities, and fluctuations of feed quality may cause mineral imbalances throughout the year, with dry season being critical due to the lignification of natural pastures and water/feed scarcity. In addition, in the wet season, heavy rains may cause nutrient leaching and, consequently, reduce the mineral contents of forage and the availability to grazing ruminants [144].

5. Conclusions

Research on tropical beef quality and composition is reported from a limited number of countries, a consequence of a lack of access to research funding, resources and infrastructure. Studies that are reported have often been based on a piecemeal approach, using limited numbers of animals and short durations. Regardless, consolidating the findings from these studies has allowed the demonstration of an axiomatic basis defining "tropical beef" as a concept.

Tropical beef is the meat obtained from cattle raised in tropical environments. The majority of the tropical cattle population remains largely uncharacterized, and production systems in the Tropics are diverse, but converge on the use of indigenous and *Bos indicus* breeds or *Bos indicus*-influenced crossbreeds under pasture feeding regimes. No one gender is used throughout tropical production systems, and while some systems allow cattle to be slaughtered at ≤2 years of age, generally, animals are ≥3 years at slaughter. These production systems generally produce lean, low-yielding carcasses and tough (>46 N), lean (≤3.6% IMF) meat, with a macronutrient composition otherwise similar to beef from animals raised elsewhere (72–74% moisture and 20–24% protein). Fatty acid profiles depend on the breed and production systems, while mineral content is influenced by the environment. Although lean and tough, tropical beef is highly acceptable to the consumers it serves and is culturally and traditionally relevant. In many countries, tropical cattle have important functions, ranging from the provision of food and income to socio-economic and cultural roles. Indeed, tropical cattle contribute to food security and income in developing regions, particularly for smallholder beef farmers for whom producing meat in a sustainable manner is an important challenge.

Funding: This research received no external funding.

Acknowledgments: The authors would like to acknowledge the invaluable contributions of Diana Sofia Andrade Chacon in the writing of this review.

Conflicts of Interest: The authors declare no conflict of interest.

References

1. Bartends, W. Climate adaptation of tropical cattle. *Annu. Rev. Anim. Biosci.* **2017**, *5*, 133–150. [CrossRef] [PubMed]
2. Tewolde, A. Evaluation and utilization of tropical breeds for efficient beef production in the Tropics: Challenges and opportunities. In Proceedings of the 3rd World Congress on Genetics Applied to Livestock Production, Lincoln, NE, USA, 16–22 July 1986; pp. 283–329.
3. Prayaga, K.C.; Henshall, J.M. Adaptability in tropical beef cattle: Genetic parameters of growth, adaptive and temperament traits in a crossbred population. *Aust. J. Exp. Agric.* **2005**, *45*, 971–983. [CrossRef]
4. Wolcott, M.L.; Johnston, D.J.; Barwick, S.A.; Iker, C.L.; Thompson, H.M.; Burrow, H.M. Genetics of meat quality and carcass traits and the impact of tenderstretching in two tropical beef genotypes. *Anim. Prod. Sci.* **2009**, *49*, 383–398. [CrossRef]
5. Meat and Livestock Australia Ltd. *Tropical Beef Production Manual*; Meat and Livestock Australia Ltd.: Sydney, Australia, 2011.
6. Burrow, H.M. Importance of adaptation and genotype x environment interactions in tropical beef breeding systems. *Animal* **2012**, *6*, 729–740. [CrossRef]
7. Carrasco-García, A.A.; Pardío-Sedas, V.T.; León-Banda, G.G.; Ahuja-Aguirre, C.; Paredes-Ramos, P.; Hernández-Cruz, B.C.; Murillo, V.V. Effect of stress during slaughter on carcass characteristics and meat quality in tropical beef cattle. *Asian Australas. J. Anim. Sci.* **2020**, *33*, 1656–1665. [CrossRef]
8. National Geographic Society. Tropics. 2011. Available online: https://www.nationalgeographic.org/encyclopedia/tropics/ (accessed on 3 March 2021).
9. State of the Tropics. *State of the Tropics 2020 Report*; James Cook University: Townsville, Australia, 2020.
10. Arnfield, J.A. Köppen Climate Classification. Encyclopedia Britannica. 2020. Available online: https://www.britannica.com/science/Koppen-climate-classification (accessed on 24 February 2021).
11. Peel, M.C.; Finlayson, B.L.; McMahon, T.A. Updated world map of the Köppen-Geiger climate classification. *Hydrol. Earth Syst. Sci.* **2007**, *11*, 1633–1644. [CrossRef]
12. FAO. How to feed the world in 2050. In Proceedings of the Expert Meeting on How to Feed the World in 2050, Rome, Italy, 24–26 June 2009.
13. Setshwaelo, L.L. Beef cattle breeding in the tropics. In *Dairy Cattle Genetics and Breeding, Adaptation, Conservation XIV, Proceedings of the 4th World Congress of Genetics Applied to Livestock Production, Edinburgh, UK, 23–27 July 1990*; Organizing Committee: Edinburgh, UK, 1990; pp. 349–359.
14. Eberhard, D.; Simons, G.; Fennig, C. *Ethnologue: Languages of the world*, 24th ed.; SIL International: Dallas, TX, USA, 2021.
15. Scholtz, M.M.; McManus, C.; Okeyo, A.M.; Theunissen, A. Opportunities for beef production in developing countries of the southern hemisphere. *Livest. Sci.* **2011**, *142*, 195–202. [CrossRef]
16. Cooke, R.F.; Daigle, C.L.; Moriel, P.; Smith, S.B.; Tedeschi, L.O.; Vendramini, J.M.B. Cattle adapted to tropical and subtropical environments: Social, nutritional, and carcass quality considerations. *J. Anim. Sci.* **2020**, *98*, skaa014. [CrossRef]
17. Elferink, E.V.; Nonhebel, S. Variations in land requirements for meat production. *J. Clean Prod.* **2007**, *15*, 1778–1786. [CrossRef]
18. Preston, T.R. A strategy for cattle production in the tropics. *World Anim. Rev.* **1977**, *21*, 11–17.
19. Preston, T.R. Sustainable systems of intensive livestock production for the humid tropics using local resources. *BSAP Occas. Publ.* **1993**, *16*, 101–105. [CrossRef]
20. Ferraz, J.B.S.; de Felício, P.E. Production systems—An example from Brazil. *Meat Sci.* **2010**, *84*, 238–243. [CrossRef]
21. Meat and Livestock Australia Ltd. Meat Standards Australia Beef Information Kit. 2011. Available online: https://www.mla.com.au/globalassets/mla-corporate/marketing-beef-and-lamb/msa_tt_beefinfokit_jul13_lr.pdf (accessed on 20 February 2021).
22. Strother, G.R.; Burns, E.C.; Smart, L.I. Resistance of purebred Brahman, Hereford, and Brahman times Hereford crossbred cattle to the lone star tick, *Amblyomma americanum* (Acarina: Ixodidae). *J. Med. Entomol.* **1974**, *11*, 559–563.
23. Frisch, J.E.; Vercoe, J.E. Food intake, eating rate, weight gains, metabolic rate and efficiency of feed utilization in *Bos taurus* and *Bos indicus* crossbred cattle. *Anim. Sci.* **1977**, *25*, 343–358. [CrossRef]
24. Turner, J.W. Genetic and biological aspects of Zebu adaptability. *J. Anim. Sci.* **1980**, *50*, 1201–1205. [CrossRef] [PubMed]
25. Murray, M.; Trail, J.C.M.; Davis, C.E.; Black, S.J. Genetic resistance to African trypanosomiasis. *J. Infect. Dis.* **1984**, *149*, 311–319. [CrossRef]
26. Felius, M. *Cattle Breeds: An Encyclopedia*; Misset: Doetinchem, The Netherlands, 1995.
27. Rege, J.E.O.; Tawah, C.L. The state of African cattle genetic resources II. Geographical distribution, characteristics and uses of present-day breeds and strains. *Anim. Genet. Resour. Inf.* **1999**, *26*, 1–25. [CrossRef]
28. Hansen, P.J. Physiological and cellular adaptations of zebu cattle to thermal stress. *Anim. Reprod. Sci.* **2004**, *82*, 349–360. [CrossRef] [PubMed]

29. Costa, D.F.A.; Quigley, S.P.; Isherwood, P.; McLennan, S.R.; Sun, X.Q.; Gibbs, S.J.; Poppi, D.P. Small differences in biohydrogenation resulted from the similar retention times of fluid in the rumen of cattle grazing wet season C3 and C4 forage species. *Anim. Feed Sci. Technol.* **2019**, *253*, 101–112. [CrossRef]
30. Mattioli, R.C.; Pandey, V.S.; Murray, M.; Fitzpatrick, J.L. Immunogenetic influences on tick resistance in African cattle with particular reference to trypanotolerant N'Dama (*Bos taurus*) and trypanosusceptible Gobra zebu (*Bos indicus*) cattle. *Acta Trop.* **2000**, *75*, 263–277. [CrossRef]
31. Renaudeau, D.; Collin, A.; Yahav, S.; de Basilio, V.; Gourdine, J.L.; Collier, R.J. Adaptation to hot climate and strategies to alleviate heat stress in livestock production. *Animal* **2012**, *6*, 707–728. [CrossRef] [PubMed]
32. Wheeler, T.L.; Cundiff, L.V.; Koch, R.M. Effect of marbling degree on beef palatability in *Bos taurus* and *Bos indicus* cattle. *J. Anim. Sci.* **1994**, *72*, 3145–3151. [CrossRef] [PubMed]
33. Hammond, A.C.; Olson, T.A.; Chase, C.C., Jr.; Bowers, E.J.; Randel, R.D.; Murphy, C.N.; Vogt, D.W.; Tewolde, A. Heat tolerance in two tropically adapted *Bos taurus* breeds, Senepol and Romosinuano, compared with Brahman, Angus, and Hereford cattle in Florida. *J. Anim. Sci.* **1996**, *74*, 295–303. [CrossRef]
34. Elzo, M.A.; Johnson, D.D.; Wasdin, J.G.; Driver, J.D. Carcass and meat palatability breed differences and heterosis effects in an Angus–Brahman multibreed population. *Meat Sci.* **2012**, *90*, 87–92. [CrossRef] [PubMed]
35. Gama, L.T.; Bressan, M.C.; Rodrigues, E.C.; Rossato, L.V.; Moreira, O.C.; Alves, S.P.; Bessa, R.J.B. Heterosis for meat quality and fatty acid profiles in crosses among *Bos indicus* and *Bos taurus* finished on pasture or grain. *Meat Sci.* **2013**, *93*, 98–104. [CrossRef] [PubMed]
36. Prayaga, K.C.; Corbet, N.J.; Johnston, D.J.; Wolcott, M.L.; Fordyce, G.; Burrow, H.M. Genetics of adaptive traits in heifers and their relationship to growth, pubertal and carcass traits in two tropical beef cattle genotypes. *Anim. Prod. Sci.* **2009**, *49*, 413–425. [CrossRef]
37. Riley, D.G.; Chase, C.C., Jr.; Coleman, S.W.; Olson, T.A. Genetic assessment of rectal temperature and coat score in Brahman, Angus, and Romosinuano crossbred and straightbred cows and calves under subtropical summer conditions. *Livest. Sci.* **2012**, *148*, 109–118. [CrossRef]
38. Riley, D.G.; Burke, J.M.; Chase, C.C., Jr.; Coleman, S.W. Genetic effects for reproductive performance of straightbred and crossbred Romosinuano and Angus cows in a temperate zone. *Livest. Sci.* **2015**, *180*, 22–26. [CrossRef]
39. Rocha, J.F.; Martínez, R.; López-Villalobos, N.; Morris, S.T. Tick burden in *Bos taurus* cattle and its relationship with heat stress in three agroecological zones in the tropics of Colombia. *Parasites Vectors* **2019**, *12*, 73. [CrossRef]
40. Pryce, J.E.; Royal, M.D.; Garnsworthy, P.C.; Mao, I.L. Fertility in the high-producing dairy cow. *Livest. Prod. Sci.* **2004**, *86*, 125–135. [CrossRef]
41. de Roos, A.P.W.; Hayes, B.J.; Spelman, R.J.; Goddard, M.E. Linkage disequilibrium and persistence of phase in Holstein–Friesian, Jersey and Angus cattle. *Genetics* **2008**, *179*, 1503–1512. [CrossRef] [PubMed]
42. Mwai, O.; Hanotte, O.; Kwon, Y.J.; Cho, S. African indigenous cattle: Unique genetic resources in a rapidly changing world. *Asian Australas. J. Anim. Sci.* **2015**, *28*, 911–921. [CrossRef] [PubMed]
43. Anderson, D.M.; Estell, R.E.; Gonzalez, A.L.; Cibils, A.F.; Torell, L.A. Criollo cattle: Heritage genetics for arid landscapes. *Rangelands* **2015**, *37*, 62–67. [CrossRef]
44. Ramírez-Restrepo, C.A.; Vera, R.R.; Rao, I.M. Dynamics of animal performance, and estimation of carbon footprint of two breeding herds grazing native neotropical savannas in eastern Colombia. *Agric. Ecosyst. Environ.* **2019**, *281*, 35–46. [CrossRef]
45. Rege, J.E.O. The state of African cattle genetic resources. I. Classification framework and identification of threatened and extinct breeds. *Anim. Genet. Resour. Inf.* **1999**, *25*, 1–26. [CrossRef]
46. Peel, D.S.; Johnson, R.J.; Mathews, K.H. *Cow-Calf Beef Production in Mexico*; USDA: Washington, DC, USA, 2010.
47. Parra-Bracamonte, G.M.; Lopez-Villalobos, N.; Morris, S.T.; Vázquez-Armijo, J.F. An overview on production, consumer perspectives and quality assurance schemes of beef in Mexico. *Meat Sci.* **2020**, *170*, 108239. [CrossRef]
48. Rubio Lozano, M.S.; Braña Varela, D.; Méndez Medina, R.D. *Carne de res Mexicana*; Folleto Técnico No. 15; Instituto Nacional de Investigaciones Forestales, Agrícolas y Pecuarias, Ajuchitlán: Colón, Querétaro, Mexico, 2012.
49. Poppi, D.P.; Quigley, S.P.; da Silva, T.A.C.C.; McLennan, S.R. Challenges of beef cattle production from tropical pastures. *Rev. Bras. Zootec.* **2018**, *47*, e20160419. [CrossRef]
50. Ash, A.; Hunt, L.; McDonald, C.; Scanlan, J.; Bell, L.; Cowley, R.; Watson, I.; Mclvor, J.; MacLeod, N. Boosting the productivity and profitability of northern Australian beef enterprises: Exploring innovation options using simulation modelling and systems analysis. *Agric. Syst.* **2015**, *139*, 50–65. [CrossRef]
51. O'Reagain, P.; Bushell, J.; Holmes, B. Managing for rainfall variability: Long-term profitability of different grazing strategies in a northern Australian tropical savanna. *Anim. Prod. Sci.* **2011**, *51*, 210–224. [CrossRef]
52. Agus, A.; Widi, T.S.M. Current situation and future prospects for beef cattle production in Indonesia—A review. *Asian Australas. J. Anim. Sci.* **2018**, *31*, 976–983. [CrossRef]
53. Teye, G.A.; Sunkwa, W.K. Carcass characteristics of tropical beef cattle breeds (West African Shorthorn, sanga and zebu) in Ghana. *Afr. J. Food Agric. Nutr. Dev.* **2010**, *10*, 2866–2883. [CrossRef]
54. Rege, J.E.O.; Aboagye, G.S.; Tawah, C.L. Shorthorn cattle of West and Central Africa IV. Production characteristics. *World Anim. Rev.* **1994**, *78*, 33–48.

55. Asizua, D.; Mpairwe, D.; Kabi, F.; Mutetikka, D.; Hvelplund, T.; Weisbjerg, M.R.; Madsen, J. Effects of grazing and feedlot finishing duration on the performance of three beef cattle genotypes in Uganda. *Livest. Sci.* **2017**, *199*, 25–30. [CrossRef]
56. Nfor, B.M.; Corazzin, M.; Fonteh, F.A.; Aziwo, N.T.; Galeotti, M.; Piasentier, E. Quality and safety of beef produced in Central African Sub-region. *Ital. J. Anim. Sci.* **2014**, *214*, 392–397. [CrossRef]
57. Deffo, V.; Pamo, E.T.; Tchotsoua, M.; Lieugomg, M.; Arene, C.J.; Nwagbo, E.C. Determination of the Critical Period for Cattle Farming in Cameroon. 2011. Available online: http://academicjournals.org/article/article1379432605_Deffo%20et%20al.pdf (accessed on 24 February 2021).
58. Salifou, C.F.A.; Dahouda, M.; Houaga, I.; Picard, B.; Hornick, J.L.; Micol, D.; Farougou, S.; Mensah, G.A.; Clinquart, A.; Youssao, A.K.I. Muscle characteristics, meat tenderness and nutritional qualities traits of Borgou, Lagunaire and Zebu Fulani bulls raised on natural pasture in Benin. *Int. J. Anim. Vet. Adv.* **2013**, *5*, 143–155. [CrossRef]
59. Ball, B.; Johnson, E.R. The influence of breed and sex on saleable beef yield. *Aust. J. Exp. Agric.* **1989**, *29*, 483–487. [CrossRef]
60. Wythes, J.R.; Shorthose, W.R.; Dodt, R.M.; Dickinson, R.F. Carcass and meat quality of *Bos indicus* x *Bos taurus* and *Bos taurus* cattle in northern Australia. *Aust. J. Exp. Agric.* **1989**, *29*, 757–763. [CrossRef]
61. Wythes, J.R.; Shorthose, W.R. Chronological age and dentition effects on carcass and meat quality of cattle in northern Australia. *Aust. J. Exp. Agric.* **1991**, *31*, 145–152. [CrossRef]
62. Newman, S.; Burrow, H.M.; Shepherd, R.K.; Bindon, B.M. Meat quality traits of grass-and grain-finished Brahman crosses for domestic and export markets. *Proc. Assoc. Adv. Anim. Breed. Genet.* **1999**, *13*, 235–238.
63. Newman, S.; Burrow, H.M.; Shepherd, R.K.; Bindon, B.M. Carcass yield traits of grass-and grain-finished Brahman crosses for domestic and export markets. *Proc. Assoc. Adv. Anim. Breed. Genet.* **1999**, *13*, 231–234.
64. Schatz, T.J.; Thomas, S.; Geesink, G. Comparison of the growth and meat tenderness of Brahman and F1 Senepol × Brahman steers. *Anim. Prod. Sci.* **2014**, *54*, 1867–1870. [CrossRef]
65. Norman, G.A.; de Felicio, P.E. Effects of breed and nutrition on the productive traits of Zebu, Charolais and crossbreed beef cattle in south-east Brazil—Part 1: Body and gross carcase composition. *Meat Sci.* **1981**, *5*, 425–438. [CrossRef]
66. Maggioni, D.; de Marques, J.A.; Rotta, P.P.; Perotto, D.; Ducatti, T.; Visentainer, J.V.; do Prado, I.N. Animal performance and meat quality of crossbred young bulls. *Livest. Sci.* **2010**, *127*, 176–182. [CrossRef]
67. Pflanzer, S.B.; de Felício, P.E. Moisture and fat content, marbling level and color of boneless rib cut from Nellore steers varying in maturity and fatness. *Meat Sci.* **2011**, *87*, 7–11. [CrossRef]
68. Bressan, M.C.; Rodrigues, E.C.; Rossato, L.V.; Ramos, E.M.; da Gama, L.T. Physicochemical properties of meat from *Bos taurus* and *Bos indicus*. *Rev. Bras. Zootec.* **2011**, *40*, 1250–1259. [CrossRef]
69. Diniz, F.B.; Villela, S.D.J.; Mourthé, M.H.F.; Paulino, P.V.R.; Boari, C.A.; Ribeiro, J.S.; Barroso, J.A.; Pires, A.V.; Martins, P.G.M.A. Evaluation of carcass traits and meat characteristics of Guzerat-crossbred bulls. *Meat Sci.* **2016**, *112*, 58–62. [CrossRef]
70. Pereira, A.S.C.; Baldi, F.; Sainz, R.D.; Utembergue, B.L.; Chiaia, H.L.J.; Magnabosco, C.U.; Manicardi, F.R.; Araujo, F.R.C.; Guedes, C.F.; Margarido, R.C.; et al. Growth performance, and carcass and meat quality traits in progeny of Poll Nellore, Angus and Brahman sires under tropical conditions. *Anim. Prod. Sci.* **2015**, *55*, 1295–1302. [CrossRef]
71. Belew, J.B.; Brook, J.C.; McKenna, D.R.; Savell, J.W. Warner-Bratzler shear evaluations of 40 bovine muscles. *Meat Sci.* **2003**, *64*, 507–512. [CrossRef]
72. Rodas-González, A.; Huerta-Leidenz, N.; Jerez-Timaure, N.; Miller, M.F. Establishing tenderness thresholds of Venezuelan beef steaks using consumer and trained sensory panels. *Meat Sci.* **2009**, *83*, 218–223. [CrossRef]
73. Rodriguez, J.; Unruh, J.; Villarreal, M.; Murillo, O.; Rojas, S.; Camacho, J.; Jaeger, J.; Reinhardt, C. Carcass and meat quality characteristics of Brahman cross bulls and steers finished on tropical pastures in Costa Rica. *Meat Sci.* **2014**, *96*, 1340–1344. [CrossRef]
74. Chávez, A.; Pérez, E.; Rubio, M.S.; Méndez, R.D.; Delgado, E.J.; Díaz, D. Chemical composition and cooking properties of beef forequarter muscles of Mexican cattle from different genotypes. *Meat Sci.* **2012**, *91*, 160–164. [CrossRef]
75. de Nadai Bonin, M.; Pedrosa, V.B.; e Silva, S.D.L.; Bünger, L.; Ross, D.; da Costa Gomes, R.; de Almeida Santana, M.H.; de Córdova Cucco, D.; de Rezende, F.M.; Ítavo, L.C.V.; et al. Genetic parameters associated with meat quality of Nellore cattle at different anatomical points of longissimus: Brazilian standards. *Meat Sci.* **2021**, *171*, 108281. [CrossRef]
76. Civit, G.S.; Méndez, M.R.D.; Iturbe, C.F.; Rosiles, R.; Casis, L.; Rubio, L.M.S. Chemical composition, meat quality and consumer acceptability in Mexican (Guadalajara, Chihuahua and Veracruz) retail beef. In Proceedings of the 51st International Congress of Meat Science and Technology, Baltimore, MA, USA, 7–12 August 2005; pp. 424–428.
77. Santrich-Vacca, D.; Cianzio, D.; Rivera, A.; Casas, A.; Macchiavelli, R. Quality and chemical composition of beef from cattle of two age groups in Puerto Rico. *J. Agric. Univ. Puerto Rico* **2013**, *97*, 57–74. [CrossRef]
78. Vazquez-Mendoza, O.V.; Aranda-Osorio, G.; Huerta-Bravo, M.; Kholif, A.E.; Elghandour, M.M.Y.; Salem, A.Z.M.; Maldonado-Simán, E. Carcass and meat properties of six genotypes of young bulls finished under feedlot tropical conditions of Mexico. *Anim. Prod. Sci.* **2017**, *57*, 1186–1192. [CrossRef]
79. Jerez-Timaure, N.; Huerta-Leidenz, N. Effects of breed type and supplementation during grazing on carcass traits and meat quality of bulls fattened on improved savannah. *Livest. Sci.* **2009**, *121*, 219–226. [CrossRef]
80. Nassu, R.T.; Tullio, R.R.; Berndt, A.; Francisco, V.C.; Diesel, T.A.; Alencar, M.M. Effect of the genetic group, production system and sex on the meat quality and sensory traits of beef from crossbred animals. *Trop. Anim. Health Prod.* **2017**, *49*, 1289–1294. [CrossRef] [PubMed]

81. Pflanzer, S.B.; de Felício, P.E. Effects of teeth maturity and fatness of Nellore (*Bos indicus*) steer carcasses on instrumental and sensory tenderness. *Meat Sci.* **2009**, *83*, 697–701. [CrossRef] [PubMed]
82. Van Cleef, E.H.C.B.; D'Áurea, A.P.; Fávaro, V.R.; van Cleef, F.O.S.; Barducci, R.S.; Almeida, M.T.C.; Machado Neto, O.R.; Ezequiel, J.M.B. Effects of dietary inclusion of high concentrations of crude glycerin on meat quality and fatty acid profile of feedlot fed Nellore bulls. *PLoS ONE* **2017**, *12*, e0179830. [CrossRef]
83. Lage, J.F.; Berchielli, T.T.; San Vito, E.; Silva, R.A.; Ribeiro, A.F.; Reis, R.A.; Dallantonia, L.R.; Simonetti, L.R.; Delevatti, L.M.; Machado, M. Fatty acid profile, carcass and meat quality traits of young Nellore bulls fed crude glycerin replacing energy sources in the concentrate. *Meat Sci.* **2014**, *96*, 1158–1164. [CrossRef] [PubMed]
84. Méndez, R.D.; Meza, C.O.; Berruecos, J.M.; Garcés, P.; Delgado, E.J.; Rubio, M.S. A survey of beef carcass quality and quantity attributes in Mexico. *J. Anim. Sci.* **2009**, *87*, 3782–3790. [CrossRef]
85. Huerta-Leidenz, N.; Jerez Timaure, N. Eating quality of meat from bovines in Venezuela: A review. *Rev. Fac. Agron. LUZ* **2020**, *37*, 169–206.
86. Mach, N.; Bach, A.; Realini, C.E.; Font i Furnols, M.; Velarde, A.; Devant, M. Burdizzo pre-pubertal castration effects on performance, behaviour, carcass characteristics, and meat quality of Holstein bulls fed high-concentrate diets. *Meat Sci.* **2009**, *81*, 329–334. [CrossRef] [PubMed]
87. Seideman, S.C.; Cross, H.R.; Oltjen, R.R.; Schanbacher, B.D. Utilization of the intact male for red meat production: A review. *J. Anim. Sci.* **1982**, *55*, 826–840. [CrossRef]
88. Huerta-Leidenz, N.O.; Rodas-González, A.; Jerez Timaure, N.; Arispe, M.; Rivero, J.M. Efecto de la clase de machos bovinos y el peso de la canal sobre el rendimiento comercial en cortes Venezolanos. *Rev. Cient.* **1999**, *9*, 33–39.
89. Field, R.A. Effect of castration on meat quality and quantity. *J. Anim. Sci.* **1971**, *32*, 849–858. [CrossRef]
90. Seideman, S.C.; Cross, H.R.; Crouse, J.D. Carcass characteristics, sensory properties and mineral content of meat from bulls and steers. *J. Food Qual.* **1989**, *11*, 497–507. [CrossRef]
91. Zorrilla-Ríos, J.M.; Lancaster, P.A.; Goad, C.L.; Horn, G.W.; Hilton, G.G.; Galindo, J.G. Quality evaluation of beef carcasses produced under tropical conditions of México. *J. Anim. Sci.* **2013**, *91*, 477–482. [CrossRef] [PubMed]
92. Delgado, E.J.; Rubio, M.S.; Iturbe, F.A.; Méndez, R.D.; Cassís, L.; Rosiles, R. Composition and quality of Mexican and imported retail beef in Mexico. *Meat Sci.* **2005**, *69*, 465–471. [CrossRef] [PubMed]
93. Duckett, S.K.; Neel, J.P.S.; Lewis, R.M.; Fontenot, J.P.; Clapham, W.M. Effects of forage species or concentrate finishing on animal performance, carcass and meat quality. *J. Anim. Sci.* **2013**, *91*, 1454–1467. [CrossRef]
94. Leheska, J.M.; Thompson, L.D.; Howe, J.C.; Hentges, E.; Boyce, J.; Brooks, J.C.; Shriver, B.; Hoover, L.; Miller, M.F. Effects of conventional and grass-feeding systems on the nutrient composition of beef. *J. Anim. Sci.* **2008**, *86*, 3575–3585. [CrossRef]
95. Huerta-Leidenz, N.; Hernández, O.; Rodas-González, A.; Ordóñez, V.; Pargas, J.H.L.; Rincón, E.; del Villar, A.; Bracho, B. Body weight and carcass dressing as affected by sex class, breed type, muscle thickness, age and provenance of Venezuelan cattle. *NACAMEH* **2013**, *7*, 75–96. [CrossRef]
96. Uzcátegui Bracho, S.; Arenas de Moreno, L.; Jerez-Timaure, N.; Huerta-Leidenz, N.; Byers, F.M. Bull production factors affecting proximate and mineral composition of cooked *longissimus* steaks. In Proceedings of the 50th International Congress of Meat Science and Technology, Helsinki, Finland, 8–13 August 2004.
97. Ngapo, T.M.; Braña Varela, D.; Rubio Lozano, M. Mexican consumers at the point of meat purchase. Beef choice. *Meat Sci.* **2017**, *134*, 34–43. [CrossRef]
98. Bautista-Martínez, Y.; Hernández-Mendo, O.; Crosby-Galván, M.M.; Joaquin-Cancino, S.; Albarrán, M.R.; Salinas-Chavira, J.; Granados-Rivera, L.D. Physicochemical characteristics and fatty acid profile of beef in Northeastern Mexico: Grazing vs feedlot systems. *CyTA J. Food* **2020**, *18*, 147–152. [CrossRef]
99. Valero Leal, K.; Huerta-Leidenz, N.; Uzcátegui Bracho, S.; de Moreno, L.A.; Ettiene, G.; Buscema, I. Comparative analyses of proximal and fatty acid composition of grass-fed meat from water buffaloes (*Bubalus bubalis*) and zebu-influenced beef cattle at 24 months of age. In Proceedings of the 46th International Congress of Meat Science and Technology, Buenos Aires, Argentina, 27 August–1 September 2000; pp. 74–75.
100. Uzcátegui, B.S.; Huerta-Leidenz, N.; de Moreno, L.A.; Colina, G.; Jerez-Timaure, N. Contenido de humedad, lípidos totales y ácidos grasos del músculo longissimus crudo de bovinos en Venezuela. *Arch. Latinoam. Nutr.* **1999**, *49*, 171–180. [PubMed]
101. Uzcátegui-Bracho, S.; Arenas de Moreno, L.; Jerez-Timaure, N.; Huerta-Leidenz, N.; Giuffrida-Mendoza, M.; Ortega, J. Principal component analysis to characterise the chemical composition of beef according to age and gender. In Proceedings of the 52nd International Congress of Meat Science and Technology, Dublin, Ireland, 13–18 August 2006; pp. 155–156.
102. Arenas de Moreno, L.; Vidal, A.; Huerta-Sánchez, D.; Navas, Y.; Uzcátegui-Bracho, S.; Huerta-Leidenz, N. Análisis comparativo proximal y de minerales entre carnes de iguana, pollo y res. *Arch. Latinoam. Nutr.* **2000**, *4*, 409–415.
103. Uzcátegui-Bracho, S.; Giuffrida-Mendoza, M.; Arenas de Moreno, L.; Jerez-Timaure, N. Contenido proximal, lípidos y colesterol de las carnes de res, cerdo y pollo obtenidas de expendios carniceros de la zona sur de Maracaibo. *Rev. Venez. de Tecnol. y Soc.* **2010**, *3*, 13–29.
104. Riyanto, J.; Nuhriawangsa, A.M.P.; Pramono, A.; Widyawati, S.D.; Purnika, D. The chemical quality of silverside Simental Ongole crossbred meat at various roasting temperatures. In Proceedings of the International Conference on Science and Applied Science, Surakarta City, Indonesia, 12 May 2018; AIP Publishing LLC.: Melville, NY, USA, 2014; p. 020042.

105. Das Graças Padre, R.; Aricetti, J.A.; Gomes, S.T.M.; de Goes, R.H.d.T.B.; Moreira, F.B.; do Prado, I.N.; Visentainer, J.V.; de Souza, N.E.; Matsushita, M. Analysis of fatty acids in *Longissimus* muscle of steers of different genetic breeds finished in pasture systems. *Livest. Sci.* **2007**, *110*, 57–63. [CrossRef]
106. Ojong, W.B.; Saccà, E.; Corazzin, M.; Sepulcri, A.; Piasentier, E. Body and meat characteristics of young bulls from Zebu Goudali of Cameroon and its crosses with the Italian Simmental. *Ital. J. Anim. Sci.* **2018**, *17*, 240–249. [CrossRef]
107. Ruiz, M.R.; Matsushita, M.; Visentainer, J.V.; Hernandez, J.A.; Ribeiro, E.L.d.A.; Shimokomaki, M.; Reeves, J.J.; de Souza, N.E. Proximate chemical composition and fatty acid profiles of *Longissimus thoracis* from pasture fed LHRH immunocastrated, castrated and intact *Bos indicus* bulls. *S. Afr. J. Anim. Sci.* **2005**, *35*, 13–18.
108. Aricetti, J.A.; Rotta, P.P.; Prado, R.M.D.; Perotto, D.; Moletta, J.L.; Matsushita, M.; Prado, I.N.D. Carcass characteristics, chemical composition and fatty acid profile of longissimus muscle of bulls and steers finished in a pasture system bulls and steers finished in pasture systems. *Asian Australas. J. Anim. Sci.* **2008**, *21*, 1441–1448. [CrossRef]
109. Giuffrida-Mendoza, M.; de Moreno, L.A.; Huerta-Leidenz, N.; Uzcátegui-Bracho, S.; Valero-Leal, K.; Romero, S.; Rodas-González, A. Cholesterol and fatty acid composition of *longissimus thoracis* from water buffalo (*Bubalus bubalis*) and Brahman-influenced cattle raised under savannah conditions. *Meat Sci.* **2015**, *106*, 44–49. [CrossRef] [PubMed]
110. Ilavarasan, R.; Abraham, R.J.J.; Appa Rao, V. Influence of age on meat quality characteristics and proximate composition of cattle meat of Tamil Nadu. *J. Environ. Biol. Sci.* **2016**, *30*, 185–187.
111. Arenas de Moreno, L.; Giuffrida-Mendoza, M.; Bulmes, L.; Uzcátegui-Bracho, S.; Huerta-Leidenz, N.; Jerez-Timaure, N. Effect of strategic supplementation, implant regime and gender on the proximate and mineral composition of raw and cooked bovine meat. *Rev. Cient.* **2008**, *18*, 65–72.
112. Uzcátegui-Bracho, S.; Rodas-González, A.; Hennig, K.; Arenas de Moreno, L.; Leal, M.; Vergara-López, J.; Huerta-Leidenz, N. Efecto de la suplementación a pastoreo sobre la composición proximal, mineral y contenido de colesterol del músculo longissimus dorsi de novillos Criollo Limonero. In Proceedings of the XX Reunión de ALPA, Cusco, Peru, 22–25 October 2007.
113. Uzcátegui-Bracho, S.; Rodas-González, A.; Hennig, K.; Arenas de Moreno, L.; Leal, M.; Vergara-López, J.; Jerez-Timaure, N. Composición proximal, mineral y contenido de colesterol del músculo *Longissimus dorsi* de novillos Criollo Limonero suplementados a pastoreo. *Rev. Cient.* **2008**, *18*, 589–594.
114. Montero-Lagunes, M.; Juárez-Lagunes, F.I.; García-Galindo, H.S. Perfil de ácidos grasos en carne de toretes Europeo x Cebú finalizados en pastoreo y en corral. *Rev. Mex. Cienc. Pecu.* **2011**, *2*, 137–149.
115. Huerta-Sánchez, D.; Villa, V.; Arenas de Moreno, L.; Huerta-Leidenz, N.; Guiffrida, M.; Rodas-González, A. Comparison of imported vs. domestic beef cuts for restaurant use in Venezuela, II Marbling level, proximate and mineral composition. In Proceedings of the 50th International Congress of Meat Science and Technology, Helsinki, Finland, 8–13 August 2004; pp. 1091–1094.
116. Huerta-Montauti, D.; Villa, V.; Arenas de Moreno, L.; Rodas-González, A.; Giuffrida-Mendoza, M.; Huerta-Leidenz, N. Proximate and mineral composition of imported versus domestic beef cuts for restaurant use in Venezuela. *J. Muscle Foods* **2007**, *18*, 237–252. [CrossRef]
117. Bressan, M.C.; Rodrigues, E.C.; de Paula, M.d.L.; Ramos, E.M.; Portugal, P.V.; Silva, J.S.; Bessa, R.B.; Telo da Gama, L. Differences in intramuscular fatty acid profiles among *Bos indicus* and crossbred *Bos Taurus* × *Bos indicus* bulls finished on pasture or with concentrate feed in Brazil. *Ital. J. Anim. Sci.* **2016**, *15*, 10–21. [CrossRef]
118. Wheeler, T.L.; Miller, R.K.; Savell, J.W.; Cross, H.R. Palatability of chilled and frozen beef steaks. *J. Food Sci.* **1990**, *55*, 301–304. [CrossRef]
119. Miller, M.F.; Cross, H.R.; Lunt, D.K.; Smith, S.B. Lipogenesis in acute and 48-hour cultures of bovine intramuscular and subcutaneous adipose tissue explants. *J. Anim. Sci.* **1991**, *69*, 162–170. [CrossRef]
120. Campbell, E.M.G.; Sanders, J.O.; Lunt, D.K.; Gill, C.A.; Taylor, J.F.; Davis, S.K.; Riley, D.G.; Smith, S.B. Adiposity, lipogenesis, and fatty acid composition of subcutaneous and intramuscular adipose tissues of Brahman and Angus crossbred cattle. *J. Anim. Sci.* **2016**, *94*, 1415–1425. [CrossRef]
121. Huerta-Léidenz, N.; Ríos, G. La castración del bovino a diferentes estadios de su crecimiento. II. Las características de la canal. Una revisión. *Rev. Fac. Agron. LUZ* **1993**, *10*, 163–187.
122. Dikeman, M.E. Effects of metabolic modifiers on carcass traits and meat quality. *Meat Sci.* **2007**, *77*, 121–135. [CrossRef] [PubMed]
123. Wood, J.D.; Enser, M.; Fisher, A.V.; Nute, G.R.; Sheard, P.R.; Richardson, R.I.; Hughes, S.I.; Whittington, F.M. Fat deposition, fatty acid composition and meat quality: A review. *Meat Sci.* **2008**, *78*, 343–358. [CrossRef]
124. Webb, E.C.; Erasmus, L.J. The effect of production system and management practices on the quality of meat products from ruminant livestock. *S. Afr. J. Anim. Sci.* **2013**, *43*, 413–423. [CrossRef]
125. Nfor, B.M.; Corazzin, M.; Fonteh, F.A.; Sepulcri, A.; Aziwo, N.T.; Piasentier, E. Fatty acid profile of zebu beef cattle from the Central African sub-region. *S. Afr. J. Anim. Sci.* **2014**, *44*, 148–154. [CrossRef]
126. Vlaeminck, B.; Fievez, V.; Cabrita, A.R.J.; Fonseca, A.J.M.; Dewhurst, R.J. Factors affecting odd-and branched-chain fatty acids in milk: A review. *Anim. Feed Sci. Technol.* **2006**, *131*, 389–417. [CrossRef]
127. Itoh, M.; Johnson, C.B.; Cosgrove, G.P.; Muir, P.D.; Purchas, R.W. Intramuscular fatty acid composition of neutral and polar lipids for heavy-weight Angus and Simmental steers finished on pasture or grain. *J. Sci. Food Agric.* **1999**, *79*, 821–827. [CrossRef]
128. Simopoulos, A.P. Omega-6/omega-3 essential fatty acid ratio and chronic diseases. *Food Rev. Int.* **2004**, *20*, 77–90. [CrossRef]

129. Piasentier, E.; Bovolenta, S.; Moioli, B.; Orrù, L.; Valusso, R.; Corazzin, M. Fatty acid composition and sensory properties of Italian Simmental beef as affected by gene frequency of Montbéliarde origin. *Meat Sci.* **2009**, *83*, 543–550. [CrossRef]
130. Corazzin, M.; Bovolenta, S.; Sepulcri, A.; Piasentier, E. Effect of whole linseed addition on meat production and quality of Italian Simmental and Holstein young bulls. *Meat Sci.* **2012**, *90*, 99–105. [CrossRef] [PubMed]
131. Ripoll, G.; Blanco, M.; Albertí, P.; Panea, B.; Joy, M.; Casasús, I. Effect of two Spanish breeds and diet on beef quality including consumer preferences. *J. Sci. Food Agric.* **2014**, *94*, 983–992. [CrossRef]
132. Enser, M.; Richardson, R.I.; Wood, J.D.; Gill, B.P.; Sheard, P.R. Feeding linseed to increase the *n*-3 PUFA of pork: Fatty acid composition of muscle, adipose tissue, liver and sausages. *Meat Sci.* **2000**, *55*, 201–212. [CrossRef]
133. Yang, A.; Lanari, M.C.; Brewster, M.; Tume, R.K. Lipid stability and meat colour of beef from pasture- and grain-fed cattle with or without vitamin E supplement. *Meat Sci.* **2002**, *60*, 41–50. [CrossRef]
134. Madron, M.S.; Peterson, D.G.; Dwyer, D.A.; Corl, B.A.; Baumgard, L.H.; Beermann, D.H.; Bauman, D.E. Effect of extruded full-fat soybeans on conjugated linoleic acid content of intramuscular, intermuscular, and subcutaneous fat in beef steers. *J. Anim. Sci.* **2002**, *80*, 1135–1143. [CrossRef] [PubMed]
135. Realini, C.E.; Duckett, S.K.; Brito, G.W.; Dalla Rizza, M.; De Mattos, D. Effect of pasture vs. concentrate feeding with or without antioxidants on carcass characteristics, fatty acid composition, and quality of Uruguayan beef. *Meat Sci.* **2004**, *66*, 567–577. [CrossRef]
136. Giuffrida de Mendoza, M.; Arenas de Moreno, L.; Huerta-Leidenz, N.; Uzcátegui-Bracho, S.; Beriain, M.J.; Smith, G.C. Occurrence of conjugated linoleic acid in *longissimus dorsi* muscle of water buffalo (*Bubalus bubalis*) and zebu-type cattle raised under savannah conditions. *Meat Sci.* **2005**, *69*, 93–100. [CrossRef] [PubMed]
137. Kelly, M.L.; Kolver, E.S.; Bauman, D.E.; Van Amburgh, M.E.; Muller, L.D. Effect of intake of pasture on concentrations of conjugated linoleic acid in milk of lactating cows. *J. Dairy Sci.* **1998**, *81*, 1630–1636. [CrossRef]
138. Lobato, J.F.P.; Freitas, A.K.; Devincenzi, T.; Cardoso, L.L.; Tarouco, J.U.; Vieira, R.M.; Dillenburg, D.R.; Castro, I. Brazilian beef produced on pastures: Sustainable and healthy. *Meat Sci.* **2014**, *98*, 336–345. [CrossRef]
139. Montoya, C.; García, J.F.; Barahona, R. Contenido de ácidos grasos en carne de bovinos cebados en diferentes sistemas de produccíon en el trópico Colombiano. *Vitae* **2015**, *22*, 205–214. [CrossRef]
140. Eichhorn, J.M.; Coleman, L.J.; Wakayama, E.J.; Blomquist, G.J.; Bailey, C.M.; Jenkins, T.G. Effects of breed type and restricted versus ad libitum feeding on fatty acid composition and cholesterol content of muscle and adipose tissue from mature bovine females. *J. Anim. Sci.* **1986**, *63*, 781–794. [CrossRef]
141. Manner, W.; Maxwell, R.J.; Williams, J.E. Effects of dietary regimen and tissue site on bovine fatty acid profiles. *J. Anim. Sci.* **1984**, *59*, 109–121. [CrossRef]
142. Aboujaoude, C.; Pereira, A.S.C.; Feitosa, F.L.B.; de Lemos, M.V.A.; Chiaia, H.L.J.; Berton, M.P.; Peripolli, E.; de Oliveira Silva, R.M.; Ferrinho, A.M.; Mueller, L.F.; et al. Genetic parameters for fatty acids in intramuscular fat from feedlot-finished Nelore carcasses. *Anim. Prod. Sci.* **2018**, *58*, 234–243. [CrossRef]
143. Kelly, M.J.; Tume, R.K.; Newman, S.; Thompson, J.M. Genetic variation in fatty acid composition of subcutaneous fat in cattle. *Anim. Prod. Sci.* **2013**, *53*, 129–133. [CrossRef]
144. Ribeiro, D.M.; Mourato, M.P.; Almeida, A.M. Assessing mineral status in edible tissues of domestic and game animals: A review with special emphasis in tropical regions. *Trop. Anim. Health Prod.* **2019**, *51*, 1019–1032. [CrossRef]
145. Fernandes, E.A.D.N.; Sarriés, G.A.; Bacchi, M.A.; Mazola, Y.T.; Gonzaga, C.L.; Sarriés, S.R.V. Trace elements and machine learning for Brazilian beef traceability. *Food Chem.* **2020**, *333*, 127462. [CrossRef] [PubMed]
146. Huerta-Leidenz, N.; Arenas de Moreno, L.; Moron-Fuenmayor, O.; Uzcátegui-Bracho, S. Composición mineral del músculo *longissimus* crudo derivado de canales producidas y clasificadas en Venezuela. *Arch. Latinoam. Nutr.* **2003**, *53*, 96–101.
147. Giuffrida-Mendoza, M.; Arenas de Moreno, L.; Uzcátegui-Bracho, S.; Rincón-Villalobos, G.; Huerta-Leidenz, N. Mineral content of *Longissimus dorsi thoracis* from water buffalo and Zebu-influenced cattle at four comparative ages. *Meat Sci.* **2007**, *75*, 487–493. [CrossRef] [PubMed]

MDPI

Review

Enhancing the Nutritional Value of Red Meat through Genetic and Feeding Strategies

Manuel Juárez [1,*,†], Stephanie Lam [1,†], Benjamin M. Bohrer [2,†], Michael E. R. Dugan [1,†], Payam Vahmani [3,†], Jennifer Aalhus [1,†], Ana Juárez [4,†], Oscar López-Campos [1,†], Nuria Prieto [1,†] and Jose Segura [1,†]

[1] Lacombe Research and Development Centre, Agriculture and Agri-Food Canada, Lacombe, AB T4L 1W1, Canada; stephanie.lam@canada.ca (S.L.); mike.dugan@canada.ca (M.E.R.D.); Jennifer.Aalhus@outlook.com (J.A.); oscar.lopezcampos@canada.ca (O.L.-C.); Nuria.PrietoBenavides@canada.ca (N.P.); jose.seguraplaza@canada.ca (J.S.)

[2] Department of Animal Sciences, The Ohio State University, Columbus, OH 43210, USA; bohrer.13@osu.edu

[3] Department of Animal Science, University of California Davis, Davis, CA 95616, USA; pvahmani@ucdavis.edu

[4] Oceanographic Center of Cadiz, Spanish Institute of Oceanography (IEO), 11006 Cádiz, Spain; ana.juarez@ieo.es

* Correspondence: manuel.juarez@canada.ca; Tel.: +1-403-782-8118

† These authors contributed equally to this work.

Abstract: Consumption of red meat contributes to the intake of many essential nutrients in the human diet including protein, essential fatty acids, and several vitamins and trace minerals, with high iron content, particularly in meats with high myoglobin content. Demand for red meat continues to increase worldwide, particularly in developing countries where food nutrient density is a concern. Dietary and genetic manipulation of livestock can influence the nutritional value of meat products, providing opportunities to enhance the nutritional value of meat. Studies have demonstrated that changes in livestock nutrition and breeding strategies can alter the nutritional value of red meat. Traditional breeding strategies, such as genetic selection, have influenced multiple carcass and meat quality attributes relevant to the nutritional value of meat including muscle and fat deposition. However, limited studies have combined both genetic and nutritional approaches. Future studies aiming to manipulate the composition of fresh meat should aim to balance potential impacts on product quality and consumer perception. Furthermore, the rapidly emerging fields of phenomics, nutrigenomics, and integrative approaches, such as livestock precision farming and systems biology, may help better understand the opportunities to improve the nutritional value of meat under both experimental and commercial conditions.

Keywords: beef; lamb; pork; trace elements; micronutrients; fatty acids; genomics; heritability

Citation: Juárez, M.; Lam, S.; Bohrer, B.M.; Dugan, M.E.R.; Vahmani, P.; Aalhus, J.; Juárez, A.; López-Campos, O.; Prieto, N.; Segura, J. Enhancing the Nutritional Value of Red Meat through Genetic and Feeding Strategies. *Foods* **2021**, *10*, 872. https://doi.org/10.3390/foods10040872

Academic Editor: Huerta-Leidenz Nelson and Markus F. Miller

Received: 17 March 2021
Accepted: 13 April 2021
Published: 16 April 2021

1. Introduction

Meat consumption has played a substantial role in human evolution. The ability of early humans to use basic rudimentary tools, such as stones and sticks to control and use fire to cook meat products procured from hunting, as well as their access to bone marrow, led to the consumption of nutrient dense foods with higher energy and protein content. This lessened the need for large jaws/teeth and a bulky digestive system and consequently, resulted in a larger endocranial cavity allowing for brain growth to occur [1,2]. In regions of the world where humans have thrived, animal products, especially meat products, have been an important component in the human diet due to its high level of biologically available nutrients, including protein, iron, zinc and B-complex vitamins, especially B_{12} [3,4].

Despite the importance of red meat in human evolution and its high nutritional value in the human diet, the consumption of red meat (defined in this review as livestock meat from beef, pork and lamb) has received significant negative attention in both scientific studies and the popular press. This includes the increasing number of studies suggesting the association between red meat consumption and negative effects on human health [5–7] and the environment [8,9]. However, human dietary studies often have no supporting evidence of a cause-and-effect relationship between red meat consumption and negative effects on human health. Furthermore, factors other than red meat consumption, including meat processing and cooking methods, lifestyle factors [10–12], and adiposity index of the individuals [13] are known to influence negative effects attributed to red meat consumption, such as inflammation. However, as studies emerge, more evidence is revealing that the removal of red meat from the human diet may lead to negative health effects, such as lower bone mineral density and higher bone fracture rates [14]. In support, studies have shown the importance of red meat nutrients for the pregnant and aging population to maintain healthy vitamin and mineral status and skeletal muscle mass, respectively [15]. Nevertheless, negative attention towards the consumption of red meat is increasing, especially the perception of a strong association between the consumption of red meat products and the development of some forms of cancer [16]. Parallel concerns on the consumption of red meat and its impact on climate change and greenhouse gas emissions for most nations in the world is also increasing [8,9], leading to an emphasis on improving the sustainability of livestock production. It is known that diets high in refined foods, sugars, oils and meats are expected to contribute up to 80% of the increase in greenhouse gases originating from food production and global land clearing [9]. In addition to environmental concerns, animal welfare, food safety, and cost of meat production are becoming increasingly important in developed countries [17,18]. As a consequence, new recommendations from national and international health organizations in developed countries suggest a decrease in the consumption of the protein food group, including red meats, and an increase in the consumption of the fruits and vegetables food group which includes plant-based protein alternatives [19,20].

In recent years, a constant decrease in per capita meat consumption in developed countries has been observed; despite this, meat remains a key component of human diets around the world [21], with more than 90% of consumers regularly eating meat in developed countries [22,23]. Following this trend, meat consumption in developing countries continues to increase every year, with projections indicating a parallel increase alongside the growth in world population [24]. With different demands and increases in red meat consumption worldwide, the variability in dietary recommendations that exists across the world in different regions should also be considered, as these dietary recommendations are specific to different societal and cultural norms [25]. This has shown that different regions worldwide have varying recommendations for portions of the food groups and what is considered 'healthy' has changed over time [25]. This has emphasized the importance of understanding the potential to manipulate red meat nutrients to meet specific needs of different regions. The nutrient density and nutrient value of meat products has been reviewed and measured, showing the highly bioavailable and nutrient dense quality of beef, pork and lamb compared to other non-meat food products [3]. However, the understanding of what environmental and genetic factors influence the variability in the nutrient composition of red meat is incomplete.

Variability in the nutrient composition of red meat is known to exist due to differences in animal species (ruminant and monogastric) and breeds, as well as different strategies for feeding and rearing livestock, which has allowed for opportunities to manipulate the nutritional value. This has been studied by implementing different strategies to enhance its beneficial properties [26–28], resulting in value-added meats with enhanced composition which are now available in the marketplace. For example, research studies have focused on docosapentaenoic acid (DPA) in red meat as a valuable terrestrial source of long chain fatty acids, to better understand the value and content of naturally occurring *trans* fatty acids present in red meat [29]. In support, numerous studies have provided evidence that nutrient composition variation in red meat depends on the complex interaction between the animal's genetics, the production environment and their interaction. While environmental factors can influence meat nutritional value, such as production system, animal age, gender, or physical activity, diet has shown to have the largest impact on red meat composition and will be the main focus of this review, along with genetic studies including both breed comparisons and genetic parameter analyses.

2. Literature Review Methodology

A literature search was conducted using the following databases: Scopus, Biological Abstracts, CAB Abstracts and FSTA. The search terms included: 'beef' AND/OR 'bovine' AND/OR 'pork' AND/OR 'swine' AND/OR 'lamb' AND/OR 'ovine' AND/OR 'red meat' AND/OR '*longissimus thoracis lumborum*; muscle' AND/OR 'REA' AND/OR 'ribeye' AND 'nutrition' AND/OR 'diet' AND/OR 'supplement' AND/OR 'additive' AND/OR 'feeding program' AND/OR 'genetic' AND/OR 'heritability' AND/OR 'genomic breeding values' AND/OR 'GEBVs' AND 'minerals'AND/OR 'trace elements' AND/OR 'copper' AND/OR 'iron' AND/OR 'zinc' AND/OR 'selenium' AND/OR 'iodine' AND/OR 'sodium' AND/OR 'calcium' AND/OR 'cobalt' AND/OR 'vitamin A' AND/OR 'vitamin E' AND/OR 'vitamin B_2' AND/OR 'vitamin B_9' AND/OR 'vitamin B_{12}' AND/OR 'fat' AND/OR 'fatty acids' AND/OR '*trans*' AND/OR 'conjugated linoleic acids'. The search was filtered for publication dates during or after the year 1999 and excluded documents on processed meat, human nutrition, and clinical trials. The results (6800) were sorted by species and nutrient. In addition, the references from the articles obtained by this method were used to identify additional relevant material.

3. Manipulating the Nutritional Value of Red Meat

Dietary nutrients are categorized as micro- or macronutrients based on dietary requirements of humans. Micronutrients are further categorized into vitamins and trace elements. Water soluble vitamins include B-complex vitamins, while fat soluble vitamins include vitamins A, D, and E, among others. Trace elements can be grouped between those highly influenced by diet and liver metabolism, such as iron, copper, and zinc, and those less affected, such as iodine and selenium. Macronutrients include proteins and fats, as well as carbohydrates and water; however, the latter two are not as relevant when considering the nutritional value of red meat and therefore are not included in this review.

3.1. Micronutrients

3.1.1. Vitamins

Concentration ranges for several vitamins in beef, pork and lamb are shown in Table 1, revealing a high variability in the concentration of vitamins in red meat among different species.

Table 1. Nutrient range of vitamins in red meats.

Vitamins	Beef	Pork	Lamb
Vitamin A (μg/100 g)	5.00–11.5	2.00–6.10	7.80–8.60
Vitamin E (mg/100 g)	0.03–1.10	0.01–0.86	0.08–1.20
B_2 (mg/100 g)	0.09–0.80	0.05–1.23	0.11–0.25
B_{12} (μg/100 g)	0.40–3.10	0.30–1.10	0.60–2.50

Although retinoic acid and retinal are the most physiologically active forms of vitamin A, other forms also include free retinol, retinyl esters, β-carotene, and carotenals. Vitamin A is essential for maintaining normal vision, the immune system and reproductive function. Vitamin A deficiency is common in undeveloped countries, especially among children and women at reproductive age, but is rarely seen in more developed countries [30]. It is mainly stored in the liver, but several studies have also shown potential for manipulation of its content in milk, plasma and subcutaneous fat through dietary supplementation of livestock [31–33].

Domínguez et al. [33] studied the effect of feeding chestnuts to Celta pigs during the finishing phase on retinol concentration in adipose tissue. Their results show a retinol concentrations ranging from 0.63 to 0.76 μg/g retinol in various tissue including rump fat, subcutaneous *biceps femoris*, and subcutaneous dorsal fat, and a concentration of 527 μg/g in the liver. Results in beef *longissimus* from Duckett at al. [34] showed a greater concentration in α-tocopherol, β-carotene, thiamine and riboflavin in cattle managed in a pasture-finishing system compared to counterparts managed in a high concentrate-finishing system. In addition, a study comparing different lamb rearing systems by Osorio et al. [35] revealed a highly significant difference between retinol concentration in lamb *longissimus* muscle in maternal milk rearing (10.83 μg/100 g) compared to milk replacement rearing (43.69 μg/100 g). Similarly, Blanco et al. [36] measured vitamin concentration in *longissimus thoracis* muscle of lambs and compared production systems during rearing of suckling lambs raised indoors to suckling lambs raised on pasture. The results reveal a highly significant difference in lutein, retinol, α-tocopherol, and γ-tocopherol, with higher concentrations observed in *longissimus* muscle of suckling lambs in the pasture-raised system. Taylor et al. [37] described the first heritability values (0.36, 0.03, 0.79 and 0.48) related to hepatic vitamin A concentration in beef cattle at 235, 340, 600 and 710 days of age, respectively. More recently, Kato et al. [38] also described decreasing heritability values with age of 0.37, 0.24, 0.16 and 0.07 (at <13, 14–18, 19–21 and >13 months, respectively) for vitamin A concentration in serum from calves with Japanese Black sires and Holstein dams. Nevertheless, the potential for genetic manipulation by assessing genetic parameters and heritability of vitamin A in livestock muscle has not been reported to date.

Vitamin E deficiency is caused by inadequate dietary intake or by a disorder causing fat malabsorption. Due to its important role, vitamin E is a common ingredient supplemented in animal nutrition [39]. With its importance in overall health to consumers as well as its role in shelf-life stability, there is interest to increase vitamin E concentration in red meat. Multiple studies have shown the potential to increase vitamin E in different red meat products, using it as a biological antioxidant to enhance meat color stability and slow the rate of lipid oxidation [40]. Absorption of vitamin E is known to be proportional to the vitamin E status of the animal in most species and this status varies by several factors, including fat intake, digestion, liver function or an excess of dietary zinc. Muscle deposition mainly depends on dietary compound source (synthetic, *all-rac-α*-tocopheryl acetate vs. natural, RRR-α-tocopheryl acetate), dosage, species, tissue and time of supplementation [41]. Leal et al. [42] described a difference of 1.07–1.27% for α-tocopherol accumulation in lambs depending on the dietary dose and vitamin E source. Kim et al. [43] studied the effect of supra-nutritional vitamin E (35, 300 and 700 IU) supplementation for 14, 28 and 42 days before slaughter. They reported that vitamin E supplementation for 28 days before slaughter maximizes the *longissimus thoracis et. lumborum* muscle vitamin E concentration. An enhancement of vitamin E accumulation has been described when decreasing vitamin A levels from the diet in pigs [44,45] and lambs [36]. Selenium deficiency hinders vitamin E absorption and greater vitamin E concentrations in *longissimus thoracis* muscle of pigs have been observed when organic selenium source was included in the diet [46]. Regarding genetics, most studies have reported no breed differences, and no relevant research was found providing genetic parameters for vitamin E in meat. Studies investigating changes in gene expression, protein abundance and the concentration of metabolites after vitamin E dietary inclusion are also limited. Factors such as vitamin E form and tissue type is known to affect gene expression. To better understand the underlying genetic mechanisms of the abosprtion and molecular mechanisms of vitamin E, one approach is to integrate omics technology to better understand its biological role and potential to be manipulated [47]. Currently, few studies have evaluated genetic parameters for vitamin E in muscle tissue. Ntawubizi et al. [48] described a heritability for α-tocopherol of 0.30 in plasma collected at slaughter from Pietrain (Landrace—Large White) pigs from two performance test stations.

B vitamins are water-soluble micronutrients which are best absorbed as a complex and play many important roles in the human body. Among other functions, thiamine (vitamin B_1) enables conversion of blood glucose into biological energy, having a modulatory role in the acetylcholine neurotransmitter system. Riboflavin (vitamin B_2) is involved in carbohydrate, protein and fat metabolism processes, as well as other biological mechanisms associated with fatty acid and iron metabolism and thyroid regulation. Forms of niacin (vitamin B_3), such as nicotinamide adenine dinucleotide (NAD) and NAD phosphate (NADP) are included in a vast array of processes and enzymes involved in every aspect of peripheral and brain cell function. Pantothenic acid (vitamin B_5) is part of co-enzyme A (vitamin B_6). It is involved in protein metabolism, red blood cell metabolism, hemoglobin or neurotransmitter formation and maintaining blood glucose. Biotin (vitamin B_7) is important for carbon dioxide fixation, carbohydrate and fat metabolism, and plays a key role in glucose metabolism and homeostasis. Folic acid (vitamin B_9) acts as a co-enzyme for leucopoiesis, erythropoiesis and nucleoprotein synthesis, and is necessary for the synthesis and regeneration of monoamine neurotransmitters. Cobalamin (vitamin B_{12}) is essential for the maturation of erythrocytes, cell growth and reproduction and the formation of myelin and nucleoproteins; low levels of cobalamin results in a functional folate deficiency [49]. In general, recommended amounts of these vitamins are achieved by humans through their diet and deficiency is more common in developing countries. However, in developed countries, the vegan sub-population is identified as at risk for deficiency, especially for cobalamin. This is because vitamin B_{12} is naturally present only in animal-derived products, which requires constant supplementation among individuals consuming vegan diets.

Regarding ruminants, B-complex vitamins are degraded in the rumen [50,51] and the nutritional requirements of the animal are usually met or exceeded with microbial fermentation in the rumen; therefore, dietary supplementation is not normally implemented by ruminant nutritionists. Vitamin B_{12} synthesis requires adequate levels of dietary cobalt. However, high levels of dietary starch reduce vitamin B_{12} levels in the rumen, leading to reduced vitamin B_{12} in meat from ruminants. A recent study suggested that the species of bacteria and bacterial consumption of vitamin B_{12} in the rumen may better represent overall levels, compared to bacterial production [52]. Duckett, Neel, Fontenot, and Clapham [34] compared high concentrate finishing systems to pasture finishing systems for Angus-cross steers and found nutrient potentials of 46.7 μg/100 g for B_1 and 233 μg/100 g for B_2. In addition to production systems, feed additives also have an impact on the nutrient potential of B vitamin concentration in meat. This has been observed using algae additives in pig feed, which led to nutritient potentials of 90 μg/100 g for B_6 and 0.06 μg/100 g for B_{12} [53]. Regarding monogastric animals, diets are usually supplemented with B vitamins, and while additional increases have a small impact on muscle concentration of vitamins B_9 and B_{12}, vitamin B_2 does not seem to respond to higher supplementation. To the best of our knowledge, no genetic studies stating heritability values for vitamin B concentrations in meat are currently available. Nevertheless, heritability values within the range of 0.23–0.45 were reported for B_{12} [54,55] and within the range of 0.31–0.52 for B_2 [56] concentration in cow milk.

3.1.2. Trace Elements

Meat is considered an important source of trace minerals for humans, having higher concentration compared to other foods [4] and a high content and bioavailability of copper, iron, phosphorus, magnesium and zinc [3,57].

Table 2 contains the compositional range of iron, copper, zinc, selenium and iodine in red meats including beef, pork and lamb, revealing high variation in concentrations of trace elements in red meat. Other studies have shown variability in trace element contents in meat, not only among species, but also among different muscles [58].

Although for most of the trace minerals, animal tissue concentrations are independent of intake, it has been suggested that dietary intervention [59] and genetic factors could be modified to manipulate the concentration of specific trace elements in meat with greater success than in other animal-derived food products, such as eggs or milk [60].

Table 2. Trace element nutrient composition range in red meat.

Trace Elements	Beef	Pork	Lamb
Copper (mg/100 g)	0.04–1.40	0.03–0.59	0.03–0.13
Iron (mg/100 g)	1.00–7.80	0.30–3.00	1.10–3.60
Zinc (mg/100 g)	2.30–7.70	0.40–5.00	2.10–9.40
Selenium (mg/100 g)	0.40–10.8	0.05–1.23	0.30–35.0
Iodine (μg/100 g)	0.20–20.0	0.40–17.0	30.0–46.0

Copper is a cofactor for several enzymes and influences iron metabolism in the human body. Copper deficiency is not common in human diets; however, under certain conditions, such as celiac disease, lower intestinal absorption, anemia and osteoporosis could occur from copper deficiency. In addition, copper intake is regulated by the liver, and this regulation is different among livestock species. For this reason, while copper concentration is highly dependant on dietary sources, muscle concentration does not respond to direct supplementation and relatively low dietary copper concentrations can often be impaired [61–64]. Nevertheless, Ponnampalam et al. [65] found higher copper concentration in meat from lambs fed a high energy-high protein finishing diet compared with meat from lambs fed high energy–moderate protein or moderate energy–high protein finishing diets. Zhao et al. [66] reported higher concentrations of copper in organic pork than in conventional pork, partially due to the organic pigs having more time to exercise, which enhanced the capability of conserving this element. Regarding genetic manipulation, most studies looking at heritability of copper in muscle have reported no genetic variation. However, using Genome-Wide Association Study (GWAS) and a Bayesian approach in a small population of Nellore beef cattle, Tizioto et al. [67] reported moderate heritability estimates for copper content. Several studies have shown potential to manipulate plasma copper concentration [68]; however, this effect was not reflected in muscle content, which may be due to liver regulation.

Zinc plays an important role in the immune system of humans and zinc deficiency can impact growth in children and lead to various health issues in adults [69]. Zinc deficiency is often linked to low bioavailability due to interactions with phytic acid from plant dietary sources. The gastrointestinal-pancreatic control of zinc absorption makes it difficult to manipulate through dietary means. Most studies have reported little or no effect on muscle concentrations following zinc supplementation [66,70,71], especially in pigs. Interactions between trace elements should be considered, as high dietary zinc or zinc supplementation can interfere with copper and iron absorption [72]. Based on the literature found, genetic manipulation could have certain impacts in sheep, but little to no impact in beef cattle. The biological significance of this manipulation would most likely be minimal. In contrast, the effect of genetic line has been reported on zinc content in meat from lambs and pigs and is mainly associated with the muscle fiber type distribution [73,74].

Iron is an essential trace element in human diets and iron deficiency is an important health concern globally [75], affecting approximately 20% of the population and 50% of the population in less developed countries [76]. Dietary iron can be found in heme and non-heme form. The heme form has much greater bioavailability (20–30% absorption vs. 5–10% absorption for non-heme iron) and can only be obtained by consuming animal foods, such as red meats. Other influences on heme iron content include muscle fiber types, as fiber type proportion varies across different muscles. Absorption of heme iron is also relatively independent of other dietary ingredients, while absorption of non-heme iron can be influenced by meat composition. Heme iron is a component of hemoproteins, including myoglobin and, therefore, also influences the characteristic red color for meat. Contrastingly, free iron ions released from heme and ferritin are the main catalysts for lipid peroxidation of meat, and heme iron has been linked to increased risk of developing colorectal cancer by its implications in multiple processes in the intestine, such as DNA damage of epithelial cells and colonic hyperproliferation [77]. Hence, while red meats are recommended as an excellent alternative to decrease iron deficiency due to their heme iron content, this same factor may be the reason for some consumers to decrease their consumption of red meats. Regardless, understanding the potential manipulation of the concentration of iron content in meat is interesting from both quality and nutritional points of view. This has been well studied in red and white veal and has shown that using interventions to reduce iron content is easier than to increase iron content.

Iron absorption is controlled at the gastrointestinal level and, as previously mentioned, only a small portion of dietary iron is absorbed, making dietary manipulation difficult [78,79]. In fact, most studies that have shown a significant impact on increasing iron content in meat used extreme concentrations of dietary iron that surpassed the regulated limits allowed for use in animal diets [80]. Dimov et al. [81] found higher levels of iron in the meat of calves fed a silage-free finishing diet compared with calves fed a silage finishing diet, which could be attributed to the higher copper content found in the silage-free finishing diet. Other dietary ingredients, such as zinc supplements and green tea, have been reported to decrease iron content [82]. On the other hand, most studies have reported a relatively high heritability for both total iron and myoglobin content in red meat [67,83–85]. This is due to the link between muscle fiber type and iron content, in which meat containing more red muscle fibers have more iron compared to meat containing more white muscle fibers [70]. The few studies evaluating genetic parameters for muscle fiber type reported similar heritability values for total iron [82].

Another trace element is selenium, which is incorporated into selenoproteins and have several pleiotropic effects, including antioxidant and anti-inflammatory effects on the production of active thyroid hormone. Low selenium status in humans leads to increased risk of mortality, poor immune function and cognitive decline. In contrast, high selenium status or selenium supplementation is known to have positive effects, such as antiviral effects, improved male and female reproductive function and reduced risk of autoimmune thyroid disease [86]. Selenium-deficiency and elevated iodine together can have negative health impacts, such as enhanced autoimmune reactions and accelerated deterioration of thyroid function through oxidative stress [87]. These reasons highlight the need to provide sufficient selenium in the human diet, and one approach is through red meat consumption.

Selenium supplementation is common in livestock nutrition in order to meet the dietary requirements of animals. Its positive impact on meat quality as part of the antioxidant enzyme glutathione peroxidase also makes selenium supplementation interesting in regard to avoiding meat quality defects, such as white muscle disease in calves and lambs when low concentrations of selenium are present in the soil. Although no homeostatic regulation has been described for dietary selenium, bioavailability seems to be lower in ruminants, which may be due to transformation in the rumen. However, supplemented selenium can be deposited in meat tissue, and it is possible to produce selenium-enriched meat products. Several studies have reported the effect of the selenium source on absorption and deposition, with organic selenium, such as selenium-enriched yeast, increasing selenium supplement in muscle content at higher rates than inorganic selenium supplementation [88–90]. In terms of genetics, relatively high heritability has been reported for the selenium content in cattle [67], indicating that some variation of selenium content can be attributed to genetic variation in cattle. The latter study indicates the potential for genetic manipulation and the implementation of new animal breeding programs to improve the selenium concentration in muscle tissue and enhance nutritional attributes; however, in general, there remains a lack of studies evaluating genetic parameters of minerals and further research is needed on larger animal populations and in various breeds and species.

Iodine deficiency in humans is a severe condition that can lead to health issues including goiter, especially during childhood. It has been suggested that the high levels of iodine in animal-derived food products has decreased iodine deficiency in several countries [91]. The consumption of meat allows for sufficient dietary intake of iodine in humans, but maximum levels can actually be exceeded when overconsuming milk and eggs [91]. Iodine is directly absorbed in the intestine and regulated by the thyroid. Few studies show that increasing iodine content in meats is possible through dietary supplementation, in both ruminant and monogastric animals [92,93]. In lambs, a great increase has been observed in diets with high levels of seaweed. Similar to selenium, this may be the reason there are no available data regarding genetic parameters for this trace element.

3.2. Macronutrients

3.2.1. Total Protein and Amino Acids

Proteins are essential macronutrients for human energy and nutrient requirements, as protein or amino acid deficiencies are known to cause severe health issues, especially in pregnant women, children [94] and the elderly. The consumption of indispensable (i.e., essential) amino acids, which cannot be synthesized in the human body, highlights the importance of consuming balanced protein sources from food. The amino acid balance and digestibility have been used to define protein quality in different food sources. Animal-derived products, including red meat, provide complete proteins. Animal protein is necessary, for instance, in situations where patients require high consumption of protein for tissue and musculoskeletal recovery [95,96]. Additionally, amino acid content plays an important role in the development of meat flavor compounds and sensory characteristics during cooking, which highlight the importance of amino acid composition for consumer acceptance of red meats.

With muscle protein being a functional tissue, studying the change in protein content of a muscle must consider changes in fat or moisture content which could therefore influence protein content. Thus, most of the studies reviewed have evaluated genetic and dietary approaches and their influence on changes in individual amino acid content. Descriptive studies have measured total protein of muscle tissue derived from red meat livestock species [3,97,98]; however, there is a lack of studies evaluating total protein or amino acid content in red meats and how these components can be modified to enhance its nutritional value using nutritional or genetic approaches [99]. Drazbo et al. [100] found that total protein in pork *longissimus dorsi* muscle was different when feeding a diet with protein and amino acid levels reduced by 15% relative to the standard levels. Specific pork loin essential amino acid composition was also different when supplemented with dietary additives like ginseng [101]. Muscle histidine and valine concentration have been observed to be lower when fed 5:1 and 10:1 PUFA ratio diets compared to 1:1 and 2.5:1 PUFA ratio (n-6:n-3) diets [102], suggesting lower concentrations of these amino acids are observed when fed a high saturated fatty acid profile. However, no effect on total protein was observed from diets with 2.0% supplemented palm oil and 0.5% or 1.0% CLA [103]. A study observed cattle fed grass silage had higher free amino acid levels compared to animals fed a concentrate diet, and many individual amino acid concentrations were also significantly different [104]. A study on lambs weaned at different ages revealed no difference in crude protein percent; however, looking at individual amino acids, essential amino acids were higher when weaning occurred at an earlier age [105], suggesting potential shifts in fiber type composition and therefore amino acid composition. Additionally, in the latter study, environmental factors such as production system or weaning system could have affected amino acid composition. A study using diet manipulation revealed protein percent in lamb leg was significantly impacted by feeding olive cake [106]; this suggests the increase in fat and decrease in moisture of the diet influences protein content.

Regarding genetic approaches to manipulate total protein or amino acid content in red meat, existing studies have revealed heritability of both total protein and individual amino acids ranges from low to high for beef and pork [107–109]. A prior study based on targeted single nucleotide polymorphisms (SNPs) and total significant SNPs revealed moderate to high heritability estimates of 0.42 and 0.26, respectively, for total protein content in pork [109]. In addition, pork breeds have also been compared to evaluate the different genetic backgrounds associated with total protein and amino acid profiles [110]. Studies have also estimated heritability of different amino acids in beef, revealing estimates of 0.34, 0.17, 0.66, 0.40 and 0.33 for alanine, glutamine, taurine, anserine and inosine, respectively [108]. Similarly, Ahlberg et al. [107] revealed high heritability estimates of total protein content in beef *semitendinosus* and *longissimus* muscles of 0.75 and 0.70, respectively. One study revealed that the *longissimus dorsi* mean amino acid content was significantly different between lamb breeds for arginine, glutamine and tyrosine [111]. The studies to date suggest a potential for manipulating individual amino acid concentrations content using diet, as well as genetic approaches, due to the moderate to high heritability of these components; however, further research is needed, specifically to evaluate the potential to manipulate total protein content.

3.2.2. Total Fat and Fatty Acid Composition

Meat lipids continue to remain as the nutrient component with the highest potential for modification, in both content and composition, presenting opportunities for value added production and health promotion [112]. For instance, low-fat of n-3 enriched meats are considered functional foods for overweight individuals, since their consumption improves the body fat index, n-3 levels and the n-6:n-3 ratio, without impacting the Healthy Eating Index or intake levels of energy or other macronutrients [10].

Numerous studies have reported values for total fat, as well as different groups of fatty acids in meat. This area of research has been highly researched for the last 20 years and continues to attract attention from researchers in the area of animal and meat sciences (Table 3). The potential for manipulation of lipids is clear when the ranges in the literature were considered, with clear differences between monogastrics and ruminants. In monogastrics such as swine and poultry, meat fatty acid composition is reflective of their diets, whereas, in ruminants, dietary unsaturated fatty acids undergo extensive biohydrogenation by the rumen bacteria and are transformed into saturated fatty acids [113]. This phenomenon limits the ability to increase the content of these fatty acids in ruminant meats through feeding polyunsaturated fatty acids (PUFA) sources such as oilseeds and fish oil [114]. Conversely, during ruminal biohydrogenation of PUFA, several intermediates are produced, and a portion of them passes from the rumen and subsequently finds its way into meat after post-ruminal absorption. Specific biohydrogenation intermediates such as conjugated linoleic acids (CLA) and vaccenic acid (VA, *trans*-11 18:1) have been associated with several health benefits including anti-inflammatory and anti-diabetic effects [115].

Table 3. Nutrient range of fatty acids in red meats.

Lipids	Beef	Pork	Lamb
Total IMF (g/100 g meat)	0.60–26.90	1.60–17.00	2.50–18.10
SFA (g/100 g IMF)	33.70–49.10	32.80–41.00	46.20–50.40
MUFA (g/100 g IMF)	24.70–56.10	39.60–49.10	32.10–45.30
Trans (g/100 g IMF)	1.50–4.00	–	3.00–6.30
PUFA (g/100 g IMF)	2.80–29.00	3.80–26.20	3.60–8.10
n-3 (g/100 g IMF)	0.38–10.40	1.20–13.40	1.50–3.50
LC n-3 (g/100 g IMF)	0.25–4.90	0.19–1.90	0.77–1.40
n-6 (g/100 g IMF)	2.80–20.20	8.70–12.80	2.10–4.60
CLA (g/100 g IMF)	0.10–1.80	0.04–3.60	0.57–1.50

Key targets for manipulation include increasing n-3 PUFA across species, and specifically in ruminants increasing contents of "healthy" PUFA biohydrogenation intermediates including CLA and VA [112,114]. In addition, a primary target has been to reduce saturated fatty acid (SFA) content as well as increasing levels of oleic acid (*cis* 9-18:1) [116]. Feeding grains, oilseeds, forages, grass or DDGS, among other feedstuffs, has a large impact on intramuscular fat (IMF) and the proportions of the different fatty acid groups. For example, meat from grass-fed ruminants tend to present lower IMF and higher proportions of n-3 PUFA, CLA and VA compared to concentrate-fed ruminants [27]. Feeding concentrate based diets, however, have been associated with decreased PUFA/SFA ratios, but, over the finishing period, there is increased conversion of SFA to monounsaturated fatty acids (MUFA), and relative rates are influenced by breed [117]. On the other hand, manipulation of PUFA biohydrogenation intermediates may have more to do with interactions between diet, rumen microbiology, and management than host genetics. Indeed, large increases in VA and CLA in steers has been linked to feeding management, for example, feeding a PUFA rich supplement (flaxseed co-extruded with peas) before feeding hay, instead of feeding a hay and supplement mix, has led to a substantial increase in VA and CLA [118,119]; these differences are related to shifts in the rumen microbial population [120].

While many studies have reported the potential for manipulation of VA in beef and lamb, it is important to look at the total amount of *trans* fatty acids and the relative proportions of *trans*-10-18:1 and *trans*-11-18:1, as *trans*-10 has been associated with detrimental effects on blood lipid profiles through upregulation of hepatic triacylglycerol and cholesterol synthesis [115,121]. Again, host genetics are not considered as a primary factor in the accumulation and proportions of *trans* fatty acids, but different forages, grass or vitamin supplementations can lead to shifts in *trans* fatty acid amounts and isomer proportions [27]. In beef and lamb, feeding sources of PUFA leads to a large number of biohydrogenation intermediates including several conjugated and non-conjugated 18:2 and 18:3 isomers for which the roles in the human body are still unclear [115]. CLA has also received more attention recently due to the potential health benefits of nitro-fatty acids in humans [122], but the role of host genetics in these processes is not yet clear. In fact, in ruminants, dietary ingredients and additives that modify the rumen microbiome may have a larger effect than direct supplementation of fatty acid supplements. On the other hand, dietary supplementation of PUFA has a large impact on pork fat composition, leading to large increases in PUFA and n-3, even long chain n-3, especially when using marine sources, such as fish oil and algae [28]. In cattle, however, some limitations exist regarding long chain PUFA deposition due to their preferential incorporation into phospholipids [123]. Further manipulation of pork IMF can also be achieved through altering lean deposition by reducing the protein or lysine content in diets (i.e., causing lean to fat repartitioning), or by adding CLA into pig diets [124].

The relationship between total fat content and relative proportions of fatty acids is important to consider, as higher IMF corresponds to lower relative PUFA content, due to the smaller contribution of membrane phospholipids [125]. For this reason, when reporting fatty acid profiles in meat, it is important to provide either total IMF or use mg per 100 g of meat as the unit, instead of the percentage of fatty acids in total fat. Similarly, consumers do not eat denuded muscles, but commercial cuts, which combine lean with seam fat and subcutaneous fat. Thus, while the manipulation of fatty acid profile in IMF may be more limited, it is possible to obtain a larger impact when considering changes in the whole primal, including all fat depots [126]. In general, dietary effects observed in IMF tend to have a larger impact on larger fat depots. This is important when trying to enhance the lipid profile in order to reach certain health claims. In fact, an alternative for fresh meat products, such as ground meat, could be accomplished by either supplementing the diets of a small percentage of the animals or selecting carcasses with a naturally higher concentration of certain beneficial fatty acids, and then mixing the fat from those carcasses with lean from the regular population. Manipulation of fatty acid profiles also has to take into consideration effects on meat and fat quality (taste, oxidative stability, fat softness, etc.), as enhancing the healthfulness of the fatty acid profile will be of limited value if overall quality is negatively affected. Thus, studies investigating manipulation of fatty acid profiles need to be linked with complimentary studies on meat or meat product quality, including sensory evaluation [112].

Inter- and intra-breed differences are well known in terms of total IMF variability, with very obvious cases of genetic groups with higher marbling [127,128], while populations selected for other traits correlated to total fat, such as lean meat yield, have seen a decrease in IMF as a negative side effect [129–131] Studies show a medium to high heritability not only for total IMF, but also for the majority of fatty acid groups which can be endogenously synthesized in both ruminants and monogastrics [132,133]. According to GWAS studies in different species, both total IMF and fatty acid composition in meat are influenced by key regulatory genes with major effects and multiple genes with smaller effects, and have shown moderate to high heritability estimates for IMF and low to medium heritability for specific fatty acids [134–139]. However, despite the potential to include IMF fatty acids in breeding programs, antagonistic genetic relationships with performance have usually minimized the emphasis on selection for these traits. A recent study [140] reported a series of genetic markers that could be used to manipulate IMF without impacting backfat thickness, opening new opportunities for animal selection. In terms of manipulating meat fatty acid composition, the influence of fatness on the lipid profile (decrease of relative proportion of PUFA with higher levels of IMF) must be taken into consideration [133]. Nevertheless, multiple studies have found SNPs for a number of candidate genes regulating intramuscular fatty acid metabolism [141]. Within the last few years, numerous studies have focused on alternatives to traditional genetic selection. This includes the use of transgenic animals (by nuclear transfer of modified DNA to an embryo) which can increase the endogenous production of certain beneficial fatty acids, such as omega-3 fatty acids [142–146]. This approach is also possible for feedstuffs, with crops being genetically modified to produce long-chain omega-3 fatty acids usually only available from marine sources [147]. Although these strategies present great potential for IMF and fatty acid manipulation, ethical and safety concerns still need to be addressed [148,149]. Moreover, consumer perception of genetically modified organisms and animal welfare could limit the large-scale implementation of these strategies [150].

4. Considerations

Enhancing the nutritional value of red meats continues to attract much attention from the scientific community and support from the industry. Research indicates great potential for the dietary manipulation of certain nutrients in red meats, such as vitamin E, selenium, total IMF, or fatty acid profile. In general, studies suggest greater potential for genetic selection for desirable IMF and fatty acid composition; genetic potential also exists for changing total protein when considering individual amino acids. Further research is needed to understand the genetic potential to manipulate iron content and therefore muscle fiber type composition. Fat content and lipid profile represented the fraction with the highest potential for manipulation either through diet or genetic selection [112]. However, most studies have used approaches that independently evaluate the impact of either genetics or nutritional strategies, and few studies have explored the interactions between these two major factors [151–153]. Furthermore, recent research on microbiome manipulation have shown an impact on meat composition [154]. Holistic approaches, such as livestock precision farming, systems biology, livestock phenomics, and nutrigenomics, have the ability to integrate genetic, environmental and phenotypic information, leading to better understanding of the biological system as a whole and unlocking the true potential for manipulating meat nutritional attributes [155,156].

Establishing the justification to modify the nutritional content of red meat and understanding its consequences should also be considered due to ethical concerns. Among the reasons commonly described to justify enhancing meat nutritional value, the importance of balancing currently deficient diets in developed countries, especially for populations at risk, should be mentioned, as well as the need for a more complete nutrient-dense product for the diets of developing countries [157,158]. Furthermore, new research findings and dietary recommendations from public and private institutions will continue to shape our understanding of the impact of different foods on human health; therefore, providing alternative approaches to modify the nutritional value of meat will allow the industry to address future challenges in this area [159]. However, the main justification identified in research studies is the production of value-added differentiated products [160,161]. Other sectors based on animal products, such as the egg and dairy industry, have developed strong marketplace differentiation based on nutritional enhancement claims [158]. With the exception of further processed meats, this strategy has reached lower success in the meat sector compared to other industries. While some fresh meats are using enhanced nutritional value as their differentiation strategy (e.g., omega-3 pork), credence attributes have become the most common form of differentiation to appeal to consumers with higher standards for aspects such as animal welfare, environmental concerns or the impact of livestock production on antimicrobial resistance [162,163]. Ethical aspects and sustainability concerns could be raised regarding the use of highly valuable feedstuffs, such as feed ingredients sourced from marine resources, for animal feed [164]. Additionally, changes in diet or genetic selection should consider potential impacts on animal welfare, as well as consumer perception of certain practices. As an example, the current perception from a large part of the population regarding genetically engineered foods could negate any commercial benefit from the use of transgenic animals or feedstock to enhance the nutritional value of meat [165]. In addition to the consideration of societal perspectives, limitations when using genetic selection should be considered, including the selection for desirable traits which may be negatively or positively correlated with other undesirable traits. Furthermore, although nutritional enhancement claims can be attractive for some consumers, it is well known that modifying meat composition can lead to changes in appearance, firmness, shelf-life and palatability; all of which could influence the consumer's acceptance [166,167]. Surveys have shown that, while consumers may choose to consume chicken for its perception as a healthy alternative, the drivers leading to the purchase of red meats are mainly related to the eating experience [163]. Since palatability will ultimately determine consumer satisfaction [168], research is needed to evaluate quality attributes and how they are affected when using different strategies to modify the nutritional value of meat.

5. Conclusions

Traditional and novel omics research indicates the potential to manipulate fresh meat composition and therefore enhance its nutritional value. Enhancing the nutritional value of meat could be maximized by combining nutritional strategies and genetic selection, opening opportunities to develop added-value products. Certain limitations need to be considered due to complex metabolic processes and the influence of genetics on certain nutrients. Furthermore, to achieve sustainable animal production, holistic consideration of dietary and genetic strategies and its effect on animal welfare, environmental impact, product quality, and consumer perception must be considered. Traditional red meat will continue to serve as a nutrient dense food which is widely consumed due to the co-evolution of positive human sensory perceptions and thus will continue to provide health benefits to a large percentage of the world population.

Author Contributions: Conceptualization, M.J.; Writing—Original Draft Preparation, M.J., S.L., M.E.R.D., P.V., A.J., O.L.-C., N.P. and J.S.; Writing—Review and Editing, M.J., B.M.B., J.A., and S.L.; and Funding Acquisition, M.J. All authors have read and agreed to the published version of the manuscript.

Funding: This research received no external funding.

Institutional Review Board Statement: Not applicable.

Informed Consent Statement: Not applicable.

Data Availability Statement: This study did not report any data.

Acknowledgments: The authors acknowledge financial support from the Agriculture and Agri-Food Canada.

Conflicts of Interest: The authors declare no conflict of interest.

References

1. Mann, N. Meat in the human diet: An anthropological perspective. *J. Nutr. Diet.* **2007**, *64*, S102–S107. [CrossRef]
2. Stanford, C.B.; Bunn, H.T. *Meat-Eating and Human Evolution*; Oxford University Press: Oxford, UK, 2001.
3. Bohrer, B.M. Review: Nutrient density and nutritional value of meat products and non-meat foods high in protein. *Trends Food Sci. Technol.* **2017**, *65*, 103–112. [CrossRef]

4. Pereira, P.M.; Vicente, A.F. Meat nutritional composition and nutritive role in the human diet. *Meat Sci.* **2013**, *93*, 586–592. [CrossRef]
5. Lim, J.H.; Kim, Y.S.; Lee, J.E.; Youn, J.; Chung, G.E.; Song, J.H.; Yang, S.Y.; Kim, J.S. Dietary pattern and its association with right-colonic diverticulosis. *J. Gastroenterol. Hepatol.* **2021**, *36*, 144–150. [CrossRef]
6. Al-Shaar, L.; Satija, A.; Wang, D.D.; Rimm, E.B.; Smith-Warner, S.A.; Stampfer, M.J.; Hu, F.B.; Willett, W.C. Red meat intake and risk of coronary heart disease among US men: Prospective cohort study. *BMJ* **2020**, *371*, m4141. [CrossRef] [PubMed]
7. Garam, J.; Hannah, O.; Gitanjali, M.S.; Dahyun, P.; Min-Jeong, S. Impact of dietary risk factors on cardiometabolic and cancer mortality burden among Korean adults: Results from nationally representative repeated cross-sectional surveys 1998–2016. *NRP* **2020**, *14*, 384–400. [CrossRef]
8. Tilman, D.; Clark, M. Global diets link environmental sustainability and human health. *Nature* **2014**, *515*, 518–522. [CrossRef]
9. McMichael, A.J.; Powles, J.W.; Butler, C.D.; Uauy, R. Energy and health 5 food, livestock production, energy, climate change, and health. *Lancet* **2007**, *370*, 1253–1263. [CrossRef]
10. Celada, P.; Sánchez-Múniz, F.J. Are meat and meat product consumptions harmful? Their relationship with the risk of colorectal cancer and other degenerative diseases. *J. An. Real Acad. Farm.* **2016**, *82*, 68–90.
11. Ferguson, L.R. Meat and cancer. *Meat Sci.* **2010**, *84*, 308–313. [CrossRef] [PubMed]
12. Wyness, L.; Weichselbaum, E.; O'Connor, A.; Williams, E.; Benelam, B.; Riley, H.; Stanner, S. Red meat in the diet: An update. *Nutr. Bull.* **2011**, *36*, 34–77. [CrossRef]
13. Mazidi, M.; Kengne, A.P.; George, E.S.; Siervo, M. The association of red meat intake with inflammation and circulating intermediate biomarkers of type 2 diabetes is mediated by central adiposity. *Br. J. Nutr.* **2019**, 1–20. [CrossRef]
14. Iguacel, I.; Miguel-Berges, M.; Gómez-Bruton, A.; Moreno, L.; Julián, C. Veganism, vegetarianism, bone mineral density, and fracture risk: A systematic review and meta-analysis. *Nutr. Rev.* **2019**, *77*, 1–18. [CrossRef]
15. Wyness, L. The role of red meat in the diet: Nutrition and health benefits. *Proc. Nutr. Soc.* **2016**, *75*, 227–232. [CrossRef]
16. Bouvard, V.; Loomis, D.; Guyton, K.; Grosse, Y.; Ghissassi, F.E.; Benbrahim-Tallaa, L.; Guha, N.; Mattock, H.; Straif, K. Carcinogenicity of consumption of red and processed meat. *Lancet Oncol.* **2015**, *16*, 1599–1600. [CrossRef]
17. Grunert, K.G.; Sonntag, W.I.; Glanz-Chanos, V.; Forum, S. Consumer interest in environmental impact, safety, health and animal welfare aspects of modern pig production: Results of a cross-national choice experiment. *Meat Sci.* **2018**, *137*, 123–129. [CrossRef] [PubMed]
18. Alonso, M.E.; González-Montaña, J.R.; Lomillos, J.M. Consumers' Concerns and Perceptions of Farm Animal Welfare. *Animals* **2020**, *10*, 385. [CrossRef] [PubMed]
19. EU Science Hub. *Food-Based Dietary Guidelines in Europe*; Report, 2021; EU Science Hub European Commission: Brussels, Belgium, 2021.
20. Health Canada. *Eating Well with Canada's Food Guide*; Health Canada: Toronto, ON, Canada, 2016.
21. Ritchie, H. and Roser, M. *Meat and Dairy Production*; Our World Data: Oxford, UK, 2019.
22. Reinhart, R. *Few Americans Vegetarian or Vegan*; Gallup: Washington, DC, USA, 2018.
23. Massow, M. *Vegan Virtue Signalling Complicates Trend Interpretation*; foodFOCUS: Waalwijk, The Netherlands, 2019.
24. FAO. *The Future of Food and Agriculture—Trends and Challenges*; Report, 2017; FAO: Rome, Italy, 2017.
25. Fernandez, M.; Raheem, D.; Ramos, F.; Carrascosa, C.; Saraiva, A.; Raposo, A. Highlights of Current Dietary Guidelines in Five Continents. *Int. J. Environ. Res. Public Health* **2021**, *18*, 2814. [CrossRef]
26. Chikwanha, O.C.; Vahmani, P.; Muchenje, V.; Dugan, M.E.R.; Mapiye, C. Nutritional enhancement of sheep meat fatty acid profile for human health and wellbeing. *Food Res. Int.* **2018**, *104*, 25–38. [CrossRef]
27. Vahmani, P.; Rolland, D.; Mapiye, C.; Dunne, P.; Aalhus, J.; Juárez, M.; McAllister, T.; Prieto, N.; Dugan, M. Increasing desirable polyunsaturated fatty acid concentrations in fresh beef intramuscular fat. *CAB Rev.* **2017**, *12*, 1–17. [CrossRef]
28. Dugan, M.E.R.; Vahmani, P.; Turner, T.D.; Mapiye, C.; Juárez, M.; Prieto, N.; Beaulieu, A.D.; Zijlstra, R.T.; Patience, J.F.; Aalhus, J.L. Pork as a Source of Omega-3 (n-3) Fatty Acids. *Clin. Med.* **2015**, *4*, 1999–2011. [CrossRef]
29. Dugan, M.; Aldai, N.; Aalhus, J.; Rolland, D.; Kramer, J. Review:Trans-forming beef to provide healthier fatty acid profiles. *Can. J. Anim. Sci.* **2011**, *91*, 545–556. [CrossRef]
30. Green, A.S.; Fascetti, A.J. Meeting the Vitamin A Requirement: The Efficacy and Importance of β-Carotene in Animal Species. *Sci. World J.* **2016**, *2016*, 7393620. [CrossRef]
31. Radu, C.; Lobon, S.; Molino, F.; Sanz, A.; Joy, M.; Ferrer, J.; Blanco, M. The concentration of liposoluble vitamins in the milk of the ewe and the meat of the sucking lamb according to the food received. In Proceedings of the XVI Jornadas Sobre Produccion Animal, Zaragoza, Spain, 19–20 May 2015; Volume 19, pp. 170–172.
32. Jin, Q.; Cheng, H.; Wan, F.; Bi, Y.; Liu, G.; Liu, X.; Zhao, H.; You, W.; Liu, Y.; Tan, X. Effects of feeding β-carotene on levels of β-carotene and vitamin A in blood and tissues of beef cattle and the effects on beef quality. *Meat Sci.* **2015**, *110*, 293–301. [CrossRef] [PubMed]
33. Domínguez, R.; Martinez, S.; Gomez, M.; Carballo, J.; Franco, I. Fatty acids, retinol and cholesterol composition in various fatty tissues of Celta pig breed: Effect of the use of chestnuts in the finishing diet. *J. Food Compos. Anal.* **2015**, *37*, 104–111. [CrossRef]
34. Duckett, S.K.; Neel, J.P.S.; Fontenot, J.P.; Clapham, W.M. Effects of winter stocker growth rate and finishing system on: III. Tissue proximate, fatty acid, vitamin, and cholesterol content. *J. Anim. Sci.* **2009**, *87*, 2961–2970. [CrossRef] [PubMed]

35. Osorio, M.T.; Zumalacárregui, J.M.; Cabeza, E.A.; Figueira, A.; Mateo, J. Effect of rearing system on some meat quality traits and volatile compounds of suckling lamb meat. *Small Rumin. Res.* **2008**, *78*, 1–12. [CrossRef]
36. Blanco, M.; Lobón, S.; Bertolín, J.R.; Joy, M. Effect of the maternal feeding on the carotenoid and tocopherol content of suckling lamb tissues. *Arch. Anim. Nutr.* **2019**, *73*, 472–484. [CrossRef] [PubMed]
37. Taylor, R.L.; Pahnish, O.F.; Roubicek, C.B. Hepatic and Blood Concentrations of Carotene and Vitamin A in Unsupplemented Range Cattle2. *J. Anim. Sci.* **1968**, *27*, 1477–1486. [CrossRef]
38. Kato, Y.; Ito, M.; Hirooka, H. Genetic parameters of serum vitamin A and total cholesterol concentrations and the genetic relationships with carcass traits in an F 1 cross between Japanese Black sires and Holstein dams. *J. Anim. Sci.* **2011**, *89*, 951–958. [CrossRef]
39. Azzi, A. Tocopherols, tocotrienols and tocomonoenols: Many similar molecules but only one vitamin E. *Redox Biol.* **2019**, *26*, 101259. [CrossRef]
40. Idamokoro, E.M.; Falowo, A.B.; Oyeagu, C.E.; Afolayan, A.J. Multifunctional activity of vitamin E in animal and animal products: A review. *Anim. Sci. J.* **2020**, *91*, e13352. [CrossRef]
41. Leal, L.N.; Beltrán, J.A.; Alonso, V.; Bello, J.M.; den Hartog, L.A.; Hendriks, W.H.; Martín-Tereso, J. Dietary vitamin E dosage and source affects meat quality parameters in light weight lambs. *J. Sci. Food Agric.* **2018**, *98*, 1606–1614. [CrossRef]
42. Leal, L.N.; Jensen, S.K.; Bello, J.M.; Den Hartog, L.A.; Hendriks, W.H.; Martín-Tereso, J. Bioavailability of α-tocopherol stereoisomers in lambs depends on dietary doses of all-rac- or RRR-α-tocopheryl acetate. *Animal* **2019**, *13*, 1874–1882. [CrossRef]
43. Kim, J.C.; Jose, C.G.; Trezona, M.; Moore, K.L.; Pluske, J.R.; Mullan, B.P. Supra-nutritional vitamin E supplementation for 28 days before slaughter maximises muscle vitamin E concentration in finisher pigs. *Meat Sci.* **2015**, *110*, 270–277. [CrossRef] [PubMed]
44. Olivares, A.; Rey, A.I.; Daza, A.; Lopez-Bote, C.J. High dietary vitamin A interferes with tissue alpha -tocopherol concentrations in fattening pigs: A study that examines administration and withdrawal times. *Animal* **2009**, *3*, 1264–1270. [CrossRef] [PubMed]
45. Ayuso, M.; Óvilo, C.; Fernández, A.; Nuñez, Y.; Isabel, B.; Daza, A.; López-Bote, C.J.; Rey, A.I. Effects of dietary vitamin A supplementation or restriction and its timing on retinol and α-tocopherol accumulation and gene expression in heavy pigs. *JAFST* **2015**, *202*, 62–74. [CrossRef]
46. Calvo, L.; Segura, J.; Toldrá, F.; Flores, M.; Rodríguez, A.I.; López-Bote, C.J.; Rey, A.I. Meat quality, free fatty acid concentration, and oxidative stability of pork from animals fed diets containing different sources of selenium. *Food Sci. Technol. Int.* **2017**, *23*, 716–728. [CrossRef]
47. Kim, H.K.; Han, S.N. Vitamin E: Regulatory role on gene and protein expression and metabolomics profiles. *IUBMB Life* **2019**, *71*, 442–455. [CrossRef]
48. Ntawubizi, M.; Raes, K.; De Smet, S. Genetic parameter estimates for plasma oxidative status traits in slaughter pigs. *J. Anim. Sci.* **2019**, *98*. [CrossRef]
49. Kennady, V.; Virmani, M.; Malik, R.K.; Rajalakshmi, K.; Kasthuri, S. The Role of B Vitamins in Livestock Nutrition. *J. Vet. Med. Res.* **2018**, *5*, 1162.
50. Santschi, D.; Chiquette, J.; Berthaiume, R.; Martineau, R.; Matte, J.; Mustafa, A.; Girard, C. Effects of the forage to concentrate ratio on B-vitamin concentrations in different ruminal fractions of dairy cows. *Can. J. Anim. Sci.* **2005**, *85*, 389–399. [CrossRef]
51. Girard, C.L.; Santschi, D.E.; Stabler, S.P.; Allen, R.H. Apparent ruminal synthesis and intestinal disappearance of vitamin B12 and its analogs in dairy cows1. *J. Dairy Sci.* **2009**, *92*, 4524–4529. [CrossRef]
52. Franco-López, J.; Duplessis, M.; Bui, A.; Reymond, C.; Poisson, W.; Blais, L.; Chong, J.; Gervais, R.; Rico, D.; Cue, R.; et al. Correlations between the Composition of the Bovine Microbiota and Vitamin B 12 Abundance. *mSystems* **2020**, *5*. [CrossRef]
53. Promeyrat, A.; Vautier, A.; Lhommeau, T.; Roux, A.l.; Biannic, O. Adding algae to pig feed: Influence on meat quality and composition of meat and offal. In Proceedings of the Journées de la Recherche Porcine, Online, 1–4 February 2020; Volume 52, p. G19.
54. Rutten, M.J.; Bouwman, A.C.; Sprong, R.C.; van Arendonk, J.A.M.; Visker, M.H. Genetic Variation in Vitamin B-12 Content of Bovine Milk and Its Association with SNP along the Bovine Genome. *PLoS ONE* **2013**, *8*, e62382. [CrossRef] [PubMed]
55. Duplessis, M.; Pellerin, D.; Cue, R.I.; Girard, C.L. Short communication: Factors affecting vitamin B12 concentration in milk of commercial dairy herds: An exploratory study. *J. Dairy Sci.* **2016**, *99*, 4886–4892. [CrossRef] [PubMed]
56. Poulsen, N.A.; Rybicka, I.; Larsen, L.B.; Buitenhuis, A.J.; Larsen, M.K. Short Communication: Genetic variation of riboflavin content in bovine milk. *J. Dairy Sci.* **2015**, *98*, 3496–3501. [CrossRef]
57. Menezes, E.A.; Oliveira, A.F.; França, C.J.; Souza, G.B.; Nogueira, A.R.A. Bioaccessibility of Ca, Cu, Fe, Mg, Zn, and crude protein in beef, pork and chicken after thermal processing. *Food Chem.* **2018**, *240*, 75–83. [CrossRef]
58. Lombardi-Boccia, G.; Lanzi, S.; Aguzzi, A. Aspects of meat quality: Trace elements and B vitamins in raw and cooked meats. *J. Food. Compos. Anal.* **2005**, *18*, 39–46. [CrossRef]
59. Flynn, A.; Cashman, K. Nutritional aspects of minerals in bovine and human milks. *J. Adv. Dairy Chem.* **1997**, *3*, 257–302.
60. Rooke, J.A.; Flockhart, J.F.; Sparks, N.H. The potential for increasing the concentrations of micro-nutrients relevant to human nutrition in meat, milk and eggs. *J. Agric. Sci.* **2010**, *148*, 603–614. [CrossRef]
61. Norouzian, M.A.; Ghiasi, S.E. Carcass performance and meat mineral content in Balouchi lamb fed pistachio by-products. *Meat Sci.* **2012**, *92*, 157–159. [CrossRef] [PubMed]
62. Cheng, J.; Fan, C.; Zhang, W.; Zhu, X.; Yan, X.; Wang, R.; Jia, Z. Effects of dietary copper source and level on performance, carcass characteristics and lipid metabolism in lambs. *AAAP* **2008**, *21*, 685–691. [CrossRef]

63. Wilkinson, J.; Hill, J.; Livesey, C. Accumulation of potentially toxic elements in the body tissues of sheep grazed on grassland given repeated applications of sewage sludge. *Anim. Sci.* **2001**, *72*, 179–190. [CrossRef]
64. Cheng, Y.F.; Chen, Y.P.; Du, M.F.; Chao, W.; Zhou, Y.M. Evaluation of Dietary Synbiotic Supplementation on Growth Performance, Muscle Antioxidant Ability and Mineral Accumulations, and Meat Quality in Late-finishing Pigs. *Kafkas Univ. Vet. Fak. Derg.* **2018**, *24*. [CrossRef]
65. Ponnampalam, E.N.; Plozza, T.; Kerr, M.G.; Linden, N.; Mitchell, M.; Bekhit, A.E.D.A.; Jacobs, J.L.; Hopkins, D.L. Interaction of diet and long ageing period on lipid oxidation and colour stability of lamb meat. *Meat Sci.* **2017**, *129*, 43–49. [CrossRef]
66. Zhao, Z.; Barcus, M.; Kim, J.; Lum, K.L.; Mills, C.; Lei, X.G. High dietary selenium intake alters lipid metabolism and protein synthesis in liver and muscle of pigs. *J. Nutr.* **2016**, *146*, 1625–1633. [CrossRef]
67. Tizioto, P.C.; Taylor, J.F.; Decker, J.E.; Gromboni, C.F.; Mudadu, M.A.; Schnabel, R.D.; Coutinho, L.L.; Mourão, G.B.; Oliveira, P.S.N.; Souza, M.M.; et al. Detection of quantitative trait loci for mineral content of Nelore *Longissimus Dorsi* Muscle. *Genet. Sel. Evol.* **2015**, *47*, 1–9. [CrossRef]
68. Wiener, G. Review of genetic aspects of mineral metabolism with particular reference to copper in sheep. *Livest. Prod. Sci.* **1979**, *6*, 223–232. [CrossRef]
69. Prasad, A.S. Zinc in human health: Effect of zinc on immune cells. *Mol. Med.* **2008**, *14*, 353–357. [CrossRef]
70. Ponnampalam, E.N.; Vahedi, V.; Giri, K.; Lewandowski, P.; Jacobs, J.L.; Dunshea, F.R. Muscle antioxidant enzymes activity and gene expression are altered by diet-induced increase in muscle essential fatty acid (α-linolenic acid) concentration in sheep used as a model. *Nutrients* **2019**, *11*, 723. [CrossRef]
71. Kessler, J.; Morel, I.; Dufey, P.A.; Gutzwiller, A.; Stern, A.; Geyer, H. Effect of organic zinc sources on performance, zinc status and carcass, meat and claw quality in fattening bulls. *Livest. Prod. Sci.* **2003**, *81*, 161–171. [CrossRef]
72. Kelleher, S.L.; Lönnerdal, B. Zinc supplementation reduces iron absorption through age-dependent changes in small intestine iron transporter expression in suckling rat pups. *J. Nutr.* **2006**, *136*, 1185–1191. [CrossRef]
73. Babicz, M.; Kasprzyk, A. Comparative analysis of the mineral composition in the meat of wild boar and domestic pig. *Ital. J. Anim. Sci.* **2019**, *18*, 1013–1020. [CrossRef]
74. Ponnampalam, E.N.; Kerr, M.G.; Butler, K.L.; Cottrell, J.J.; Dunshea, F.R.; Jacobs, J.L. Filling the out of season gaps for lamb and hogget production: Diet and genetic influence on carcass yield, carcass composition and retail value of meat. *Meat Sci.* **2019**, *148*, 156–163. [CrossRef]
75. López, M.A.; Martos, F.C. Iron availability: An updated review. *Int. J. Food. Sci. Nutr.* **2004**, *55*, 597–606. [CrossRef] [PubMed]
76. Martínez-Navarrete, N.; Camacho, M.; Martınez-Lahuerta, J.; Martınez-Monzó, J.; Fito, P. Iron deficiency and iron fortified foods—A review. *Int. Food Res. J.* **2002**, *35*, 225–231. doi:10.1016/S0963-9969(01)00189-2. [CrossRef]
77. Ishikawa, S.; Tamaki, S.; Ohata, M.; Arihara, K.; Itoh, M. Heme induces DNA damage and hyperproliferation of colonic epithelial cells via hydrogen peroxide produced by heme oxygenase: A possible mechanism of heme-induced colon cancer. *Mol. Nutr. Food Res.* **2010**, *54*, 1182–1191. [CrossRef] [PubMed]
78. Apple, J.; Wallis-Phelps, W.; Maxwell, C.; Rakes, L.; Sawyer, J.; Hutchison, S.; Fakler, T. Effect of supplemental iron on finishing swine performance, carcass characteristics, and pork quality during retail display. *J. Anim. Sci.* **2007**, *85*, 737–745. [CrossRef] [PubMed]
79. Ponnampalam, E.N.; Behrendt, R.; Kerr, M.G.; Raeside, M.C.; McDonagh, M.B. The influence of the level of ewe gestation nutrition and lamb finishing diet on long-chain polyunsaturated fat concentration, antioxidant and mineral status, and colour stability of meat. *Anim. Prod. Sci.* **2018**, *58*, 1481–1487. [CrossRef]
80. O'Sullivan, M.G.; Byrne, D.V.; Stagsted, J.; Andersen, H.J.; Martens, M. Sensory colour assessment of fresh meat from pigs supplemented with iron and vitamin E. *Meat Sci.* **2002**, *60*, 253–265. [CrossRef]
81. Dimov, K.; Kalev, R.; Penchev, P. Effect of finishing diet with excluded silage on amino-acid, fatty-acid, and mineral composition of meat (*M. Longisimus Dorsi*) Calves. *Bulg. J. Agric. Sci.* **2012**, *18*, 288–295.
82. Zembayashi, M.; Lunt, D.K.; Smith, S.B. Dietary tea reduces the iron content of beef. *Meat Sci.* **1999**, *53*, 221–226. doi:10.1016/S0309-1740(99)00058-3. [CrossRef]
83. Hermesch, S.; Jones, R.M. Genetic parameters for haemoglobin levels in pigs and iron content in pork. *Animal* **2012**, *6*, 1904–1912. [CrossRef]
84. Mateescu, R.G.; Garrick, D.J.; Tait, R.G., Jr.; Garmyn, A.J.; Duan, Q.; Liu, Q.; Mayes, M.S.; Van Eenennaam, A.L.; VanOverbeke, D.L.; Hilton, G.G.; et al. Genome-wide association study of concentrations of iron and other minerals in *Longissimus* Muscle Angus Cattle. *J. Anim. Sci.* **2013**, *91*, 3593–3600. [CrossRef] [PubMed]
85. King, D.; Shackelford, S.; Wheeler, T. Relative contributions of animal and muscle effects to variation in beef lean color stability. *J. Anim. Sci.* **2011**, *89*, 1434–1451. [CrossRef]
86. Rayman, M.P. Selenium and human health. *Lancet* **2012**, *379*, 1256–1268. [CrossRef]
87. Rostami, R.; Nourooz-Zadeh, S.; Mohammadi, A.; Khalkhali, H.R.; Ferns, G.; Nourooz-Zadeh, J. Serum Selenium Status and Its Interrelationship with Serum Biomarkers of Thyroid Function and Antioxidant Defense in Hashimoto's Thyroiditis. *Antioxidants* **2020**, *9*, 1070. [CrossRef]
88. Mehdi, Y.; Dufrasne, I. Selenium in cattle: A review. *Molecules* **2016**, *21*, 545. [CrossRef]

89. Falk, M.; Bernhoft, A.; Framstad, T.; Salbu, B.; Wisløff, H.; Kortner, T.M.; Kristoffersen, A.B.; Oropeza-Moe, M. Effects of dietary sodium selenite and organic selenium sources on immune and inflammatory responses and selenium deposition in growing pigs. *J. Trace Elem. Med. Biol.* **2018**, *50*, 527–536. [CrossRef]
90. Paiva, F.A.; Saran Netto, A.; Correa, L.B.; Silva, T.H.; Guimaraes, I.C.S.B.; Del Claro, G.R.; Cunha, J.A.; Zanetti, M.A. Organic selenium supplementation increases muscle selenium content in growing lambs compared to inorganic source. *Small Rumin. Res.* **2019**, *175*, 57–64. [CrossRef]
91. Flachowsky, G.; Berk, A.; Meyer, U. Iodine transfer from feed into meat and other food of animal origin. *Fleischwirtschaft* **2007**, *87*, 83–87.
92. Meyer, U.; Weigel, K.; Schöne, F.; Leiterer, M.; Flachowsky, G. Effect of dietary iodine on growth and iodine status of growing fattening bulls. *Livest Sci.* **2008**, *115*, 219–225. [CrossRef]
93. Schöne, F.; Zimmermann, C.; Quanz, G.; Richter, G.; Leiterer, M. A high dietary iodine increases thyroid iodine stores and iodine concentration in blood serum but has little effect on muscle iodine content in pigs. *Meat Sci.* **2006**, *72*, 365–372. [CrossRef]
94. Millward, D.J. Macronutrient intakes as determinants of dietary protein and amino acid adequacy. *J. Nutr.* **2004**, *134*, 1588S–1596S. [CrossRef]
95. Howard, E.E.; Pasiakos, S.M.; Fussell, M.A.; Rodriguez, N.R. Skeletal muscle disuse atrophy and the rehabilitative role of protein in recovery from musculoskeletal injury. *Adv. Nutr.* **2020**, *11*, 989–1001. [CrossRef] [PubMed]
96. Prelack, K.; Dylewski, M.; Sheridan, R.L. Practical guidelines for nutritional management of burn injury and recovery. *Burns* **2007**, *33*, 14–24. [CrossRef] [PubMed]
97. Purchas, R.W.; Wilkinson, B.H.P.; Carruthers, F.; Jackson, F. A comparison of the nutrient content of uncooked and cooked lean from New Zealand beef and lamb. *J. Food Compos. Anal.* **2014**, *35*, 75–82. [CrossRef]
98. Greenfield, H.; Arcot, J.; Barnes, J.A.; Cunningham, J.; Adorno, P.; Stobaus, T.; Tume, R.K.; Beilken, S.L.; Muller, W.J. Nutrient composition of Australian retail pork cuts 2005/2006. *Food Chem.* **2009**, *117*, 721–730. [CrossRef]
99. Huang, S.; Wang, L.M.; Sivendiran, T.; Bohrer, B.M. Review: Amino acid concentration of high protein food products and an overview of the current methods used to determine protein quality. *Crit. Rev. Food Sci. Nutr.* **2018**, *58*, 2673–2678. [CrossRef]
100. Drazbo, A.; Sobotka, W.; Matusevicius, P. The effect of production system, dietary protein levels and amino acid supplementation on performance, carcass traits and meat quality in growing-finishing pigs. *Vet. Ir. Zootech.* **2012**, *57*, 3–9. doi:10.1186/s40104-019-0381-2. [CrossRef]
101. Jin, S.K.; Kim, I.S.; Kim, S.J.; Jeong, K.J.; Lee, J.R. Fatty Acid, Amino Acid Composition and Sensory Traits of Pork from Pigs Fed Artificial Culture Medium of Wild Ginseng. *Korean J. Food Sci. Anim. Resour.* **2006**, *26*, 349–355.
102. Li, F.; Duan, Y.; Li, Y.; Tang, Y.; Geng, M.; Oladele, O.A.; Kim, S.W.; Yin, Y. Effects of dietary n-6:n-3 PUFA ratio on fatty acid composition, free amino acid profile and gene expression of transporters in finishing pigs. *BJN* **2015**, *113*, 739–748. [CrossRef] [PubMed]
103. Intarapichet, K.O.; Maikhunthod, B.; Thungmanee, N. Physicochemical characteristics of pork fed palm oil and conjugated linoleic acid supplements. *Meat Sci.* **2008**, *80*, 788–794. [CrossRef] [PubMed]
104. Koutsidis, G.; Elmore, J.S.; Oruna-Concha, M.J.; Campo, M.M.; Wood, J.D.; Mottram, D.S. Water-soluble precursors of beef flavour: I. Effect of diet and breed. *Meat Sci.* **2008**, *79*, 124–130. [CrossRef]
105. Chai, J.; Diao, Q.; Zhao, J.; Wang, H.; Deng, K.; Qi, M.; Nie, M.; Zhang, N. Effects of rearing system on meat quality, fatty acid and amino acid profiles of Hu lambs. *Anim. Sci.* **2018**, *89*, 1178–1186. [CrossRef]
106. Mioč, B.; Pavič, V.; Vnučec, I.; Prpić, Z.; Kostelić, A.; Sušić, V. Effect of olive cake on daily gain, carcass characteristics and chemical composition of lamb meat. *Czech. J. Anim. Sci.* **2008**, *52*, 31–36. [CrossRef]
107. Ahlberg, C.M.; Schiermiester, L.N.; Howard, T.J.; Calkins, C.R.; Spangler, M.L. Genome wide association study of cholesterol and poly- and monounsaturated fatty acids, protein, and mineral content of beef from crossbred cattle. *Meat Sci.* **2014**, *98*, 804–814. [CrossRef]
108. Sakuma, H.; Saito, K.; Kohira, K.; Ohhashi, F.; Shoji, N.; Uemoto, Y. Estimates of genetic parameters for chemical traits of meat quality in Japanese black cattle. *Anim. Sci.* **2017**, *88*, 203–212. [CrossRef]
109. Lee, Y.S.; Jeong, H.; Taye, M.; Kim, H.J.; Ka, S.; Ryu, Y.C.; Cho, S. Genome-wide Association Study (GWAS) and Its Application for Improving the Genomic Estimated Breeding Values (GEBV) of the Berkshire Pork Quality Traits. *AAAP* **2015**, *28*, 1551–1557. [CrossRef]
110. Salejda, A.; Krasnowska, G.; Blaszczuk, M. Analysis of quality properties in raw meat and fats from fatteners breeding in Wielkopolska. *Acta Sci. Pol. Technol. Aliment.* **2009**, *8*, 11–19.
111. Peraza-Mercado, G.; Jaramillo-Lopez, E.; Alarcon-Rojo, A.D. Fatty acid and amino acid composition of Pelibuey and Polypay x Rambouillet lambs. *AEJAES* **2010**, *8*, 197–205.
112. Mapiye, C.; Aldai, N.; Turner, T.; Aalhus, J.; Rolland, D.; Kramer, J.; Dugan, M. The labile lipid fraction of meat: From perceived disease and waste to health and opportunity. *Meat Sci.* **2012**, *92*, 210–220. [CrossRef]
113. Doreau, M.; Meynadier, A.; Fievez, V.; Ferlay, A. Chapter 19—Ruminal metabolism of fatty acids: Modulation of polyunsaturated, conjugated, and *trans* fatty acids in meat and milk. In *Handbook of Lipids in Human Function;* Elsevier: Amsterdam, The Netherlands, 2016; pp. 521–542. [CrossRef]
114. Vahmani, P.; Mapiye, C.; Prieto, N.; Rolland, D.C.; McAllister, T.A.; Aalhus, J.L.; Dugan, M.E.R. The scope for manipulating the polyunsaturated fatty acid content of beef: A review. *J. Anim. Sci. Biotechnol.* **2015**, *6*, 29. [CrossRef]

115. Vahmani, P.; Ponnampalam, E.N.; Kraft, J.; Mapiye, C.; Bermingham, E.N.; Watkins, P.J.; Proctor, S.D.; Dugan, M.E.R. Bioactivity and health effects of ruminant meat lipids. Invited Review. *Meat Sci.* **2020**, *165*, 108114. [CrossRef]
116. Smith, S.B.; Lunt, D.K.; Smith, D.R.; Walzem, R.L. Producing high-oleic acid beef and the impact of ground beef consumption on risk factors for cardiovascular disease: A review. *Meat Sci.* **2020**, *163*, 108076. [CrossRef] [PubMed]
117. Huerta-Leidenz, N.O.; Cross, H.R.; Savell, J.W.; Lunt, D.K.; Baker, J.F.; Smith, S.B. Fatty acid composition of subcutaneous adipose tissue from male calves at different stages of growth. *J. Anim. Sci.* **1996**, *74*, 1256–1264. [CrossRef] [PubMed]
118. Vahmani, P.; Aalhus, J.; Rolland, D.; McAllister, T.; Prieto, N.; Block, H.; Proctor, S.; Guan, L.; Dugan, M. Sequential feeding of lipid supplement enriches beef adipose tissues with 18: 3n-3 biohydrogenation intermediates. *Lipids* **2017**, *52*, 641–649. [CrossRef] [PubMed]
119. Vahmani, P.; Rolland, D.C.; McAllister, T.A.; Block, H.C.; Proctor, S.D.; Guan, L.L.; Prieto, N.; Lopez-Campos, O.; Aalhus, J.L.; Dugan, M.E.R. Effects of feeding steers extruded flaxseed on its own before hay or mixed with hay on animal performance, carcass quality, and meat and hamburger fatty acid composition. *Meat Sci.* **2017**, *131*, 9–17. [CrossRef] [PubMed]
120. Petri, R.M.; Vahmani, P.; Yang, H.E.; Dugan, M.E.; McAllister, T.A. Changes in rumen microbial profiles and subcutaneous fat composition when feeding extruded flaxseed mixed with or before hay. *Front Microbiol.* **2018**, *9*, 1055. [CrossRef] [PubMed]
121. Vahmani, P.; Meadus, W.J.; Duff, P.; Rolland, D.C.; Dugan, M.E.R. Comparing the lipogenic and cholesterolgenic effects of individual *Trans*-18:1 Isomers Liver Cells. *Eur. J. Lipid Sci. Technol.* **2017**, *119*, 1600162. [CrossRef]
122. Rom, O.; Khoo, N.K.H.; Chen, Y.E.; Villacorta, L. Inflammatory signaling and metabolic regulation by nitro-fatty acids. *Nitric Oxide* **2018**, *78*, 140–145. [CrossRef]
123. Dugan, M.E.; Mapiye, C.; Vahmani, P. Chapter 4—Polyunsaturated Fatty Acid Biosynthesis and Metabolism in Agriculturally Important Species. In *Polyunsaturated Fatty Acid Metabolism*; Elsevier: Amsterdam, The Netherlands, 2018; pp. 61–86. [CrossRef]
124. Dugan, M.E.R.; Aalhus, J.L.; Uttaro, B. Nutritional manipulation of pork quality: Current opportunities. In *Advances in Pork Production*; University of Alberta: Edmonton, AB, Canada, 2004; Volume 15, pp. 237–243.
125. Wood, J.D.; Enser, M.; Fisher, A.V.; Nute, G.R.; Sheard, P.R.; Richardson, R.I.; Hughes, S.I.; Whittington, F.M. Fat deposition, fatty acid composition and meat quality: A review. *Meat Sci.* **2008**, *78*, 343–358. [CrossRef] [PubMed]
126. Turner, T.; Mapiye, C.; Aalhus, J.; Beaulieu, A.; Patience, J.; Zijlstra, R.; Dugan, M. Flaxseed fed pork: N-3 fatty acid enrichment and contribution to dietary recommendations. *Meat Sci.* **2014**, *96*, 541–547. [CrossRef]
127. Straadt, I.K.; Aaslyng, M.D.; Bertram, H.C. Sensory and consumer evaluation of pork loins from crossbreeds between Danish Landrace, Yorkshire, Duroc, Iberian and Mangalitza. *Meat Sci.* **2013**, *95*, 27–35. [CrossRef] [PubMed]
128. Gotoh, T.; Joo, S.T. Characteristics and health benefit of highly marbled Wagyu and Hanwoo beef. *Korean J. Food Sci. Anim. Resour.* **2016**, *36*, 709. [CrossRef]
129. Shackelford, S.; Koohmaraie, M.; Cundiff, L.; Gregory, K.; Rohrer, G.; Savell, J. Heritabilities and phenotypic and genetic correlations for bovine postrigor calpastatin activity, intramuscular fat content, Warner-Bratzler shear force, retail product yield, and growth rate. *J. Anim. Sci.* **1994**, *72*, 857–863. [CrossRef] [PubMed]
130. Warner, R.D.; Greenwood, P.L.; Pethick, D.W.; Ferguson, D.M. Genetic and environmental effects on meat quality. *Meat Sci.* **2010**, *86*, 171–183. [CrossRef] [PubMed]
131. Wood, J.; Richardson, R.; Nute, G.; Fisher, A.; Campo, M.; Kasapidou, E.; Sheard, P.; Enser, M. Effects of fatty acids on meat quality: A review. *Meat Sci.* **2004**, *66*, 21–32. [CrossRef]
132. Pitchford, W.; Deland, M.; Siebert, B.; Malau-Aduliand, A.; Bottema, C. Genetic variation in fatness and fatty acid composition of crossbred cattle. *J. Anim. Sci.* **2002**, *80*, 2825–2832. [CrossRef]
133. De Smet, S.; Raes, K.; Demeyer, D. Meat fatty acid composition as affected by fatness and genetic factors: A review. *Anim. Res.* **2004**, *53*, 81–98.:2004003. [CrossRef]
134. Viterbo, V.S.; Lopez, B.I.M.; Kang, H.; Kim, H.; Song, C.W.; Seo, K.S. Genome wide association study of fatty acid composition in Duroc swine. *AAAP* **2018**, *31*, 1127. [CrossRef]
135. Pewan, S.B.; Otto, J.R.; Huerlimann, R.; Budd, A.M.; Mwangi, F.W.; Edmunds, R.C.; Holman, B.W.B.; Henry, M.L.E.; Kinobe, R.T.; Adegboye, O.A. Genetics of omega-3 long-chain polyunsaturated fatty acid metabolism and meat eating quality in Tattykeel Australian White lambs. *Genes* **2020**, *11*, 587. [CrossRef] [PubMed]
136. Cesar, A.S.; Regitano, L.C.; Mourão, G.B.; Tullio, R.R.; Lanna, D.P.; Nassu, R.T.; Mudado, M.A.; Oliveira, P.S.; do Nascimento, M.L.; Chaves, A.S. Genome-wide association study for intramuscular fat deposition and composition in Nellore cattle. *BMC Genet.* **2014**, *15*, 1–15. [CrossRef] [PubMed]
137. Luo, W.; Cheng, D.; Chen, S.; Wang, L.; Li, Y.; Ma, X.; Song, X.; Liu, X.; Li, W.; Liang, J.; et al. Genome-wide association analysis of meat quality traits in a porcine Large White—Minzhu intercross population. *Int. J. Biol. Sci.* **2012**, *8*, 580–595. [CrossRef] [PubMed]
138. Zhang, Y.; Zhang, J.; Gong, H.; Cui, L.; Zhang, W.; Ma, J.; Chen, C.; Ai, H.; Xiao, S.; Huang, L. Genetic correlation of fatty acid composition with growth, carcass, fat deposition and meat quality traits based on GWAS data in six pig populations. *Meat Sci.* **2019**, *150*, 47–55. [CrossRef]
139. Srikanth, K.; Lee, E.; Kwon, A.; Jang, G.; Chung, H. Association of a single nucleotide polymorphism in the calneuron 1 gene on meat quality and carcass traits in hanwoo (*Bos Taurus Coreanae*). *J. Anim. Plant Sci.* **2018**, *28*, 651–655.

140. Zhang, Z.; Zhang, Z.; Oyelami, F.; Sun, H.; Xu, Z.; Ma, P.; Wang, Q.; Pan, Y. Identification of genes related to intramuscular fat independent of backfat thickness in Duroc pigs using single-step genome-wide association. *Anim. Genet.* **2021**, *52*, 108–113. [CrossRef]
141. Li, C.; Aldai, N.; Vinsky, M.; Dugan, M.; McAllister, T. Association analyses of single nucleotide polymorphisms in bovine stearoyl-CoA desaturase and fatty acid synthase genes with fatty acid composition in commercial cross-bred beef steers. *Anim. Genet.* **2012**, *43*, 93–97. [CrossRef] [PubMed]
142. Lai, L.; Kang, J.X.; Li, R.; Wang, J.; Witt, W.T.; Yong, H.Y.; Hao, Y.; Wax, D.M.; Murphy, C.N.; Rieke, A. Generation of cloned transgenic pigs rich in omega-3 fatty acids. *Nat. Biotechnol.* **2006**, *24*, 435–436. [CrossRef]
143. Tang, F.; Yang, X.; Liu, D.; Zhang, X.; Huang, X.; He, X.; Shi, J.; Li, Z.; Wu, Z. Co-expression of fat1 and fat2 in transgenic pigs promotes synthesis of polyunsaturated fatty acids. *Transgenec Res.* **2019**, *28*, 369–379. [CrossRef]
144. Richards, M.P.; Kathirvel, P.; Gong, Y.; Lopez-Hernandez, A.; Walters, E.M.; Prather, R.S. Long chain omega-3 fatty acid levels in loin muscle from transgenic (fat-1 gene) pigs and effects on lipid oxidation during storage. *Food Biotechnol.* **2011**, *25*, 103–114. [CrossRef]
145. Xin, X.B.; Yang, S.P.; Li, X.; Liu, X.F.; Zhang, L.L.; Ding, X.B.; Zhang, S.; Li, G.P.; Guo, H. Proteomics insights into the effects of MSTN on muscle glucose and lipid metabolism in genetically edited cattle. *Gen. Comp. Endocrinol.* **2020**, *291*, 113237. [CrossRef]
146. Moghadasian, M.H. Advances in dietary enrichment with N-3 fatty acids. *Crit. Rev. Food Sci. Nutr.* **2008**, *48*, 402–410. [CrossRef]
147. Venegas-Calerón, M.; Sayanova, O.; Napier, J.A. An alternative to fish oils: Metabolic engineering of oil-seed crops to produce omega-3 long chain polyunsaturated fatty acids. *Prog. Lipid Res.* **2010**, *49*, 108–119. [CrossRef]
148. Ruan, J.; Xu, J.; Chen-Tsai, R.Y.; Li, K. Genome editing in livestock: Are we ready for a revolution in animal breeding industry? *Transgenic Res.* **2017**, *26*, 715–726. [CrossRef]
149. Lamas-Toranzo, I.; Guerrero-Sánchez, J.; Miralles-Bover, H.; Alegre-Cid, G.; Pericuesta, E.; Bermejo-Alvarez, P. CRISPR is knocking on barn door. *Reprod. Domest. Anim.* **2017**, *52*, 39–47. [CrossRef]
150. Eriksson, S.; Jonas, E.; Rydhmer, L.; Röcklinsberg, H. Invited review: Breeding and ethical perspectives on genetically modified and genome edited cattle. *J. Dairy Sci.* **2018**, *101*, 1–17. [CrossRef]
151. Juárez, M.; Dugan, M.E.R.; López-Campos, Ó.; Prieto, N.; Uttaro, B.; Gariépy, C.; Aalhus, J.L. Relative contribution of breed, slaughter weight, sex, and diet to the fatty acid composition of differentiated pork. *Can. J. Anim. Sci.* **2017**, *97*, 395–405. [CrossRef]
152. Juárez, M.; Horcada, A.; Alcalde, M.J.; Valera, M.; Mullen, A.M.; Molina, A. Estimation of factors influencing fatty acid profiles in light lambs. *Meat Sci.* **2008**, *79*, 203–210. [CrossRef]
153. Juárez, M.; Basarab, J.A.; Baron, V.S.; Valera, M.; Larsen, I.L.; Aalhus, J.L. Quantifying the relative contribution of ante- and post-mortem factors to the variability in beef texture. *Animal* **2012**, *6*, 1878–1887. [CrossRef]
154. Krause, T.R.; Lourenco, J.M.; Welch, C.B.; Rothrock, M.J.; Callaway, T.R.; Pringle, T.D. The relationship between the rumen microbiome and carcass merit in Angus steers. *J. Anim. Sci.* **2020**, *98*, 1–12. [CrossRef] [PubMed]
155. Banhazi, T.M.; Lehr, H.; Black, J.; Crabtree, H.; Schofield, P.; Tscharke, M.; Berckmans, D. Precision livestock farming: An international review of scientific and commercial aspects. *Int. J. Agric. Biol. Eng.* **2012**, *5*, 1–9. [CrossRef]
156. Juárez, M. Linking livestock phenomics and precision livestock farming. *J. Anim. Sci.* **2020**, *98*, 124–124. [CrossRef]
157. Smith, J.; Sones, K.; Grace, D.; MacMillan, S.; Tarawali, S.; Herrero, M. Beyond milk, meat, and eggs: Role of livestock in food and nutrition security. *J. Anim. Front.* **2013**, *3*, 6–13. [CrossRef]
158. McAfee, A.J.; McSorley, E.M.; Cuskelly, G.J.; Moss, B.W.; Wallace, J.M.W.; Bonham, M.P.; Fearon, A.M. Red meat consumption: An overview of the risks and benefits. *Meat Sci.* **2010**, *84*, 1–13. [CrossRef]
159. Leroy, F.; Cofnas, N. Should dietary guidelines recommend low red meat intake? *Crit. Rev. Food Sci. Nutr.* **2020**, *60*, 2763–2772. [CrossRef]
160. Scollan, N.; Hocquette, J.F.; Nuernberg, K.; Dannenberger, D.; Richardson, I.; Moloney, A. Innovations in beef production systems that enhance the nutritional and health value of beef lipids and their relationship with meat quality. *Meat Sci.* **2006**, *74*, 17–33. [CrossRef]
161. Zhang, H.; Aalhus, J.L.; Gariepy, C.; Uttaro, B.; Lopez-Campos, O.; Prieto, N.; Dugan, M.E.R.; Jin, Y.; Juárez, M. Effects of pork differentiation strategies in Canada on pig performance and carcass characteristics. *Can. J. Anim. Sci.* **2016**, *96*, 512–523. [CrossRef]
162. Markus, S.B.; Aalhus, J.L.; Janz, J.A.M.; Larsen, I.L. A survey comparing meat quality attributes of beef from credence attribute-based production systems. *Can. J. Anim. Sci.* **2011**, *91*, 283–294. [CrossRef]
163. ALMA. *Canadian Consumer Retail Meat Study Final Report*; Report 2016; ALMA: Edmonton, AB, Canada, 2016.
164. De Smet, S.; Vossen, E. Meat: The balance between nutrition and health. A review. *Meat Sci.* **2016**, *120*, 145–156. [CrossRef] [PubMed]
165. Wunderlich, S.; Gatto, K.A. Consumer perception of genetically modified organisms and sources of information. *Adv. Nutr.* **2015**, *6*, 842–851. [CrossRef] [PubMed]
166. Juárez, M.; Dugan, M.E.R.; Aldai, N.; Aalhus, J.L.; Patience, J.F.; Zijlstra, R.T.; Beaulieu, A.D. Increasing omega-3 levels through dietary co-extruded flaxseed supplementation negatively affects pork palatability. *Food Chem.* **2011**, *126*, 1716–1723. [CrossRef]
167. Mir, P.; Ivan, M.; He, M.; Pink, B.; Okine, E.; Goonewardene, L.; McAllister, T.; Weselake, R.; Mir, Z. Dietary manipulation to increase conjugated linoleic acids and other desirable fatty acids in beef: A review. *Can. J. Anim. Sci.* **2003**, *83*, 673–685. [CrossRef]
168. Henchion, M.M.; McCarthy, M.; Resconi, V.C. Beef quality attributes: A systematic review of consumer perspectives. *Meat Sci.* **2017**, *128*, 1–7. [CrossRef]

 foods

MDPI

Article

Effect of Dietary Brown Seaweed (*Macrocystis pyrifera*) Additive on Meat Quality and Nutrient Composition of Fattening Pigs

Nancy Jerez-Timaure [1,*], Melissa Sánchez-Hidalgo [2], Rubén Pulido [1] and Jonathan Mendoza [3]

[1] Instituto de Ciencia Animal, Facultad de Ciencias Veterinarias, Universidad Austral de Chile, Valdivia 5090000, Chile; rpulido@uach.cl
[2] Escuela de Graduados, Facultad de Ciencias Veterinarias, Universidad Austral de Chile, Valdivia 5090000, Chile; melissaruth47@gmail.com
[3] I+D Patagonia Biotecnología SA., Valdivia 5090000, Chile; jonathan.mendoza.mv@gmail.com
* Correspondence: nancy.jerez@uach.cl

Abstract: The objective of this study was to evaluate the effects of dietary brown seaweed (*Macrocystis pyrifera*) additive (SWA) on meat quality and nutrient composition of commercial fattening pigs. The treatments were: Regular diet with 0% inclusion of SWA (CON); Regular diet with 2% SWA (2%-SWA); Regular diet with 4% SWA (4%-SWA). After slaughtering, five carcasses from each group were selected, and *longissimus lumborum* (LL) samples were taken for meat quality and chemical composition analysis. Meat quality traits (except redness intensity) were not affected ($p > 0.05$) by treatments. Samples from the 4%-SWA treatment showed the lowest a value than those from the 2%-SWA and CON treatments ($p = 0.05$). Meat samples from the 4%-SWA group contained 3.37 and 3.81 mg/100 g more of muscle cholesterol than CON and 2% SWA groups, respectively ($p < 0.05$). The SWA treatments affected ($p \leq 0.05$) the content of ash, Mn, Fe, and Cu. The LL samples from 4%-SWA had the highest content of ash; however, they showed 0.13, 0.45, and 0.23 less mg/100 g of Mn, Fe, and Zn, respectively, compared to samples from CON ($p \leq 0.05$). Fatty acids composition and macro minerals content (Na, Mg, and K) did not show variation due to the SWA treatments. Further studies are needed to understand the biological effects of these components on adipogenesis, cholesterol metabolism, and mineral deposition in muscle.

Keywords: pig; seaweed; pork quality; fatty acids; minerals; proximal composition; *Macrocystis pyrifera*

Citation: Jerez-Timaure, N.; Sánchez-Hidalgo, M.; Pulido, R.; Mendoza, J. Effect of Dietary Brown Seaweed (*Macrocystis pyrifera*) Additive on Meat Quality and Nutrient Composition of Fattening Pigs. *Foods* **2021**, *10*, 1720. https://doi.org/10.3390/foods10081720

Academic Editors: Nelson O. Huerta-Leidenz, Markus F. Mille and Juana Fernández-López

Received: 4 May 2021
Accepted: 20 July 2021
Published: 26 July 2021

1. Introduction

Pork meat is generally recognized as an excellent source of proteins, B-vitamins, minerals, especially heme iron, trace elements, and other bioactive compounds [1]. Pork has been the most widely consumed meat in the world up to 2018, accounting for 40.1% of the total global meat intake [2]. The nutritional composition of meat and its organoleptic attributes depends on many factors such as genetics, age, sex condition, production system, diet, and location of cuts/muscles [3]. Pork quality has been the primary concern for producers, researchers, meat packers, processors, retailers, and ultimate consumers [4]. However, in the last decades, consumers have been demanding healthier and nutritious foods [5]. This demand has motivated a growing interest in the study of natural supplements in pig nutrition that could enhance the productive performance and health of the animals but also improve meat quality and nutrient composition.

Currently, there is a worldwide increase in the use of seaweed as a supplement or additive in animal production, especially on sustainable production systems [6,7]. Due to the increasing interest in organic foods, seaweed represents an alternative to avoid chemical or synthetic ingredients in the diet of animals [7,8].

Among the huge variety of seaweeds, brown seaweed is widely distributed throughout the globe and is also very abundant on the Chilean coast [9]. There are approximately 1500–2000 species of brown algae worldwide, and some species, such as *Macrocystis pyrifera* (giant kelp), play an important role in the ecosystem, growing up to 20 m and forming underwater kelp forests [10]. Their big size and easy harvesting management offer important advantages for animal consumption [11].

In general, brown seaweed has a lower protein content and lipids, but it contains higher levels of minerals, fatty acids, and essential amino acids and bioactive components than red and green seaweeds [11]. Moreover, brown seaweed contains vitamins A, B1, B12, D, E, and C, and folic acid, riboflavin, niacin, and pantothenic acid [12] and are an excellent source of antioxidants and bioactive compounds [13,14].

The inclusion of seaweed additives in the diet of lambs [15], beef cattle [16], fish [17], laying hens [18], and rabbits [19] has been studied. Additionally, the use of seaweed extracts in growing and fattening pigs has shown an improved growth performance and feed efficiency [9,20–22]. Extensive reviews about the inclusion of seaweeds in monogastric production have been published [21,23,24]. However, fewer studies have investigated the effects of dietary seaweed on pork carcass/meat quality traits [22,25–28]; but none of them has reported the effects of these natural additives on pork meat composition and quality.

The magnitude of the associate response to the inclusion of seaweed in the animals' diet on growth performance and carcass traits depends on the type of seaweed used (red, brown, green, and yellow), the bioactive components present in the extract, and the proportion and frequency used in the diet [11]. The objective of this study is to evaluate the effects of the inclusion of brown seaweed (*Macrocystis pyrifera*) additive in the diet of fattening pigs on pork meat quality and nutrient composition.

2. Materials and Methods

2.1. Sampling and Seaweed Additive Description

The study was conducted with 240 pigs (castrated males and females) of approximately 14 weeks of age (52.5 ± 2.8 kg) in an intensive production unit located in the Ñuble Region (Chile). Before the fattening phase, animals were separated into 12 groups of 20 pigs each. Groups were assigned to one of the following treatments: Control Group: Regular diet with 0% of seaweed additive (SWA); 2%-SWA: Regular diet + 20 kg of SWA per 1000 kg of concentrate; 4%-SWA: Regular diet + 40 kg of SWA per 1000 kg of concentrate. Pigs were balanced by weight and sex condition. The SWA is a lyophilized product of *Macrocystis pyrifera*, that maintains its chemical-physical characteristics and bioactive compounds. The composition of the regular diet and the chemical composition of the SWA are shown in Table 1.

Table 1. Composition of the regular diet and chemical composition of the seaweed additive (SWA).

Diet Ingredient	%	Chemical Composition of the Seaweed Additive	% *
Triticale	70.00	Dry matter	92.67
Wheat bran	8.10	Ash	25.60
Soya meal	20.20	Crude protein	8.87
Salt	0.42	Crude fibre	2.87
Calcium carbonate	0.60	Neutral-detergent fibre	7.06
Phosphate	0.17	Acid-detergent fibre	11.78
Oil	1.01	Ether extract	0.29
Vitamin-mineral premix [1]	0.10	Nitrogen free extract	55.04
Lisin	0.08		
Methionine	0.04		
Threonine	0.02		
quantum blue [2]	0.01		
Enocase [3]	0.02		

* based on dry matter. [1]: contains vitamins: A, D3, E, K3, B2, and B12; and minerals: Mn, Cu, I, Zn, Fe, Se, and Ca. [2]: an enhanced E. coli phytase. [3]: an enzyme preparation with endo-1,4-β-xylanase.

Five female pigs per treatment/replicate were selected for the *postmortem* evaluation to avoid the sex variation factor. The average final live weight was 106.7 ± 2.2 kg. The animals were transported to a slaughterhouse plant facility located 30 km away from the commercial farm and slaughtered after 12 h of lairage. The carcasses were chilled for 48 h *postmortem* at 4 °C. The entire portion of the *longissimus lumborum* (LL) muscle was removed from each left carcass side, and samples of 2.5 cm thickness were obtained. Two samples were used immediately for pH and color evaluation, and four samples of each loin were packaged and frozen at −20 °C for 30 days for the rest of the analysis.

2.2. Meat Quality Evaluation

Instrumental color was measured in fresh samples 48 h *postmortem*. A Hunter Lab Mini Scan XE Plus (Hunter Associates, Reston, VA, USA) was used with a 2.5-cm open port, Illuminant A, and 2° standard observer to objectively evaluate color. Three readings were obtained from the muscle surface, and the mean was calculated. Readings were obtained after exposing the muscles to air for 30 min (bloom). The color scale used was Hunter L, a, b. The L value represents lightness; a and b values represent redness, and yellowness, respectively. Warner–Bratzler Shear force (WBSF) and Water Holding Capacity (WHC) were evaluated in samples cooked in a convection oven (Albin Trotter model E-EMB Digital) to a final internal temperature of 70 °C following the guidelines of the American Meat Science Association [29]. The temperature was monitored using an Omega thermocouple thermometer type T (Omega Engineering, Inc., Stamford, CT, USA) inserted into the geometric center of each steak. The cooked steaks were chilled for 2 h at 2 °C, and then eight cores (1.27 cm in diameter) were removed parallel to the muscle fiber orientation. Cores were sheared once each on the Warner-Bratzler Meat Shear apparatus (GR Manufacturing Co., Manhattan, NY, USA) to get WBSF values. WHC was determined as cooking loss, which was determined by weight, expressed as a percentage compared to the original weight of the sample. A taste preference test was performed with 19 panelists (16 women and 3 men). Two steaks of each treatment were used in each session (two sessions). The tests were carried out in individual evaluation cabinets illuminated with red light. Each panelist, in each session, tasted three samples (one from each treatment) at random, and they were asked to select the best preference of the three samples.

2.3. Nutrient Composition of Meat

Moisture, protein, and fat content of meat samples were determined according to the AOAC [30]. All experiments were done in triplicate. Duplicates of 10 g of ground meat were calcined in a furnace at 550 °C for 6 h. After cooling, the residue (white ash) was subjected to an acid digestion process with 10 mL of a 20% v/v hydrochloric acid solution by heating on a hot plate for 10 min. Mineral analyses were conducted by atomic absorption and/or atomic emission [30], following the analytical methods described by Pelkin-Elmer [31]. Mineral content was expressed both as mg/100 g of fresh tissue or as % dry matter (DM).

Cholesterol content was determined by duplicate by gas chromatography according to Fletouris et al. [32]. The fatty acid composition was determined by direct fatty acid methyl esters (FAME) synthesis as described by Cantelolops et al. [33]. The FAME was analyzed by an Aligent 6890 GC system (Aligent Technologies, Santa Clara, CA, USA) equipped with a flame ionization detector and capillary column CP-sil388 (30 m length, 0.25 mm i.d., 0.20 µm film thickness) with a split injection of 1:50. Helium was used as a carrier gas. The temperature of the detector and injector was 250 °C. The initial temperature in the oven was 100 °C, and it reached 220 °C with an increasing rate of 5 °C/min. The fatty acids were identified by comparing their FAME retention times with sigma reference standards (Supelco™ 37 Component FAME mix, Sigma, St. Louis, MO, USA). FAME mix contains n3, n6, and n9 isomers. Results were reported as g/100 g of fresh muscle tissue.

2.4. Atherogenic (AI) and Thrombogenic Indexes (TI) and H/h Index

The risk of atherosclerosis and/or thrombogenesis was evaluated by using the atherogenic (AI) and thrombogenic indexes (TI). They were calculated based on the obtained fatty acid results, using the following equations [34]: AI = (C12:0 + 4 × C14:0 + C16:0)/[ΣMUFA + Σ (n − 6) + (Σ (n − 3)]; TI = (C14:0 + C16:0 + C18:0)/[0.5 × ΣMUFA + 0.5 × Σ(n − 6) + 3 × Σ(n − 3) + Σ(n − 3)/Σ(n − 6)]. The ratio between hypercholesterolemic (H) and hypocholesterolemic fatty acids was calculated as described by Monteiro et al. [35]: H/h = (C14:0 + C16:0)/ (C18:1 + C18:2 + C18:3 + C20:3 + C20:4 + C20:5 + C22:4 + C22:5 + C22:6).

2.5. Statistical Analysis

A One-way Analysis of Variance (ANOVA) was performed using a mixed model with SWA treatment as the main factor and animal as the random effect. The value $p \leq 0.05$ was used to declare the significant difference between the average scores. Tukey's multiple comparison test was used for the comparison of means. The Bonferroni correction was also performed to adjusts probability of *p*-values. χ^2 test was used for taste preference data.

3. Results

3.1. Meat Quality Traits

Meat quality traits evaluated in the *longissimus lumborum* (LL) muscle of fattening pigs fed with different levels of SWA are presented in Table 2. The pH values (measured 45 min and 24 h *postmortem*) for every treatment were in the range of normal values without being statistically different among treatments ($p > 0.05$). The ANOVA detected a significant effect of treatment ($p = 0.05$) on the a value (redness). Meat samples from the 4%-SWA treatment had less red intensity values when compared to those from the 2%-SWA and the control group ($p \leq 0.05$). No difference ($p > 0.05$) was observed for L value (lightness) and b value (yellowness). The instrumental tenderness and the cooking losses (expressed in percentage) were not affected by the SWA treatments ($p > 0.05$). The samples from the control group and 4%-SWA resulted in the best taste preference by panelists, being significantly different from the sample of the 2%-SWA group ($p \leq 0.05$). Samples from the control and the 4%-SWA groups showed a similar percentage of taste preference (Figure S1).

Table 2. Effects of the inclusion of seaweed additive (SWA) in the diet of fattening pigs on meat quality traits.

Variable	Treatment SWA			SEM	*p* Value
	Control	2%	4%		
Muscular pH, 45 min	6.08	6.34	6.21	0.08	0.13
Muscular pH, 24 h	5.61	5.64	5.63	0.07	0.93
Redness (a value)	5.88 [a]	5.61 [a]	4.95 [b]	0.24	0.05
Yellowness (b value)	9.32	9.60	9.06	0.14	0.18
Lightness (L value)	45.79	47.61	4495	0.96	0.37
Cooking loss, %	17.15	16.55	14.85	0.42	0.23
Shear force, kg	2.27	2.22	2.11	0.08	0.86

Means within a row lacking a common superscript letter differ ($p \leq 0.05$). SEM: Standard Error Mean.

3.2. Nutrient Composition of Meat

The ANOVA revealed that the SWA treatments only affected the total content of ash ($p < 0.001$). The LL samples of animals that were fed with the 4%-SWA presented a higher percentage of total ash compared to the other treatments (Table 3; $p \leq 0.05$). However, no mean differences in the total ash content were found ($p > 0.05$) when comparing the control with the 2%-SWA group. The LL samples from the 4%-WSA exhibited a decreased percentage of total lipids without being statistically different ($p = 0.07$).

Table 3. Effects of the inclusion of seaweed additive (SWA) in the diet of fattening pigs on the proximate composition of pork meat.

Variable [1]	Treatment SWA			SEM	*p* Value
	Control	2%	4%		
Moisture	73.77	74.08	74.47	0.21	0.26
Dry matter	26.92	25.92	25.52	0.26	0.26
Total ash	1.23 [a]	1.26 [a]	1.42 [b]	0.01	0.0001
Crude proteins	23.77	23.24	23.13	0.17	0.46
Total lipids	1.13	1.36	1.05	0.12	0.07

[1] values are expressed as g/100 g of fresh muscular tissue. Means within a row lacking a common superscript letter differ ($p \leq 0.05$). SEM: Standard Error Mean.

The ANOVA showed a significant effect of the SWA treatments on the total cholesterol content ($p \leq 0.05$). Figure 1 shows that meat samples from animals fed with the highest % of SWA (4%) contained 3.37 and 3.81 mg/100 g more muscular tissue cholesterol content than the control and the 2%-SWA groups, respectively ($p \leq 0.05$).

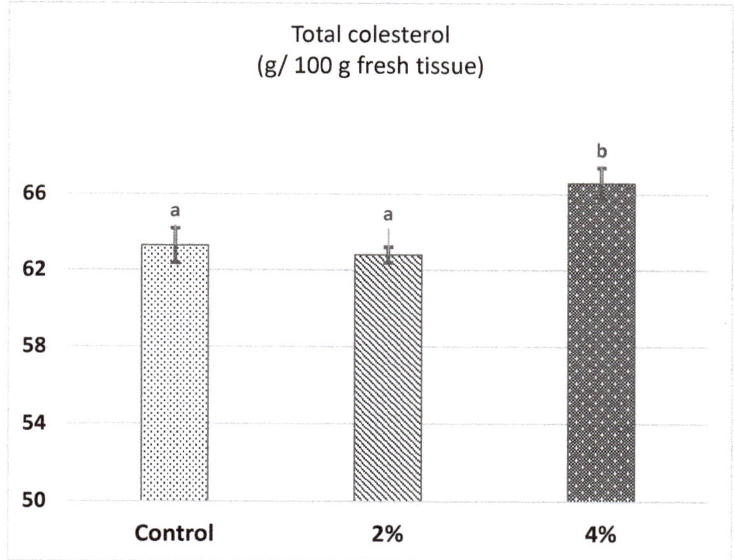

Figure 1. Mean values ± standard error means of total cholesterol in pork meat according to SWA treatments ($p = 0.02$). The different superscript letters represent significant statistical differences ($p \leq 0.05$).

The ANOVA showed that the SWA inclusion in the diet of fattening pigs did not affect the composition of saturated (SFA), monounsaturated (MUFA), and polyunsaturated fatty acids (PUFA) detected in the pork meat samples of this study. The ANOVA also found that the SWA treatments did not affect ($p > 0.05$) any of the health indexes associated with fatty acids composition: the ratio hypercholesterolemic index/hypocholesterolemic index, atherogenic index, thrombogenic index. In supplementary material (Table S1) are shown the descriptive statistics for fatty acids composition and health index in pork LL muscles.

Table 4 shows the mean values and standard error mean of several minerals evaluated in the LL samples of this study. The SWA treatments affected ($p \leq 0.05$) the content of the micro-minerals Mn, Fe, and Cu. The samples of meat with the greatest percentage of SWA (4%) had 0.13, 0.45, and 0.23 less mg/100 g of muscle in Mn, Fe, and Zn, respectively, compared with the control group ($p \leq 0.05$). The content of minerals of Na Mg and Zn were not different ($p > 0.05$) among the treatment groups; however, there was a trend ($p = 0.08$) of meat samples from the 4%-WSA treatment containing an inferior amount of K when compared to the 2%SWA and the control groups.

Table 4. Effects of the inclusion of seaweed additive (SWA) in the diet of fattening pigs on the mineral content in pork meat.

Mineral Content	Treatment SWA			SEM	*p* Value
	Control	2%	4%		
		Macrominerals [1]			
Na	44.57	50.15	44.35	1.82	0.20
Mg	23.12	25.28	24.75	1.29	0.71
K	492.67	504.62	430.03	12.50	0.08
		Microminerals [1]			
Mn	0.22 ^{b}	0.16 ^{ab}	0.09 ^{a}	0.01	0.002
Fe	1.75 ^{b}	1.18 ^{a}	1.30 ^{ab}	0.08	0.04
Cu	0.42 ^{b}	0.36 ^{b}	0.19 ^{a}	0.02	0.001
Zn	1.31	1.41	1.26	0.04	0.35

[1] values are expressed as mg/100 g of fresh muscular tissue. Means within a row lacking a common superscript letter differ ($p \leq 0.05$). SEM: Standard Error Mean.

By expressing the proximal composition of the meat on a dry basis (DM), the highly significant effect on the ash content was confirmed (Table 5; $p < 0.0001$). The samples from the animals that consumed the highest amount of SWA administered in the pigs' diet generated an increase in the amount of total ash DM of the meat. The Mn and Cu content decreased ($p = 0.002$) in the treatment with 4%-SWA, compared to control samples, but they were statistically similar to those of the 2%-SWA treatment. In addition, a trend could be evidenced in the results obtained in the Fe content ($p = 0.06$), where meat samples from 2%-SWA exhibited the lowest content of this mineral.

Table 5. Means of proximate composition and mineral content based on dry matter (DM) by treatments.

Variable	Treatment SWA			SEM	*p* Value
	Control	2%	4%		
		Proximal composition [1]			
Ash	4.71 ^{a}	4.87 ^{a}	5.55 ^{b}	0.12	<0.0001
Crude protein	90.64	89.68	90.64	0.98	0.53
Total lipids	4.31	5.26	4.13	0.91	0.43
		Macrominerals [1]			
Na	0.17	0.19	0.17	0.02	0.31
Mg	0.088	0.096	0.097	0.001	0.68
K	1.87	1.92	1.66	0.11	0.11
		Microminerals [2]			
Mn	8.30 ^{b}	6.06 ^{ab}	3.82 ^{a}	0.96	0.002
Fe	66.90	45.06	50.38	2.64	0.06
Cu	15.93 ^{b}	13.64 ^{b}	7.72 ^{a}	1.69	0.004
Zn	50.32	53.94	48.82	4.42	0.51

[1] values expressed as g/100 g of DM. [2] values expressed as mg/kg of DM. Means within a row lacking a common superscript letter differ ($p \leq 0.05$). SEM: Standard Error Mean.

4. Discussion

The effects of the inclusion of additives or supplements based on brown seaweed have been tested on growth performance, nutrients digestibility, prebiotic, antioxidant, anti-inflammatory, and immunomodulatory activities in pigs [21–24]; however, the impact of these dietary interventions on pork meat quality and nutritional composition has been less studied.

The main traits that define pork quality are pH, color, and water holding capacity [36]. The *postmortem* pH variation is an important factor that determines meat quality and has an influence on the physicochemical traits and shelf life [37,38]. Former investigations had found that the addition of seaweed in the animal diet could change carcass characteristics like marbling, color, and pH in pork [25–27,39]. Muscle pH values in this study were not affected by the inclusion of SWA and were ranged in the normal values. Muscular pH is highly correlated with the energy content of the diet [40]; however, the rate of pH decrease is influenced by multiple *antemortem* factors [41] and *postmortem* manipulation [38,42]. Michalak et al. [22] reported no statistically significant effect of a green seaweed additive (*Enteromorpha* sp.) on pH, water capacity holding, and drip loss in pork meat. Rossi et al. [19] found that sensory traits like aroma, flavor, and aroma of rabbit meat were affected by the use of 0.3 and 0.6% of dietary brown seaweed in the diet (*Laminaria* sp.). In this study, a better taste preference was observed from pork samples of animals that were fed with the highest % of SWA, without being different from those from the control groups (Figure S1).

According to the pork meat color standard [43], lightness values between 37 and 49 are considered normal. In this study, L values were not affected by treatments, and their values ranged from 44.95 to 47.61. Brown seaweed has been reported to be a rich source of natural antioxidants such as polysaccharides and polyphenols, which could improve meat color display [28,44]; also, antioxidant, antimicrobial, and immunomodulatory activities have been reported for compounds (extracts) of brown algae [8]. Moroney et al. [25] reported that spray-dried seaweed extracts that contained laminarin and fucoidan did not affect the redness values when incorporated in fresh pork via the animal diet. However, in another study, the same authors [26] reported that the SWA significantly reduced the redness intensity compared with the control. Additionally, Rajauria et al. [28] reported that an addition of 5.3% of seaweed extracts (*Laminaria* spp.) in the diet of finishing pigs significantly reduced the redness intensity compared to the control. In this study, the redness intensity was also significantly reduced in samples from animals that were fed with the highest percentage of SWA (4%).

The lower values detected in meat samples from animals that were fed with the highest % of SWA in this study could have been related to the lower iron content present in the muscle from the same treatment (4%-SWA), or could also be related to some interactions between the polysaccharides present in the SWA and the oxymyoglobin in the pork meat [14].

In this study, meat samples from 4%-SWA treatments had lower levels of iron. Ponnampalam et al. [45] reported that increased muscle heme iron concentration resulted in higher values in beef displayed for 48 to 72 h *postmortem*.

We hypothesized that samples from pigs that were fed with the highest % of SWA would increase the PUFA values and reduced cholesterol content; however, meat samples from animals that were fed a regular diet plus 4% SWA got the highest levels of total cholesterol compared to the other treatments ($p < 0.05$). Supplementation of laying hens with 1–2% of dried *Enteromorpha porifera* seaweed resulted in a reduction in cholesterol in the yolk [18]. Rossi et al. [19] reported that the inclusion of dietary levels of brown seaweed (*Laminaria* spp.) did not affect the content of cholesterol in rabbit meat. To our knowledge, there is no previous report on the effect of dietary SWA on cholesterol content in pork samples. Our results suggest that there is an important effect of the components of seaweed that impact the content of cholesterol. Muscle samples from pigs that were fed with 2%-SWA exhibited the highest % of total lipids (5.26% DM) with no significant differences with the other groups. It would be important to confirm these results with a carcass with similar intramuscular content since it is known that there is a high correlation between intramuscular fat and meat lipid composition [46]. Ruqqia et al. [47] tested 13 seaweed extract from different species for hypolipidaemic potential in normal rats, and they found that some of these extracts caused a decrease in total serum cholesterol, triglyceride, and LDL cholesterol but an increase in

HDL cholesterol. These results suggest that not all seaweed has the same effect on cholesterol metabolism. Since seaweed represents a group of organisms with diverse types of bioactive compounds, further studies are needed to understand the biological effects of the extract of *Macrocystis pyrifera* on cholesterol metabolism.

The proximal composition of the SWA used in this study is described in Table 1. Lipid content, crude protein, and ash content are lower than the values reported by Ortiz et al. [14] in the fresh seaweed (*Macrocystis piryfera*). These authors reported (in DM) 10.8% for ash, 13.2% for crude protein and 0.7% for total lipids, and 75.5% for carbohydrates. Seaweed composition depends on the harvest conditions, the habitat, and many other external conditions such as water temperature, light intensity, and nutrient concentration in the water [48].

The fatty acid composition of intramuscular tissue is affected by dietary lipid composition, de novo lipogenesis, desaturation, and the difference in the utilization of various fatty acids by the animal body [49]. In LL samples from this study, palmitic acid (C16:0) was the most abundant SFA, oleic acid (C18:1*c* $n-9$) was the most abundant MUFA, and linoleic acid (C18:2 $n-6$) the most abundant PUFA in the LL of the examined pigs. The composition of the fatty acids of the LL samples from this study is similar to those reported by Alonso et al. [50] and Parunovic et al. [51] in pork meat with no dietary seaweed inclusion. Comparing the fatty acids composition of LL samples with *Macrocystis pyrifera* seaweeds, the most abundant MUFA was also 18:1*c* $n-9$ (oleic acid) with 19.64 ± 0.08 %; the linoleic acid (18:2 $n-6$) reached values of 43.41% and the predominant SFA was also palmitic acid (C16:0; 16.17 ± 0.06) [14]. El Bahr et al. [52] reported a significant increase in the levels of EPA, DHA, total PUFA, and arachidonic acid in breast muscle of broiler chickens fed with microalgae extracts (1 g/kg diet), suggesting that high contents of methionine and lysine (present in the microalgae) were positively correlated with the increase in PUFA. *Macrocystis pyrifera* contains low contents of proteins and essential amino acids like methionine and lysine [14] compared to a microalgae, like *Arthrospira* sp. [27].

Several studies found that the long-chain ω-3 PUFA content in the muscle or adipose tissue was largely independent on the timing of feeding ω-3-PUFA-rich diets [53]; in this study, pigs were fed with the SWA for 45 days during the fattening phase; however, the effect of dietary SWA was not significant. Perhaps, it is necessary to incorporate the SWA during the growing and/or finishing phase to evaluate the impact on meat chemical composition.

Fat and fatty acids are important because of their effects on human health. In this study, the index value H/h and IA ratios turned out to be less than 1 (Supplementary Table S1), which indicated that regardless of the inclusion of SWA in the diets, all samples are categorized as healthy meats. The relation between H/h comes from the functional effects of fatty acids in cholesterol metabolism and gives a superior measure of the nutritional evaluation of fats from a nutritional standpoint [35]. The relation H/h is a suitable indicator to evaluate the risk of elevated blood cholesterol since it excludes C18:0 but includes two important hypercholesterolemic, palmitic acid (C14:0) and oleic acid (C16:0) are known to be the most important hypercholesterolemic fatty acids [35].

The balance in the relationship $\omega6$ and $\omega3$ ($\omega6/\omega3$) plays an important prevention role for severe chronic disorders and autoimmune diseases, and these authors agree to the average recommended value of 5:1. The $\omega6/\omega3$ ratio, which is currently recommended by the OMS, should be lower than 10, in *M. piryfera* this ratio is 7.42 [54].

The inclusion of dietary SWA affected the content of ash; however, in this study, microminerals such as Mn, Fe, and Cu were found in less quantity in meat samples from animals that consumed the greatest proportion of SWA (4%). All seaweeds are characterized by a higher ash content (19.3–27.8% DM) than those observed in edible plants [21], being considered by some authors [27] as an important organic source of minerals for livestock nutrition; however, there is wide variability in mineral content among seaweed species. *Macrosysty pyrifera*, for example, is rich in Mg (39 ± 2.8 mg/g), Na (36.9 ± 9.9 mg/g), K (67.5 ± 22.3 mg/g) and Fe (117 mg/kg) [11,14]. On the other hand, these algae have a relatively low content of Mn (11 mg/kg), Zn (12 mg/kg), and Cu (2 mg/kg) compared to other species of brown seaweed [11,21]. There is no scientific evidence of the direct relationship between the mineral content of the dietary seaweed additive and the mineral content in meat. The numerous bioactive compounds that are present in fed additive could affect the chemical composition of meat. Moreover, the content of polysaccharides such as alginates and agar or carrageenan could cause the formation of insoluble complexes with minerals, decreasing their bioavailability [55].

It has also been reported that *Laminaria* spp. is rich in alginates, which probably hampers the bioavailability of Ca, and that the apparent absorption values of Na and K were significantly higher in rats supplemented with *Laminaria* spp. while Mg absorption was not affected [56]. Several components in fed matrices can also exhibit retention properties in minerals, such as phenolic compounds and phytic acid, which reduce the bioavailability of Fe and Zn.

5. Conclusions

The meat of pigs fed with brown sea algae additives had a less intense red color than the control pigs. From a nutritional point of view, the meat of pigs fed with a higher percentage of seaweed additive (4%) had a small but significant total ash increase and a lower percentage of total lipids. However, the fatty acid composition of pig meat was not influenced by including seaweed additive, and some microminerals like Cu, Zn, and Mn decreased in content. In general, no harmful effects were found in this study by feeding pigs with brown sea algae extracts. Since seaweed represents a group of organisms with a diverse type of bioactive compounds, further studies are needed to understand the biological effects of this SWA on adipogenesis, cholesterol metabolism, and mineral deposition in muscle.

Supplementary Materials: The following are available online at https://www.mdpi.com/article/10.3390/foods10081720/s1, Table S1: Descriptive statistics for fatty acids composition and health index in pork *longissimus lumborum* muscle. Figure S1: Taste preference percentage by panelists.

Author Contributions: N.J.-T.: Conceptualization, supervision of experiment and laboratory analysis. Responsible for statistical analysis and revision of the structure of the manuscript. M.S.-H.: original draft preparation, participation in the results and discussion section of this manuscript. R.P.: participation in the experiment design, revision and interpretation of the literature and revision of the manuscript. J.M.: data collection, and statistical analyses, supervision of data collection. All authors searched and reviewed the literature, discussed the contents of the manuscript, and approved submission. All authors have read and agreed to the published version of the manuscript.

Funding: This research was funded by the Project CORFO INNOVA CHILE 171TE2-72098.

Institutional Review Board Statement: Animals were handled according to the criteria for animal care and welfare described in the Bioethics guide for animal experimentation approved by the Bioethics Committee of the Universidad Austral de Chile (Res. No. 23-2014).

Informed Consent Statement: Not applicable.

Data Availability Statement: Data are not available in public datasets, please contact the authors.

Acknowledgments: The authors are grateful to the Pig 'production farm "Las tres esquinas" and I+D Patagonia Biotecnología SA., for their valuable support during the experimental trial and procurement of pork samples.

Conflicts of Interest: The authors declare no conflict of interest.

References

1. Reig, M.; Aristoy, M.-C.; Todrá, F. Variability in the contents of pork meat nutrients and how it may affect food composition databases. *Food Chem.* **2013**, *140*, 479–482. [CrossRef]
2. National Pork Board. *World per Capita Pork Consumption*; National Pork Board: DesMoines, IA, USA, 2020. Available online: https://porkcheckoff.org/pork-branding/facts-statistics/ (accessed on 21 December 2020).

3. Toldrá, F.; Rubio, M.A.; Navarro, J.L.; Cabrerizo, L. Quality aspects of pork and its nutritional impact. In *Quality of Fresh and Processed Foods Advances in Experimental Medicine and Biology*; Shahidi, F., Spanier, F., Ho, C.-T., Braggins, T., Eds.; Kluwer Academic/Plenum Publications: New York, NY, USA, 2004; Volume 542, pp. 25–32. [CrossRef]
4. Valaitienė, V.; Stanytė, G.; Klementavičiūtė, J.; Jankauskas, A. Nutritional value of and element content in meat from various pig breeds. *Anim. Sci. Pap. Rep.* **2017**, *35*, 419–428.
5. Karmaus, A.L.; Jones, W. Future foods symposium on alternative proteins: Workshop proceedings. *Trends Food Sci. Technol.* **2021**, *107*, 12–129. [CrossRef]
6. Rebours, C.; Marinho-Soriano, E.; Zertuche-González, J.A.; Hayashi, L.; Vásquez, J.A.; Kradolfer, P.; Soriano, G.; Ugarte, R.; Abreu, M.H.; Bay-Larsen, I. Seaweeds: An opportunity for wealth and sustainable livelihood for coastal communities. *J. Appl. Phycol.* **2014**, *26*, 1939–1951. [CrossRef] [PubMed]
7. Rajauria, G. *Seaweeds: A Sustainable Feed for Livestock and Aquaculture*, 1st ed.; Tiwari, B.K., Troy, D., Eds.; Academic Press: Amsterdam, The Netherlands, 2015; pp. 389–413. [CrossRef]
8. Chojnacka, K.; Saeid, A.; Witkowska, Z.; Tuhy, Ł. Biologically active compounds in seaweed extracts—The prospects for the application. *Open Conf. Proc. J.* **2012**, *3* (Suppl. 1-M4), 20–28. [CrossRef]
9. Baca, S.; Cervantes, M.; Gómez, R.; Espinoza, S.; Sauer, W.; Morales, A.; Araiza, B.; Torrentera, N. Effects of adding seaweed meal to wheat diets for growing and finishing pigs. *Rev. Comput. Prod. Porcina* **2008**, *15*, 324–328.
10. Bleakley, S.; Hayes, M. Algal Proteins: Extraction, Application and Challenges concerning production. *Foods* **2017**, *6*, 33. [CrossRef]
11. Makkar, H.; Tran, G.; Heuzé, V.; Giger-Reverdin, S.; Lessire, M.; Lebas, F.; Ankers, P. Seaweeds for livestock diets: A review. *Anim. Feed Sci. Tech.* **2016**, *212*, 1–17. [CrossRef]
12. Quintral, V.; Morales, C.; Sepúlveda, M.; Schwartz. Propiedades nutritivas y saludables de algas marinas y su potencial como ingrediente funcional. *Rev. Chil. Nutr.* **2012**, *39*, 196–202. [CrossRef]
13. MacArtain, P.; Gill, C.I.R.; Brooks, M.; Campbell, R.; Rowland, I.R. Nutritional value of edible seaweeds. *Nutr. Rev.* **2007**, *65*, 535–543. [CrossRef]
14. Ortiz, J.; Uquiche, E.; Robert, P.; Romero, N.; Quitral, V.; Lantén, C. Functional and nutritional value of the Chilean seaweeds *Codium fragile*, *Gracilaria chilensis* and *Macrocystis pyrifera*. *Eur. J. Lipid. Technol.* **2008**, *111*. [CrossRef]
15. Rjiba-Ktita, S.; Chermiti, A.; Mahouachi, M. The use of seaweeds (Ruppia maritima and Chaetomorpha linum) for lamb fattening during drought periods. *Small Rumin. Res.* **2010**, *91*, 116–119. [CrossRef]
16. Morrill, J.; Sawyer, J.; Smith, S.; Miller, R.; Johnson, M.; Wickersham, T. Post-extraction algal residue in beef steer finishing diets: II. Beef flavor, fatty acid composition, and tenderness. *Algal Res.* **2017**, *25*, 578–583. [CrossRef]
17. Vadher, K.K.; Vyas, A.A.; Parmar, P.V. Effects of seaweed, *Gracilaria* sp. as feed additive in diets on growth and survival of Labeo rohita. *J. Exp. Zool. India* **2016**, *19*, 771–778.
18. Wang, S.B.; Jia, Y.H.; Wang, L.H.; Zhu, F.H.; Lin, Y.T. *Enteromorpha porifera* supplemental level: Effects on laying performance, egg quality, immune function, and microflora in feces of laying hens. *Chin. J. Anim. Nutr.* **2013**, *25*, 1346–1352.
19. Rossi, R.; Vizzarri, F.; Chiapparini, S.; Ratti, S.; Casamassima, D.; Palazzo, M.; Corino, C. Effects of dietary levels of brown seaweeds and plant polyphenols on growth and meat quality parameters in growing rabbit. *Meat Sci.* **2020**, *161*, 1077987. [CrossRef] [PubMed]
20. Gardiner, G.E.; Campbell, A.J.; O'Doherty, J.V.; Pierce, E.; Lynch, P.B.; Leonard, F.C.; Stanton, C.; Ross, R.P.; Lawlor, P.G. Effect of Ascophyllum nodosum extract on growth performance, digestibility, carcass characteristics and selected intestinal microflora populations of grower-finisher pigs. *Anim. Feed Sci. Technol.* **2008**, *141*, 259–273. [CrossRef]
21. Corino, C.; Modina, S.C.; Di Giancamillo, A.; Chiapparini, S.; Rossi, R. Seaweeds in pig nutrition. *Animals* **2019**, *9*, 1126. [CrossRef] [PubMed]
22. Michalak, I.; Chojnacka, K.; Korniewicz, D. Effect of Marine Macroalga *Enteromorpha* sp. Enriched with Zn(II) and Cu(II) ions on the Digestibility, Meat Quality and Carcass Characteristics of Growing Pigs. *J. Mar. Sci. Eng.* **2020**, *8*, 347. [CrossRef]
23. Angell, A.R.; Angell, S.F.; de Nys, R.; Paul, N.A. Seaweed as a protein source for monogastric livestock. *Trends Food Sci. Technol.* **2016**, *54*, 74–84. [CrossRef]
24. Øverland, M.; Mydland, L.; Skrede, A. Marine macroalgae as sources of protein and bioactive compounds in feed for monogastric animals. *J. Sci. Food Agric.* **2019**, *99*, 13–24. [CrossRef] [PubMed]
25. Moroney, N.C.; O'Grady, M.N.; O'Doherty, J.V.; Kerry, J.P. Addition of seaweed (*Laminaria digitata*) extracts containing laminarin and fucoidan to porcine diets: Influence on the quality and shelf-life of fresh pork. *Meat Sci.* **2012**, *92*, 423–429. [CrossRef]
26. Moroney, N.C.; O'Grady, M.N.; Robertson, R.C.; Stanton, C.; O'Doherty, J.V.; Kerry, J.P. Influence of level and duration of feeding polysaccharide (laminarin and fucoidan) extracts from brown seaweed (*Laminaria digitata*) on quality indices of fresh pork 2015. *Meat Sci.* **2015**, *99*, 132141. [CrossRef] [PubMed]
27. Madeira, M.S.; Cardoso, C.; Lopes, P.A.; Coelho, D.; Afonso, C.; Bandarra, N.M.; Prates, J.A.M. Microalgae as feed ingredients for livestock production and meat quality: A review. *Livestock Sci.* **2017**, *205*, 111–121. [CrossRef]
28. Rajauria, G.; Draper, J.; McDonnell, M.; O'Doherty, J.V. Effect of dietary seaweed extracts, galactooligosaccharide and vitamin E supplementation on meat quality parameters in finisher pigs. *Innov. Food Sci. Emerg. Technol.* **2016**, *37*, 269–275. [CrossRef]
29. AMSA (American Meat Science Association). *Research Guidelines for Cookery, Sensory Evaluation, and Instrumental Tenderness Measurements of Meat*, 2nd ed.; American Meat Science Association: Champaign, IL, USA, 2016.

30. AOAC. *Official Methods of Analysis, Vol. II. Association of Official Analytical Chemist*, 15th ed.; Association of Official Analytical Chemist: Washington, DC, USA, 1990.
31. Pelkin-Elmer. *Analytical Methods for Atomic Absorption Spectrophotometry*; ACS Publications: Norwalk, CT, USA, 1994.
32. Fletouris, N.; Botsoglou, I.; Psomas, A.; Mantis, I. Rapid Determination of Cholesterol in Milk and Milk Products by Direct Saponification and Capillary Gas Chromatography. *Int. J. Dairy Sci.* **1998**, *81*, 2833–2840. [CrossRef]
33. Cantelolops, D.; Reid, A.P. Determination of Lipids in Infant Formula Powder by Direct Extraction Methylation of Lipids and Fatty Acid Methyl Esters (FAME) Analysis by Gas Chromatography. *J. AOAC Int.* **1999**, *82*, 1128–1139. [CrossRef]
34. Ulbricht, T.L.; Southgate, D.A.T. Coronary heart disease: Seven dietary factors. *Lancet* **1991**, *338*, 985–992. [CrossRef]
35. Monteiro, A.C.G.; Santos-Silva, J.; Bessa, R.J.B.; Navas, D.R.; Lemos, J.P.C. Fatty acid composition of intramuscular fat of bulls and steers. *Livestock Sci.* **2006**, *99*, 13–19. [CrossRef]
36. Rosenvold, K.; Andersen, H.J. Factors of significance for pork quality—A review. *Meat Sci.* **2003**, *64*, 219–366. [CrossRef]
37. Brewer, M.S.; Mc-Keith, F.K. Consumer-rated quality characteristics as related to purchase intent of fresh pork. *J. Food Sci.* **1999**, *64*, 171–174. [CrossRef]
38. Jacob, R.; Hopkins, D. Techniques to reduce the temperature of beef muscle early in the *postmortem* period—A review. *Anim. Prod. Sci.* **2014**, *54*, 482–493. [CrossRef]
39. Rossi, R.; Pastorelli, G.; Cannata, S.; Tavaniello, S.; Maiorano, G.; Corino, C. Effect of long-term dietary supplementation with plant extract on carcass characteristics meat quality and oxidative stability in pork. *Meat Sci.* **2013**, *95*, 542–548. [CrossRef] [PubMed]
40. Gallo, C.; Apaoblaza, A.; Pulido, R.; Jerez-Timaure, N. Efectos de una suplementación energética en base a maíz roleado sobre las características de calidad de la canal y la incidencia de corte oscuro en novillos. *Arch. Med. Vet.* **2013**, *45*, 237–245. [CrossRef]
41. Gallo, C.; Tadich, N. Bienestar animal y calidad de carne durante los manejos previos al faenamiento en bovinos. *REDVET* **2008**, *9*, 9–10.
42. Ponnampalam, E.; Hopkins, D.; Bruce, H.; Li, D.; Baldi, G.; Bekhit, A. Causes and contributing factors to "dark cutting" meat: Current trends and future directions: A review. *Compr. Rev. Food Sci. Food Saf.* **2017**, *16*, 400–430. [CrossRef]
43. NPPC; National Pork Producers Council. Pork Composition and Quality Assessment Procedures. Des Moines, IA, USA. 2000. Available online: https://porkgateway.org/resource/pork-composition-and-quality-assessment-procedures.pdf (accessed on 11 July 2020).
44. Moroney, N.C.; O'Grady, M.N.; O'Doherty, J.V.; Kerry, J.P. Effect of a brown seaweed (Laminaria digitata) extract containing laminarin and fucoidan on the quality and shelf-life of fresh and cooked minced pork patties. *Meat Sci.* **2013**, *94*, 304–311. [CrossRef]
45. Ponnampalam, E.N.; Butler, K.L.; McDonagh, M.B.; Jacobs, J.L.; Hopkins, D.L. Relationship between muscle antioxidant status, forms of iron, polyunsaturated fatty acids, and functionality (retail colour) of meat in lambs. *Meat Sci.* **2012**, *90*, 297–303. [CrossRef]
46. Catillo, G.; Zappaterra, M.; Lo Fiego, D.P.; Steri, R.; Davoli, R. Relationships between EUROP carcass grading and backfat fatty acid composition in Italian Large White heavy pigs. *Meat Sci.* **2021**, *171*, 108291. [CrossRef]
47. Ruqqia, K.; Sultana, V.; Ara, A.; Ehteshamul-Haque, S.; Athar, A. Hypolipidaemic potential of seaweeds in normal, triton-induced and high-fat diet-induced hyperlipidaemic rats. *J. Appl. Phycol.* **2015**, *27*, 571–579. [CrossRef]
48. Rodriguez-Montesinos, Y.E.; Hernandez-Carmona, G. seasonal and geographic variations of *Macrocystis pyrifera* chemical composition at the western coast of Baja California. *Cienc. Mar.* **1991**, *17*, 91–107. [CrossRef]
49. Wood, J.D.; Enser, M.; Fisher, A.V.; Nute, G.R.; Sheard, P.R.; Richardson, R.I.; Hughes, S.I.; Whittington, F.M. Fat deposition, fatty acid composition and meat quality: A review. *Meat Sci.* **2008**, *78*, 343–358. [CrossRef] [PubMed]
50. Alonso, V.; Campo, M.M.; Español, S.; Roncalés, P.; Beltrán, J.A. Effect of crossbreeding and gender on meat quality and fatty acid composition. *Meat Sci.* **2009**, *81*, 209–217. [CrossRef]
51. Parunovic, N.; Petrovic, M.; Djordjevic, V.; Petronijevic, R.; Lakicevic, B.; Petrovic, Z.; Savic, R. Cholesterol content and fatty acids composition of Mangalitsa pork meat. *Procedia Food Sci.* **2015**, *5*, 215–218. [CrossRef]
52. El Barh, S.; Shousha, S.; Shehab, A.; Khattab, W.; Ahmen-Farid, O.; Sabike, I.; El-Garhy, O.; Albokhadaim, I.; Albosadah, K. Effect of dietary microalgae on growth performance profiles of amino and fatty acids, antioxidant status, and meat quality of broiler chickens. *Animals* **2020**, *10*, 761. [CrossRef] [PubMed]
53. Martínez-Ramírez, H.R.; Kramer, J.K.G.; de Lange, C.F.M. Retention of n-3 polyunsaturated fatty acids in trimmed loin and belly is independent of timing of feeding ground flaxseed to growing-finishing female pigs. *J. Anim. Sci.* **2014**, *92*, 238–249. [CrossRef]
54. Enser, M.; Hallett, K.; Hewett, B.; Fursey, G.A.J.; Wood, J.D. Fatty acid content and composition of English beef, lamb, and pork at retail. *Meat Sci.* **1996**, *42*, 443–456. [CrossRef]
55. Circuncisão, A.R.; Catarino, M.D.; Cardoso, S.M.; Silva, A.M.S. Minerals from macroalgae origin: Health benefits and risks for consumers. *Mar Drugs* **2018**, *16*, 400. [CrossRef]
56. Raes, K.; Knockaert, D.; Struijs, K.; Van Camp, J. Role of processing on bio accessibility of minerals: Influence of localization of minerals and anti-nutritional factors in the plant. *Trends Food Sci. Technol.* **2014**, *37*, 32–41. [CrossRef]

MDPI

Article

Carcass and Primal Composition Predictions Using Camera Vision Systems (CVS) and Dual-Energy X-ray Absorptiometry (DXA) Technologies on Mature Cows

José Segura, Jennifer L. Aalhus, Nuria Prieto, Ivy L. Larsen, Manuel Juárez and Óscar López-Campos *

Lacombe Research and Development Centre, Agriculture and Agri-Food Canada, Lacombe, AB T4L 1W1, Canada; jose.seguraplaza@canada.ca (J.S.); jennifer.aalhus@outlook.com (J.L.A.); nuria.prietobenavides@canada.ca (N.P.); ivyleelarsen@gmail.com (I.L.L.); manuel.juarez@canada.ca (M.J.)
* Correspondence: oscar.lopezcampos@canada.ca

Abstract: This study determined the potential of computer vision systems, namely the whole-side carcass camera (HCC) compared to the rib-eye camera (CCC) and dual energy X-ray absorptiometry (DXA) technology to predict primal and carcass composition of cull cows. The predictability (R^2) of the HCC was similar to the CCC for total fat, but higher for lean (24.0%) and bone (61.6%). Subcutaneous fat (SQ), body cavity fat, and retail cut yield (RCY) estimations showed a difference of 6.2% between both CVS. The total lean meat yield (LMY) estimate was 22.4% better for CCC than for HCC. The combination of HCC and CCC resulted in a similar prediction of total fat, SQ, and intermuscular fat, and improved predictions of total lean and bone compared to HCC/CCC. Furthermore, a 25.3% improvement was observed for LMY and RCY estimations. DXA predictions showed improvements in R^2 values of 26.0% and 25.6% compared to the HCC alone or the HCC + CCC combined, respectively. These results suggest the feasibility of using HCC for predicting primal and carcass composition. This is an important finding for slaughter systems, such as those used for mature cattle in North America that do not routinely knife rib carcasses, which prevents the use of CCC.

Keywords: beef primals; computer vision system; dual energy X-ray absorptiometry; mature cows; rib-eye camera; whole-side camera

Citation: Segura, J.; Aalhus, J.L.; Prieto, N.; Larsen, I.L.; Juárez, M.; López-Campos, Ó. Carcass and Primal Composition Predictions Using Camera Vision Systems (CVS) and Dual-Energy X-ray Absorptiometry (DXA) Technologies on Mature Cows. *Foods* **2021**, *10*, 1118. https://doi.org/10.3390/foods10051118

Academic Editors: Huerta-Leidenz Nelson and Markus F. Miller

Received: 28 March 2021
Accepted: 13 May 2021
Published: 18 May 2021

1. Introduction

In Canada, ~425,000 mature cows are harvested annually, producing over ~100,000 Tm of meat [1]. Recently, the reduced availability of cattle and the increase in beef demand have increased beef prices, particularly in cull cows [1]. In the Canadian Grading System, cull cows are segregated as Canada D-grades based on a broad classification of carcass types [2]. In contrast to the top youthful grades (Canada Prime, AAA, AA, and A), where estimations for retail cut yield are routinely provided, Canada D-grades are lacking prediction of carcass yields before carcass breakdown. Because mature beef carcasses are often boned out for further processing, yield assessments of carcasses would be an important attribute to enhance fair compensation to the cattle producers. Furthermore, accurate estimations of carcass composition have been suggested to assure an efficient utilization of specific muscles from cull cow carcasses. In this sense, Roberts et al. [3] reported that, despite darker lean, many muscles from D-grade carcasses had higher intramuscular fat content than in the youthful A/AA carcasses. Given this retail performance of muscles from cull cow carcasses, opportunities may exist to better utilize specific muscles from these carcasses.

For decades, in North America, the carcass classification has been carried out by trained personnel (graders), thus implying a certain degree of subjectivity on the quantified parameters [4]. The latest improvements in technologies to estimate body/carcass composition have shown applicability on different species, genetics, production systems, etc. [5,6].

Computer vision systems (CVS) were implemented in the early 1980s as a computerized, non-destructive, non-invasive, objective, cost-effective, and automatable technology, based on image analysis that provides measurements of the beef carcass or rib-eye proportions [6]. The CVS have been recognized as useful tools to improve the grading accuracy, precision, and consistency, thus benefiting all segments of the beef production and consumption supply chain [7]. Typically, at least one of the two CVS approaches is used. Whole-side carcass image analysis, also known as hot carcass camera (HCC) system, which is designed to be integrated into the slaughter chain to work autonomously, and/or the rib-surface image analysis system, also known as cold carcass camera (CCC), which mimics the traditional visual assessment of the knife-ribbed surface of the rib-eye at the 12th thoracic vertebrae. The HCC uses a color camera and a lighting system, including structured (striped) light. The half carcass holds steady in front of a colored background and one or two images (if ambient light must be compensated) are taken to obtain 2D information, and a third image is taken with the structured light to capture 3D information of the carcass from the degree of curvature of the striped light [8,9]. Using proprietary software, the CCC provides an objective measure of rib-eye length, width, and area, and fat thickness, which are then used to predict carcass yield, as well as marbling, lean, and fat contents, and color assessments [6]. Currently, the CCC system is widely utilized by the beef industry in North America [8–11], particularly in youthful carcasses. However, unlike youthful beef, mature cull cows are generally marketed without knife-ribbing the carcass at the grade site. Hence, prediction of lean yield using rib-eye assessment or CCC is not achievable and development of alternative methods is particularly pertinent for the industry.

On the other hand, Dual-energy X-ray absorptiometry (DXA) technology is a promising indirect method to estimate carcass composition due to its relatively low cost, high reliability of data collection, and ease of use [5]. In the literature, the feasibility, accuracy, and precision of DXA technology has been reported on salmon [12], broiler chickens [13], sheep [14], swine [15], and cattle [16]. In addition, Soladoye et al. [15], Kipper et al. [17], and López-Campos et al. [18] assessed the accuracy of DXA technology on mass measurement of primal cuts from pigs and steers. Most of the published studies reported on the use of DXA in youthful populations, with information being scarce or almost lacking for mature animals, particularly in the case of cull cows. Contrary to the CVS, DXA technology is at the early stages of industry implementation.

Thus, the objective of the present study was to evaluate the potential of computer vision systems, namely the whole-side carcass camera compared to the rib-eye camera, as well as the emerging DXA technology to predict whole-carcass and primal composition (fat, lean, and bone) of mature cows. Furthermore, the combination of both computer vision systems was also explored in order to evaluate this approach as an alternative for the beef industry to further improve the prediction accuracy on primals and carcass composition of mature beef.

2. Materials and Methods

2.1. Animals

A total of 111 cull cow left carcass sides (hot carcass weight: HCW = 346 $\pm$ 33.3 kg), sourced from a commercial abattoir (n = 72) and from the AAFC-Lacombe Research and Development Centre (AAFC-Lacombe RDC) cow herd (n = 39), were used in the present study. AAFC-Lacombe RDC animals were cared for according to the Canadian Council on Animal Care Guidelines [19] (AAFC-Lacombe RDC study plan No. 201705).

2.2. Carcass Sides, Cut-Out, and CVS and DXA Scanning

Cull cows sampled from the AAFC-Lacombe RDC herd were slaughtered at the AAFC-Lacombe RDC federally inspected abattoir. Following slaughter, carcasses were dressed and split and HCW were recorded. In turn, commercial carcass sides, harvested following the Guidelines for the humane care and handling of food animals at slaughter (Canadian Food Inspection Agency, CFIA) [20], were shipped to AAFC-Lacombe RDC facilities in a refrigerated truck following guidelines for transportation of carcasses over thirty months of age (CFIA) [21]. At the time of slaughter, HCW were recorded by the personnel of the slaughter plant. In both sample populations, pictures of each carcass side were taken using a HCC unit VBS 2000, *e+v*® Technology GmbH, Oranienburg, Germany.

Raw output data of the HCC images were composed of 187 variables describing carcass dimensions: angle (W00-W99), length (L00-L19), area (F00-F13, F20-F29), carcass contour and volumes (V00-V13, V20-V29), and color (Fe00-Fe18). Following 72 h of chilling at 2 °C, left carcass sides were weighed (CCW) to determine shrink loss. Physiological maturity of the carcasses were assessed based on the extent to which caps of the spinal processes and ribs had ossified (i.e., >50% ossification, a carcass receives a D grade) in accordance with López-Campos et al. [22] and the Canadian Beef Grading Agency (CBGA) [2]. Left carcass sides were then knife-ribbed between the 12th and 13th ribs. After 20 min of atmospheric oxygen exposure, full Canadian grading data were collected by a certified grader from the CBGA. The assessments included grade fat (minimum fat thickness over the rib in 4th quadrant from the spinous process, mm), fat thickness (at the three-quarters position from the spinous process, mm), rib-eye area (REA; in cm^2 of the *Longissimus thoracis*), and marbling score, subjectively assessed using United States Department of Agriculture (USDA) beef marbling pictorial standards as reference points [23]. Muscle scores (1–4) were also determined based on *L. thoracis* length and width, measured at the grade site [24,25]. Then, rib-eye pictures from each carcass side were taken using CCC: VBG 2000 (*e+v*® Technology GmbH, Oranienburg, Germany). Each image was then processed by manufacturer software, in real-time, to produce raw output data composed of 99 variables describing a number of measurements related to measurements on the rib-eye (n = 22), fat thickness (n = 15), and muscle and fat color (n = 15), as well as other variables (n = 47, e.g., marbling assessment, back fat dressing corrections, etc.).

Estimated total lean meat yield (LMY) was calculated according to the Jones et al. [24,25] equation LMY (%) = 63.5 + 1.05 × (muscle score) − 0.76 × (grade fat). The retail cut yield (RCY) percentage was calculated using the equation RCY (%) = 51.34 − 5.78 × (fat thickness at the $\frac{3}{4}$, inches) − 0.46 × (kidney, pelvic, and heart fat percent, KPH, %) − 0.0093 × (hot carcass weight, HCW, pounds) + 0.74 × (REA, square inches) [25].

Left carcass sides were fabricated into primal cuts with carcass breakpoints identified following the Institutional Meat Purchase Specifications (IMPS) for Fresh Beef Products, Series 100 [26]. The primals collected from the left fabricated carcass side were the chuck (IMPS #113), rib (IMPS #103), brisket (IMPS #118), flank (IMPS #193, non-trimmed), foreshank (IMPS #117), loin (IMPS #172A), round (IMPS #158A), and plate (IMPS #121). Following procedures described by López-Campos et al. [18], each primal cut was scanned with a GE Lunar iDXA unit (GE Lunar, General Electric, Madison, WI, USA) using the whole-body scan option on standard mode to estimate fat, lean, and bone weights. After DXA scanning, all left primals were fully dissected into subcutaneous fat (SQ), intermuscular fat (IM), body cavity fat (BC), lean, and bone, then weighed by trained personnel. An adequate dissection processing was carried out by highly skilled meat cutters ensuring that the difference between primal weight and the sum of total bone, total lean, and total fat was not higher than a 2%.

2.3. Statistical Analyses

All the statistical analyses were performed using SAS v. 9.4 (SAS Institute Inc., Cary, NC, USA, 2014) [27]. Either CVS or DXA estimates of lean, fat, and bone weights from each primal cut, and the overall fat, lean, and bone weights were included as independent variables in a partial least square regression (PLSR) to generate prediction equations. Therefore, four different groups were defined depending on the regression estimating variables used: HCC, CCC, combination of HCC + CCC, and DXA. All models were used to predict the reference values from the manual dissections and the calculations of LMY and RCY equations.

All PLSR models were fit using an internal full leave-one-out cross-validation, to avoid overfitting in the calibration set, and the number of latent variables (LV) used to minimize predicted residual error sums of squares (PRESS) was reported for the calibrated PLSR models.

The predictive ability of the PLSR models was evaluated in terms of coefficient of determination (R^2) and the mean square prediction error (MSPE), which was decomposed into error in central tendency (ECT), error due to regression (ER), and error due to disturbances (ED) [18]. These three fractions were calculated and expressed as percentages, as suggested by Benchaar et al. [28], as a means of describing the residual error in the models. ECT indicates how the average of CVS/DXA values deviates from the average of dissection values. ER measures the deviation of the least square regression coefficient from one, which is the value that it would have been if dissection and CVS/DXA measurements were in complete agreement. The ED is the variation in dissection measurements that is not accounted for by the least square regression of CVS/DXA measurements. In fact, this error is the unexplained variance and represents the portion of MSPE that cannot be eliminated by linear correction of the predictions [29]. Finally, when expressed as a percentage of the MSPE, the ECT, ER, and ED are called bias proportion, regression proportion (deviation of the regression slope from one), and disturbance proportion, respectively [30].

3. Results

3.1. Cow Carcass Population

All the carcasses used in the present study showed ossification processes at the caps of the thoracic vertebrae ranging from 50% to 100% ossified, resulting in carcasses graded as Canada D mature type grades [2]. Values of HCW (277.3–410.2 kg), CCW (271.3–401.9 kg), grade fat (0.0–29.0 mm), fat thickness (0.0–27.9 mm), REA (60–120 cm^2), LMY (49.0–61.0%), RCY (42.9–54.5%), and marbling scores (100–733, USDA marbling score), of the carcass population (n = 111) used were within the actual range (Table 1) of the Canadian beef carcass market [1].

Table 1. Descriptive statistics of carcass characteristics of the population used to obtain the prediction equations between the camera vision system values and whole carcass and primal composition (fat, lean, and bone).

	Mean (n = 111)	SD [1]	Min	Max
HCW [2] (kg)	345.8	33.3	277.3	410.2
CCW [3] (kg)	338.7	30.0	271.3	401.9
Grade fat (mm)	9.6	8.06	0.0	29.0
Fat thickness (mm)	10.2	5.59	0.0	27.9
Rib-eye width [4]	1.8	0.78	1	3
Rib-eye length [4]	2.7	0.53	1	3
Muscle score [4]	2.4	0.99	1	4
Ribeye area (cm^2)	83.6	11.2	60.0	120.0
LMY [5] (%)	56.3	5.75	49.0	61.0
RCY [6] (%)	49.6	2.29	42.9	54.5
Marbling scores [7]	455.6	143.2	100.0	733.0
Ossification (%) [8]	92.8	13.4	50.0	100.0

[1] SD: standard deviation; [2] HCW: Hot carcass weight; [3] CCW: Cold carcass weight; [4] Rib-eye width and length, and muscle score in agreement with Jones [24] and Segura et al. [25]; [5] LMY: estimated total lean meat yield [25]; [6] RCY: retail cut yield [25]; [7] Marbling scores: Official United States Standards for Grades of Beef Carcasses (marbling scores: 0 = Devoid, 100 = Practically Devoid, 200 = Traces, 300 = Slight, 400 = Small, 500 = Modest, 600 = Moderate, 700 = Slightly Abundant, 800 = Moderately Abundant, 900 = Abundant) [23]; [8] Ossification (%): Ossification processes of the carcasses assessed on the caps of the spinal processes and ribs (i.e., >50% ossification, a carcass receives a D grade) according to López-Campos et al. [22] and the Canadian beef Grading Agency [2].

3.2. Primal Weight Estimation

Overall, CVS lean and fat predictions (Table 2) showed high R^2 values for most of the primal cuts, while R^2 values for bone were much lower. The HCC had similar performance to the CCC for fat predictions in the primal cuts, ranging from R^2 = 0.47 to 0.88 compared to R^2 = 0.51–0.92, respectively. More specifically, fat foreshank regression models showed the lowest R^2 values, while plate, rib and round showed a 14.4% improvement with the HCC compared to CCC. Likewise, lean weight predictions in the primal cuts were superior using the HCC compared to the CCC, except for the rib with R^2 ranging from a low of 0.53 in the foreshank to a high of 0.90 in the round for the HCC, and 0.32 in the foreshank to 0.69 in the rib for the CCC. Additionally, the HCC outperformed the CCC in the prediction of bone weight, showing R^2 values as high as 0.79 in the round, while the highest R^2 for the CCC bone weight was 0.38 in the chuck. Neither of the camera systems studied was able to accurately predict flank bone weight, and the CCC could not either accurately predict bone weight in the brisket, loin, rib, plate, or foreshank ($R^2 < 0.10$, LV = 1 related variable was considered for prediction equation development).

When considering the error and/or variance partitioning, no remarkable differences were found for bone estimations. Nevertheless, CCC showed ECT values 97.4% higher and ED values 5.8% lower than HCC for fat primal estimations, although no difference was observed for MSPE. In the case of lean primal estimations, MSPE value for CCC resulted 63.2% higher than for HCC, although ED values for HCC resulted only 0.7% higher than CCC, and 81.7% of the difference was due to ER instead of ECT (24.4%) when compared to CCC (Table 2).

The combination of both CVS technologies did not improve fat estimations for most of the primals; only the round predictions (R^2 = 0.88) showed some improvements compared to the HCC or CCC estimations; 3.4% and 18.2%, respectively. Conversely, lean estimations of brisket, chuck, plate, and rib showed, respectively, a 10.8%, 3.3%, 10.1%, and 16.2% higher R^2 values for HCC + CCC than for HCC. Additionally, HCC + CCC improved bone estimations in the case of brisket (R^2 = 0.42), chuck (R^2 = 0.71), and loin (R^2 = 0.76) compared to the individual CVS.

In the HCC+CCC, the contribution of ED to MSPE value was again much higher than the inputs coming from ER and ECT values (Table 2). For fat estimations, HCC + CCC showed lower ED and ER values and higher ECT values than HCC, and lower ECT and ER but higher ED values than CCC. In the case of lean estimations, HCC + CCC showed MSPE similar values to HCC, but these were lower than CCC. The ED values were lower than HCC and similar to CCC. The ER values were again higher than ECT, as observed in the HCC to CCC comparison. In addition, no remarkable differences were found for bone estimations. Interestingly, for fat estimations, the LV number for HCC + CCC was lower than for HCC or CCC.

In contrast, DXA primal estimations (Table 3), on average, had R^2 values for fat (0.95), lean (0.97), and bone (0.82) higher than those for CVS, and even outperformed the prediction equations utilizing all camera variables (HCC + CCC; Table 2). Except for the foreshank fat weight (R^2 = 0.74), DXA lean and fat weight predictions for the rest of the primals showed R^2 values between 0.94 and 0.99 and 0.96 and 0.99, respectively. Similar to the CVS, lower values of R^2 were observed for bone than for fat and/or lean variables; however, even flank bone weight (R^2 = 0.31) was predicted more accurately using DXA than by using both camera systems combined. For the other primal bone weight predictions, DXA R^2 ranged from 0.85 to 0.94, whereas the combined camera R^2 values ranged from 0.36 to 0.76. Overall, there were improvements in most tissue primal predictions using DXA when compared to the camera systems. On average, for all the primals, there was an overall proportional improvement in DXA R^2 values of 26.0%, 48.9%, and 24.8% compared to HCC, CCC, or HCC + CCC, respectively, as well as an increase in the DXA R^2 values of 16.0%, 29.0%, and 54.7% for fat, lean, and bone estimations, respectively. The MSPE showed relatively low values and was defined by ED in a percentage higher than 98.7% for fat estimations, and higher than 99.8% for lean and bone tissue estimations (Table 3).

Table 2. Partial least square regression models estimating lean, fat, and bone for individual primal cuts from computer vision system (CVS) values. Coefficient of determination (R^2), mean square prediction error (MSPE), error in central tendency (ECT), error due to regression (ER), error due to disturbances (ED), and the number of latent variables (LV) are presented for each model.

Tissue	Primal [4]	HCC [1] (n = 105)						CCC [2] (n = 102)						HCC + CCC [3] (n = 95)					
		R^2	MSPE	ECT (%)	ER (%)	ED (%)	LV	R^2	MSPE	ECT (%)	ER (%)	ED (%)	LV	R^2	MSPE	ECT (%)	ER (%)	ED (%)	LV
Fat (kg)	Brisket	0.86	0.1942	0.35	0.14	99.51	8	0.88	0.1817	6.46	0.09	93.44	10	0.80	0.2872	2.68	0.01	97.31	2
	Chuck	0.88	2.3678	0.25	0.14	99.61	8	0.91	2.1435	9.98	0.29	89.73	10	0.87	2.8515	5.95	0.01	94.04	3
	Flank	0.86	0.9713	0.15	0.18	99.67	7	0.92	0.6498	9.23	0.12	90.66	10	0.88	0.8844	5.50	0.05	94.45	3
	Loin	0.81	2.5542	0.00	0.18	99.82	9	0.91	1.3322	8.29	0.23	91.48	10	0.85	2.1950	5.90	0.02	94.08	3
	Plate	0.87	0.6728	0.00	0.10	99.90	10	0.73	1.4707	2.74	0.02	97.24	4	0.84	0.9188	5.35	0.01	94.64	3
	Rib	0.87	1.1126	0.07	0.15	99.78	10	0.78	2.0276	3.66	0.04	96.30	3	0.86	1.3099	5.21	0.02	94.77	3
	Round	0.85	0.6854	0.27	0.01	99.71	4	0.72	1.3811	4.01	0.26	95.73	2	0.88	0.6259	6.71	0.49	92.80	2
	Foreshank	0.47	0.0332	0.08	0.00	99.92	2	0.51	0.0309	0.51	0.00	99.49	4	0.50	0.0316	1.17	0.00	98.83	2
Lean (kg)	Brisket	0.67	0.2783	0.06	0.10	99.85	2	0.62	0.3286	0.55	0.76	98.69	4	0.76	0.2107	0.48	0.70	98.83	3
	Chuck	0.85	4.6386	1.10	0.04	98.86	4	0.52	14.710	1.02	0.83	98.15	3	0.88	3.9094	0.70	4.53	94.77	5
	Flank	0.82	0.3376	2.26	0.03	97.71	9	0.55	0.8112	1.57	0.37	98.06	2	0.74	0.4638	0.13	1.11	98.75	3
	Loin	0.82	1.5263	0.41	0.16	99.43	5	0.58	3.5920	0.72	0.22	99.06	3	0.82	1.5196	0.47	0.19	99.34	4
	Plate	0.75	0.5167	0.04	0.08	99.87	3	0.46	1.1186	0.32	0.25	99.43	4	0.83	0.3492	0.66	0.73	98.61	5
	Rib	0.66	1.2365	0.41	0.07	99.52	2	0.69	1.1224	0.45	1.02	98.53	3	0.79	0.7751	0.00	0.86	99.14	3
	Round	0.90	2.0669	0.59	0.26	99.15	10	0.65	7.2982	1.20	0.58	98.23	4	0.86	2.9706	1.28	0.90	97.82	4
	Foreshank	0.53	0.1596	0.14	0.03	99.83	2	0.32	0.2328	0.80	0.17	99.03	2	0.51	0.1681	0.53	0.20	99.27	2
Bone (kg)	Brisket	0.37	0.0566	0.05	0.00	99.95	2	0.01 [5]	0.0855	0.30	0.00	99.70	1	0.42	0.0526	0.25	0.00	99.75	2
	Chuck	0.68	0.4167	0.01	0.01	99.98	4	0.38	0.8187	0.14	0.08	99.78	4	0.71	0.3886	0.46	0.24	99.31	3
	Flank	0.09 [5]	0.0086	0.03	0.01	99.97	1	0.03 [5]	0.0091	0.00	0.01	99.99	1	0.09 [5]	0.0086	0.06	0.05	99.89	1
	Loin	0.64	0.1272	0.05	0.00	99.95	4	0.03 [5]	0.3185	0.01	0.01	99.99	1	0.76	0.0848	0.03	0.25	99.72	6
	Plate	0.62	0.0598	0.02	0.01	99.97	2	0.09 [5]	0.1329	0.01	0.01	99.99	1	0.62	0.0595	0.09	0.02	99.89	2
	Rib	0.36	0.1358	0.14	0.01	99.85	2	0.04 [5]	0.1896	0.08	0.04	99.88	1	0.36	0.1369	0.59	0.33	99.08	2
	Round	0.79	0.2256	0.17	0.12	99.71	5	0.36	0.6723	0.05	0.07	99.88	4	0.75	0.2622	0.64	0.10	99.26	3
	Foreshank	0.60	0.0504	0.00	0.05	99.94	2	0.02 [5]	0.1156	0.13	0.04	99.83	1	0.55	0.0574	0.28	0.32	99.40	2

[1] HCC = hot carcass (whole-side) camera: regression models obtained using HCC variables. [2] CCC = cold carcass (rib-surface) camera: regression models obtained using CCC variables. [3] HCC + CCC = regression models obtained using the variables from both CVS. [4] Primals according to Institutional Meat Purchase Specifications (IMPS) for Fresh Beef Products, Series 100 [26]. [5] No statistically significant regression model ($p > 0.05$) was obtained. LV = 1 was considered to establish a prediction equation.

Table 3. Partial least square regression models estimating fat, lean, and bone for individual primal cuts from dual-energy X-ray absorptiometry (DXA) values (n = 111). Coefficient of determination (R^2), mean square prediction error (MSPE), error in central tendency (ECT), error due to regression (ER), error due to disturbances (ED), and the number of latent variables (LV) are presented for each model.

Tissue	Primal [1]	R^2	MSPE	ECT (%)	ER (%)	ED (%)	LV
Fat (kg)	Brisket	0.99	0.0143	0.521	0.021	99.46	10
	Chuck	0.99	0.3074	0.335	0.019	99.65	10
	Flank	0.98	0.1540	0.097	0.049	99.85	6
	Loin	0.98	0.2395	0.219	0.032	99.75	10
	Plate	0.98	0.1039	1.054	0.254	98.69	10
	Rib	0.98	0.1384	0.940	0.095	98.96	10
	Round	0.96	0.1734	0.253	0.001	99.75	10
	Foreshank	0.74	0.0160	0.096	0.025	99.88	4
Lean (kg)	Brisket	0.99	0.0128	0.088	0.094	99.82	10
	Chuck	0.99	0.4146	0.023	0.135	99.84	10
	Flank	0.97	0.0519	0.004	0.066	99.93	10
	Loin	0.95	0.3825	0.003	0.042	99.96	6
	Plate	0.95	0.0964	0.041	0.024	99.93	7
	Rib	0.98	0.0569	0.006	0.126	99.87	10
	Round	0.99	0.2775	0.123	0.093	99.78	10
	Foreshank	0.94	0.0205	0.047	0.013	99.94	9
Bone (kg)	Brisket	0.89	0.0096	0.141	0.013	99.85	5
	Chuck	0.92	0.1081	0.009	0.019	99.97	8
	Flank	0.31	0.0066	0.043	0.007	99.95	3
	Loin	0.88	0.0420	0.106	0.005	99.89	9
	Plate	0.94	0.0088	0.044	0.017	99.94	9
	Rib	0.85	0.0313	0.006	0.009	99.98	5
	Round	0.92	0.0875	0.038	0.008	99.95	6
	Foreshank	0.86	0.0179	0.026	0.004	99.97	4

[1] Primals according to Institutional Meat Purchase Specifications (IMPS) for Fresh Beef Products, Series 100 [26].

3.3. Overall Carcass Tissue Composition and Yield Estimations

Overall, relatively high R^2 values (>0.75) were obtained between the estimations with the different technologies and the actual dissection values and yield equation estimates of LMY and RCY. Particularly, high relationships ($R^2 > 0.80$) were observed between the estimations with DXA and HCC and the actual dissection values (Table 4). With the exception of LMY ($R^2 = 0.66$ vs. 0.85), the HCC had similar or higher predictions for overall total carcass composition than the CCC (Table 4). In particular, the HCC predicted fat weights similar to ($R^2 = 0.92$ vs. 0.93) and lean weights ($R^2 = 0.89$ vs. 0.67) and bone weights ($R^2 = 0.82$ vs. 0.31) better than the CCC camera. In fact, the HCC performed similar to DXA for all the total carcass composition estimates ($R^2 > 0.80$), and only dropped in prediction accuracy for the LMY and RCY ($R^2 = 0.66$ and 0.68 for the HCC and $R^2 = 0.81$ and 0.86 for the DXA). Adding the CCC variables to the prediction (HCC + CCC) resulted in very similar prediction accuracies to those of DXA for all overall total carcass composition, including the estimates of LMY and RCY.

Table 4. Partial least square regression models estimating total fat, lean, and bone amounts, and total subcutaneous (SQ), body cavity (BC), and intermuscular (IM) fat amounts for whole carcass sides and total lean meat yield (LMY) and retail cut yield (RCY) from dual-energy X-ray absorptiometry (DXA) and computer vision system (CVS) values. Coefficient of determination (R^2), mean square prediction error (MSPE), error in central tendency (ECT), error due to regression (ER), error due to disturbances (ED), and the number of latent variables (LV) are presented for each model.

	HCC [1] (*n* = 105)						CCC [2] (*n* = 102)						HCC + CCC [3] (*n* = 95)						DXA (*n* = 111)					
	R^2	MSPE	ECT (%)	ER (%)	ED (%)	LV	R^2	MSPE	ECT (%)	ER (%)	ED (%)	LV	R^2	MSPE	ECT (%)	ER (%)	ED (%)	LV	R^2	MSPE	ECT (%)	ER (%)	ED (%)	LV
Fat (kg)	0.92	29.407	0.130	0.249	99.62	10	0.93	30.104	11.66	0.427	87.91	10	0.91	35.532	9.073	0.011	90.92	3	0.99	2.5943	1.107	0.000	98.89	7
Lean (kg)	0.89	36.092	1.066	0.165	98.77	5	0.67	104.53	1.305	1.276	97.42	4	0.93	23.044	1.403	5.401	93.20	6	0.99	3.1380	0.046	0.180	99.77	8
Bone (kg)	0.82	2.4731	0.000	0.039	99.96	5	0.31	9.2266	0.061	0.151	99.79	1	0.84	2.1539	0.323	1.360	98.32	5	0.92	1.0459	0.029	0.013	99.96	5
SQ (kg)	0.88	6.5924	0.219	0.086	99.69	8	0.82	9.7306	4.224	0.007	95.77	3	0.88	6.5623	4.704	0.063	95.23	3	0.95	2.5014	0.038	0.025	99.94	10
BC (kg)	0.81	0.5213	0.404	0.136	99.46	10	0.75	0.6954	1.960	0.307	97.73	7	0.75	0.6965	4.453	0.084	95.46	4	0.81	0.5184	0.111	0.024	99.87	5
IM (kg)	0.91	10.189	0.145	0.269	99.59	10	0.91	12.097	10.30	0.462	89.24	10	0.90	12.709	8.903	0.026	91.07	3	0.98	1.7734	0.742	0.022	99.24	7
LMY (%)	0.66	7.3418	3.603	0.034	96.36	5	0.85	3.1867	4.719	0.113	95.17	5	0.90	2.2255	8.180	0.069	91.75	6	0.81	3.9807	0.176	0.482	99.34	5
RCY (%)	0.68	1.7008	0.641	0.001	99.36	10	0.65	1.8364	1.321	0.054	98.63	4	0.86	0.7776	6.983	0.589	92.43	6	0.86	0.7566	0.027	0.003	99.97	6

[1] HCC = hot carcass (whole-side) camera: regression models obtained using HCC variables. [2] CCC = cold carcass (rib-surface) camera: regression models obtained using CCC variables. [3] HCC + CCC = regression models obtained using the variables from both CVS systems.

In general, DXA estimations showed lower MSPE values than any CVS (2.0% vs. 14.6% on average, respectively), and, particularly, higher MSPE values were observed for the CCC procedure than for the HCC or HCC + CCC ones (21.4%, 11.8%, and 10.5% on average, respectively). Besides, IM, total fat, and total lean estimates showed the highest MSPE values, whereas BC fat showed the lowest. Implicating MSPE components, 88% for ED input was observed for the estimation of total fat using CCC variables, while values higher than 90% were observed for the others.

4. Discussion

Canadian mature cow grades (D grades) are assigned to one of the D_1, D_2, D_3, or D_4 grades depending on variables such as muscling (excellent, medium, or deficient), fat color (white or yellow), and fat measure (lower than, equal to, or higher than 15 mm) [2]. In the present study, similar numbers of carcasses for each grade were considered (26.2%, 25.0%, 23.8%, and 25.0%, respectively for D_1, D_2, D_3, and D_4). The ranges of the HCW, CCW, grade fat and fat thickness, LMY and RCY, REA, and marbling scores of the research carcass population used in the present study were representative of those found in the Canadian beef market [1].

The technologies used in the present study provided estimation values of the total amount of tissue and an overall description of the composition of the whole carcass and primal cuts without requiring the destructive procedure of dissection.

In the literature, most of the studies considering the use of CVS in beef carcass classification have focused on the quantification of LMY, RCY, and/or the total amount of fat, lean, and bone using CCC systems. Among others, Farrow et al. [31], Lu and Tan [32], McEvers et al. [10], and Shackelford et al. [33] used several variables obtained from the analysis of rib-eye images to define different regression equations to improve the accuracy, precision, and robustness of total tissue amount, LMY, or RCY estimations. The results reported by these authors (R^2 = 0.43–0.91) are within the range of those observed in the present study. All the authors agreed that CVS-related equations were an improvement on current prediction systems.

In agreement with the present study, Borggaard et al. [34] described similar R^2 values for total fat and RCY (%) using a BCC-2 camera and a HCC system, but carried out the statistical analysis by means of principal component analysis (PCA) and neural networks. Likewise, Pabiou et al. [35] used the VBS 2000 carcass grading unit (HCC) to predict carcass cut yields in cattle. Diverging from our study, HCC and CCW variables were used in the estimation models, and it was stated that stepwise regression showed slightly better R^2 values than the PLSR procedure, thus explaining 71%, 72%, and 75% of the variance for RCY (%), total fat (%), and total bone (%), respectively.

Vote et al. [36] compared BCSys (HCC) and CVS BeefCam (CCC) to study their potential as grading systems for Uruguayan beef carcasses. They reported higher RCY R^2 values for CVS estimations than for values from the USDA equation (values coming from graders). In agreement with the present study, for total fat and bone estimations, higher R^2 values were shown when using HCC or HCC + CCC technologies than when the CCC system was considered. In addition, in Vote et al. [36], bone amount estimations resulted in lower R^2 values than fat amount estimations, and HCW was also included in the models.

RCY and LMY values are commonly obtained from equations in which rib-eye and fat thickness measurements are considered [25]. Because the equations are built from these variables, it is not surprising that the CCC predicts RCY and LMY better than HCC, as the linear measures of rib-eye and rib-eye area along with the fat thickness obtained with the CCC are likely improving these estimates. Nevertheless, in the case of cull cows, it is possible that the industry could be more interested in lean to fat ratios, in which case, HCC predictions outperformed CCC. Hence, cows could be graded accurately in terms of lean/fat ratio using a camera system that does not require knife-ribbing.

The literature regarding the estimation of cattle primal composition is scarce. Using a dual CVS system (HCC + CCC), Cannell et al. [37] tested a total of 296 carcasses: 158 steers

(103 light (HCW $\leq$ 339 kg) and 55 heavy (HCW > 340 kg)) and 138 heifers (51 light and 87 heavy), and described, in agreement with our results, R^2 > 0.65 for primal fabrication yields on average using a selection of HCC and CCC variables (higher coefficient of correlation), HCW, and stepwise regression (higher R^2). Using a similar statistical approach, the HCC system, including the CCW variable (VBS 2000), Pabiou et al. [35] defined four cut-out groups according to their retail value (low, medium, high, and very high value) and obtained higher R^2 values (0.84, 0.65, and 0.87) than in the present study for wholesale primal weights for steers, heifers, and bulls, respectively. In turn, Craigie et al. [11] used VBS 2000 technology (HCC system) and described R^2 > 0.80 for the estimation of saleable (retail cut) sirloin weight, considering HCW in the regression models.

In other species, Rius-Vilarrasa et al. [38], using VSS 2000 (HCC system for lambs) and PLSR statistical analysis, reported R^2 values of 0.86 for breast and 0.96 for leg primals. Lorenzo et al. [39] reported R^2 values between 0.53 and 0.89 for the prediction of foal carcass composition and wholesale cut yields using HCC. Nevertheless, CCW was also considered as a describing variable in the prediction models, whereas HCW was used in the present study. The CCW has been described as a good estimator in the case of lamb carcasses [40]; however, its suitability has been questioned for cattle [35].

Kipper et al. [17] assessed the accuracy of the methodology using the concepts of trueness, defined in our case as the degree of agreement between the dissection and the instrumental estimation values, and precision, as indicative of the degree of internal agreement (dispersion). In addition, the trueness was considered to be the sum of ECT and ER; precision was associated with ED and overall accuracy was related to MSPE [17]. Paying attention to error parameters, higher ECT values in CCC and HCC + CCC than in HCC were detected in fat estimates, whereas the opposite behavior was observed for ED. Therefore, the similar values of R^2 and high values of ED imply that the three instrumental approaches could be considered highly accurate and precise, the PLSR analysis being suitable for estimation. However, CCC and HCC + CCC fat estimations showed lower trueness than HCC estimations (Table 2).

In agreement with the present results, the feasibility of DXA technology in assessing carcass composition has been stated for broiler chickens [13], pigs [15], and sheep [14], and good R^2 values have also been described for calves [16,18]. Aligning with our results, in all these studies, higher R^2 values were described for total fat and total lean estimations than for total bone estimations.

López-Campos et al. [18] described similar results for the estimation of fat, lean, and bone mass of primals using DXA with youthful cattle. The basis of the DXA technology lies in the different absorption ratios from a low and a high energy X-ray beams when interacting with the tissues. The software estimates the mass of two different tissues at each scanned voxel; therefore, it is possible to differentiate between fat and lean when no bone is present but, where the sample matrix contains bone, fat, and lean, the mass fraction can only be established as bone and soft tissue, with the individual measurements of fat and lean obtained from other regions of the scan. Therefore, the higher the amount of bone detected, the more difficult the differentiation between fat and lean. In addition, a medical DXA unit has been used in the present study, thus being calibrated for the measurement of human bone mineral content and bone mineral density, but not for the whole bone content of livestock.

Again, the fact that ED explained more than 99% of the MSPE values would imply that the differences among R^2 values were highly related to the dispersion (precision) and poorly related to the trueness. Therefore, the external factors such as calibration method, software analysis, the defining variables considered for the estimation models (HCW, CCW, marbling, color, gender, etc.), and the meat cutters' decision making (tissue differentiation and cutting) would be the defining variables of dispersion. Accordingly, the low R^2 value of flank bone might be a consequence of the low amount of bone included in this primal, thus implying a high variability in both DXA estimations and weight measurements. The foreshank showed the highest bone to soft tissue ratio, although the different contributions to lean or fat estimates remain unclear: higher and lower accuracy of lean and fat estimates of foreshank, respectively.

Regarding bone prediction, similar results to those from this study were described in pigs [41] and chickens [42]. Kipper et al. [41] and Schallier et al. [42] described better R^2 values when the predicted amount of bone was correlated with ash content. This implies that the presence of small pieces of lean or fat that were adhered to the bone decreases the accuracy and precision of the analysis, whereas it increases the error between actual and predicted values.

Similarly to DXA, computed tomography (CT) is a technology also based on X-ray attenuation. Prieto et al. [43] described lower R^2 values for IM and total fat than for SQ and total lean predictions (0.77, 0.89, 0.94, and 0.99, respectively) using spiral CT. Concurring with the present study, Navajas et al. [44] described lower R^2 values for carcass total bone than for fat and/or lean estimates when using CT technology (R^2 = 077, 0.92, and 0.96). In addition, Navajas et al. [45] described R^2 values of 0.92, 0.99, and 0.97, respectively, for fat, lean, and bone for the primal estimations.

To date, DXA has been limited by practical constraints for deployment in the industry (horizontal table scans, operation at room temperature, and rate of scan in minutes rather than seconds). However, Scott Technologies Ltd. (New Zealand) has developed an upright DXA scanner, capable of scanning at a rate of 540 lamb carcasses per hour while maintaining performance accuracy. This technology adaptation was originally used to mark anatomical features to program robotic cutting. The technology is now being envisioned as a means of lean yield prediction in beef and lamb plants in Australia and New Zealand.

May et al. [46] reported that estimated yield differences could be attributed partially to differences in seam fat deposition (different fat deposition along the carcass). Likewise, in practice, the fabrication of the boneless, closely trimmed round, loin, rib, and chuck retail cuts is performed manually by meat cutters, thus implying another subjective source of variability. Although these factors might introduce variations in the cutability, the present results suggest that both CVS and DXA technologies have the potential to estimate beef carcass traits such as total or retail cut yield performance.

Finally, it is worth mentioning that, based on its performance, DXA might be seen as the gold standard candidate technology for carcass composition estimation. Currently, DXA technology is still under development and it is also being used as a means of envisioning bone location for robotic carcass fabrication. The costs and other operational factors are limiting its industrial implementation. However, if a facility had the capabilities to set up both camera systems, and knife rib cows at the 12/13th, combining the HCC and CCC data, could result in prediction accuracies very similar to DXA. This approach would be of benefit to the plants in determining which carcasses would be profitable for specific fabrication lines.

5. Conclusions

The results of the present study suggest that tissue composition (fat, lean, and bone), either from primal cuts or full carcass sides, and yield percentages of LMY and RCY of mature cows can be accurately predicted by CVS or DXA technologies by applying partial least square regression statistical analysis.

Although DXA results showed higher accuracy, precision, and robustness than results from CVS technologies, DXA technology is still in development for cattle and would require further design adjustments for full implementation and integration into commercial slaughter plants with moving carcass lines.

On the contrary, CVS technologies (HCC and CCC cameras) are widely implemented in North America. In the present study, predictions using HCC variables led to similar or even better results (higher R^2 and lower MSPE values) than those from CCC. The implementation of HCC technology for the carcass composition estimations of mature cows has the benefit that knife ribbing of the carcasses is not required, not even for RCY (%) or LMY (%) estimations. Furthermore, the combination of both CVS technologies led to significant improvements in the tissue predictions of primal cuts and total carcass composition, particularly for lean/fat ratios, suggesting this approach as an alternative for the enhancement of the prediction accuracy on primals and carcass composition of cull cows.

Author Contributions: Data curation, I.L.L.; formal analysis, J.S. and I.L.L.; funding acquisition, J.L.A.; investigation, J.S., N.P. and Ó.L.-C.; project administration, Ó.L.-C.; supervision, Ó.L.-C.; writing—original draft, J.S.; writing—review and editing, J.L.A., N.P., M.J. and Ó.L.-C. All authors have read and agreed to the published version of the manuscript.

Funding: This study was funded by Agriculture and Agri-Food Canada (AAFC) as part of their AgriInnovation Program Stream B, grant number AIP-P335.

Institutional Review Board Statement: The study was conducted according to the guidelines of the Declaration of Helsinki, and approved by the AAFC-Lacombe Research and Development Centre Animal Care committee. The code for this study was 201702.

Acknowledgments: The authors express their gratitude to the AAFC-Lacombe Research and Development Centre (AB, Canada) Beef Unit and Meat Centre staff for animal care and management, animal slaughter and carcass fabrication, and technical collection and compilation of the research data. José Segura Plaza gratefully acknowledges the support from the Canada Sustainable Beef and Forage Science Cluster, through funding provided by the Canadian Cattlemen's Association and Agriculture and Agri-Food Canada. The authors also sincerely and gratefully acknowledge the cooperation and assistance of the Canadian Cattlemen's Association.

Conflicts of Interest: The authors declare no conflict of interest.

References

1. Canfax. Annual Report.Canfax/Canfax Research Services. 2020. Available online: www.canfax.ca (accessed on 25 March 2021).
2. Canadian Beef Grading Association (CBGA). 2020. Available online: www.beefgradingagency.ca (accessed on 25 March 2021).
3. Roberts, J.; Rodas-González, A.; Juárez, M.; López-Campos, Ó.; Larsen, I.L.; Aalhus, J.L. Muscle profiling of retail characteristics within the Canadian cull cow grades. *Can. J. Anim. Sci.* **2017**, *97*, 562–573. [CrossRef]
4. Aalhus, J.L.; López-Campos, Ó.; Prieto, N.; Rodas-González, A.; Dugan, M.E.R.; Uttaro, B.; Juárez, M. Review: Canadian beef grading—Opportunities to identify carcass and meat quality traits valued by consumers. *Can. J. Anim. Sci.* **2014**, *94*, 545–556. [CrossRef]
5. Scholz, A.M.; Buenger, L.; Kongsro, J.; Baulain, U.; Mitchell, A.D. Non-invasive methods for the determination of body and carcass composition in livestock: Dual-energy X-ray absorptiometry, computed tomography, magnetic resonance imaging and ultrasound: Invited review. *Animal* **2015**, *9*, 1250–1264. [CrossRef]
6. López-Campos, Ó.; Prieto, N.; Juárez, M.; Aalhus, J.L. New technologies available for livestock carcass classification and grading. *CAB Rev.* **2019**, *14*, 1–10. [CrossRef]
7. Woerner, D.R.; Belk, K.E. The History of Instrument Assessment of Beef: A Focus on the Last Ten Years. Cattlemen's Beef Board and National Cattlemen's Beef Association. 2008. Available online: www.beefresearch.org/CMDocs/BeefResearch/The%20History%20of%20Instrument%20Assessment%20of%20Beef.pdf (accessed on 25 March 2021).
8. Allen, P. Automated Grading of Beef Carcasses. In *Improving the Sensory and Nutritional Quality of Fresh Meat*; Kerry, J.P., Ledward, D., Eds.; Series in Food Science Technology and Nutrition; Woodhead Publishing: Cambridge, UK, 2009; pp. 479–492.
9. Craigie, C.R.; Navajas, E.A.; Purchas, R.W.; Maltin, C.A.; Buenger, L.; Hoskin, S.O.; Ross, D.W.; Morris, S.T.; Roehe, R. A review of the development and use of video image analysis (VIA) for beef carcass evaluation as an alternative to the current EUROP system and other subjective systems. *Meat Sci.* **2012**, *92*, 307–318. [CrossRef] [PubMed]
10. McEvers, T.J.; Hutcheson, J.P.; Lawrence, T.E. Quantification of Saleable Meat Yield Using Objective Measurements Captured by Video Image Analysis Technology. *J. Anim. Sci.* **2012**, *90*, 3294–3300. [CrossRef] [PubMed]
11. Craigie, C.R.; Ross, D.W.; Maltin, C.A.; Purchas, R.W.; Buenger, L.; Roehe, R.; Morris, S.T. The relationship between video image analysis (VIA), visual classification, and saleable meat yield of sirloin and fillet cuts of beef carcasses differing in breed and gender. *Livest. Sci.* **2013**, *158*, 169–178. [CrossRef]
12. Lovett, B.A.; Firth, E.C.; Plank, L.D.; Symonds, J.E.; Preece, M.A.; Herbert, N.A. Investigating a relationship between body composition and spinal curvature in farmed adult New Zealand king salmon (Oncorhynchus tshawytscha): A novel application of dual-energy X-ray absorptiometry. *Aquaculture* **2019**, *502*, 48–55. [CrossRef]
13. Goncalves, C.A.; Sakomura, N.K.; da Silva, E.P.; Baraldi Artoni, S.M.; Suzuki, R.M.; Gous, R.M. Dual energy X-ray absorptiometry is a valid tool for assessing in vivo body composition of broilers. *Anim. Prod. Sci.* **2019**, *59*, 993–1000. [CrossRef]
14. Hunter, T.E.; Suster, D.; Dunshea, F.R.; Cummins, L.J.; Egan, A.R.; Leury, B.J. Dual energy X-ray absorptiometry (DXA) can be used to predict live animal and whole carcass composition of sheep. *Small Rumin. Res.* **2011**, *100*, 143–152. [CrossRef]

15. Soladoye, O.P.; López-Campos, Ó.; Aalhus, J.L.; Gariepy, C.; Shand, P.; Juárez, M. Accuracy of dual energy X-ray absorptiometry (DXA) in assessing carcass composition from different pig populations. *Meat Sci.* **2016**, *121*, 310–316. [CrossRef] [PubMed]
16. Scholz, A.M.; Nuske, S.; Forster, M. Body composition and bone mineralization measurement in calves of different genetic origin by using dual-energy X-ray absorptiometry. *Acta Diabetol.* **2003**, *40* (Suppl. 1), S91–S94. [CrossRef] [PubMed]
17. Kipper, M.; Marcoux, M.; Andretta, I.; Pomar, C. Assessing the accuracy of measurements obtained by dual-energy X-ray absorptiometry on pig carcasses and primal cuts. *Meat Sci.* **2019**, *14*, 79–87. [CrossRef] [PubMed]
18. López-Campos, Ó.; Roberts, J.C.; Larsen, I.L.; Prieto, N.; Juárez, M.; Dugan, M.E.R.; Aalhus, J.L. Rapid and non-destructive determination of lean fat and bone content in beef using dual energy X-ray absorptiometry. *Meat Sci.* **2018**, *146*, 140–146. [CrossRef] [PubMed]
19. Canadian Council of Animal Care. *Guidelines on: The Care and Use of Farm Animals in Research, Teaching and Testing;* Canadian Council on Animal Care: Ottawa, ON, Canada, 2009.
20. The Safe Food for Canadians and Health of Animals Acts of CFIA Guidelines (A). Available online: https://inspection.canada.ca/food-safety-for-industry/food-specific-requirements-and-guidance/meat-products-and-food-animals/guidelines-humane-care-and-handling/eng/1525374637729/1525374638088 (accessed on 25 March 2021).
21. The Safe Food for Canadians and Health of Animals Acts of CFIA Guidelines (B). Available online: https://inspection.canada.ca/food-safety-for-industry/food-specific-requirements-and-guidance/meat-products-and-food-animals/srm/eng/1369768468665/1369768518427 (accessed on 25 March 2021).
22. López-Campos, Ó.; Aalhus, J.L.; Prieto, N.; Larsen, I.L.; Juárez, M.; Basarab, J.A. Effects of production system and growth promotants on the physiological maturity scores in steers. *Can J. Anim. Sci.* **2014**, *94*, 607–617. [CrossRef]
23. USDA. *Official United States Standards for Grades of Beef Carcasses;* Agricultural Marketing Service, United States Department of Agriculture: Washington, DC, USA, 1989. Available online: https://www.ams.usda.gov/grades-standards/carcass-beef-grades-and-standards (accessed on 5 May 2021).
24. Jones, S.D.M. Evaluation of the Grade Ruler Approach for the Yield Grading of Beef Carcasses. Published as Appendix 1. Available online: https://cdnsciencepub.com/doi/suppl/10.1139/cjas-2020-0035/suppl_file/cjas-2020-0035suppla.pdf (accessed on 25 March 2021).
25. Segura, J.; Aalhus, J.L.; Prieto, N.; Larsen, I.; Dugan, M.E.R.; López-Campos, Ó. Development and validation of the Canadian retail cut beef yield grades. *Can. J. Anim. Sci.* **2021**, *101*, 196–200. [CrossRef]
26. USDA. Institutional meat purchasing specifications for fresh beef. 2014. Available online: https://www.ams.usda.gov/grades-standards/imps (accessed on 25 March 2021).
27. SAS Institute Inc. *SAS 9.4 for Windows;* SAS Institute Inc: Cary, NC, USA, 2014.
28. Benchaar, C.; Rivest, J.; Pomar, C.; Chiquette, J. Prediction of methane production from dairy cows using existing mechanistinc models and regression equations. *J. Anim. Sci.* **1998**, *76*, 617–627. [CrossRef]
29. Theil, H. *Applied Economic Forecasting;* North-Holland Publishing Company: Amsterdam, The Netherlands, 1966.
30. Mercier, J.; Pomar, C.; Marcoux, M.; Goulet, F.; Thériault, M.; Castonguay, F.W. The use of dual-energy X-ray absorptiometry to estimate the dissected composition of lamb carcasses. *Meat Sci.* **2006**, *73*, 249–257. [CrossRef]
31. Farrow, R.L.; Loneragan, G.H.; Pauli, J.W.; Lawrence, T.E. An exploratory observational study to develop an improved method for quantifying beef carcass salable meat yield. *Meat Sci.* **2009**, *82*, 143–150. [CrossRef]
32. Lu, W.; Tan, J. Analysis of image-based measurements and USDA characteristics as predictors of beef lean yield. *Meat Sci.* **2004**, *66*, 483–491. [CrossRef]
33. Shackelford, S.D.; Wheeler, T.L.; Koohmaraie, M. Coupling of image analysis and tenderness classification to simultaneously evaluate carcass cutability, longissimus area, subprimal cut weights, and tenderness of beef. *J. Anim. Sci.* **1998**, *76*, 2631–2640. [CrossRef]
34. Borggaard, C.; Madsen, N.T.; Thodberg, H.H. In-line image analysis in the slaughter industry, illustrated by Beef Carcass Classification. *Meat Sci.* **1996**, *43* (Suppl. 1), 151–163. [CrossRef]
35. Pabiou, T.; Fikse, W.F.; Cromie, A.R.; Keane, M.G.; Nasholm, A.; Berry, D.P. Use of digital images to predict carcass cut yields in cattle. *Livest. Sci.* **2011**, *137*, 130–140. [CrossRef]
36. Vote, D.J.; Bowling, M.B.; Cunha, B.C.N.; Belk, K.E.; Tatum, J.D.; Montossi, F.; Smith, G.C. Video image analysis as a potential grading system for Uruguayan beef carcasses. *J. Anim. Sci.* **2009**, *87*, 2376–2390. [CrossRef] [PubMed]
37. Cannell, R.C.; Belk, K.E.; Tatum, J.D.; Wise, J.W.; Chapman, P.L.; Scanga, J.A.; Smith, G.C. Online evaluation of a commercial video image analysis system (Computer Vision System) to predict beef carcass red meat yield and for augmenting the assignment of USDA yield grades. *J. Anim. Sci.* **2002**, *80*, 1195–1201. [CrossRef] [PubMed]
38. Rius-Vilarrasa, E.; Buenger, L.; Maltin, C.; Matthews, K.R.; Roehe, R. Evaluation of Video Image Analysis (VIA) technology to predict meat yield of sheep carcasses on-line under UK abattoir conditions. *Meat Sci.* **2009**, *82*, 94–100. [CrossRef]
39. Lorenzo, J.M.; Güedes, C.M.; Agregan, R.; Sarries, M.V.; Franco, D.; Silva, S.R. Prediction of foal carcass composition and wholesale cut yields by using video image analysis. *Animal* **2018**, *12*, 174–182. [CrossRef]
40. Einarsson, E.; Eythorsdottir, E.; Smith, C.R.; Jonmundsson, J.V. The ability of video image analysis to predict lean meat yield and EUROP score of lamb carcasses. *Animal* **2014**, *8*, 1170–1177. [CrossRef]
41. Kipper, M.; Marcoux, M.; Andretta, I.; Pomar, C. Repeatability and reproducibility of measurements obtained by dual-energy X-ray absorptiometry on pig carcasses. *J. Anim. Sci.* **2018**, *96*, 2027–2037. [CrossRef]

42. Schallier, S.; Li, C.; Lesuisse, J.; Janssens, G.P.J.; Everaert, N.; Buyse, J. Dual-energy X-ray absorptiometry is a reliable non-invasive technique for determining whole body composition of chickens. *Poult. Sci.* **2019**, *98*, 2652–2661. [CrossRef]
43. Prieto, N.; Navajas, E.A.; Richardson, R.I.; Ross, D.W.; Hyslop, J.J.; Simm, G.; Roehe, R. Predicting beef cuts composition, fatty acids and meat quality characteristics by spiral computed tomography. *Meat Sci.* **2010**, *86*, 770–779. [CrossRef] [PubMed]
44. Navajas, E.A.; Richardson, R.I.; Fisher, A.V.; Hyslop, J.J.; Ross, D.W.; Prieto, N.; Simm, G.; Roehe, R. Predicting beef carcass composition using tissue weights of a primal cut assessed by computed tomography. *Animal* **2010**, *4*, 1810–1817. [CrossRef] [PubMed]
45. Navajas, E.A.; Glasbey, C.A.; Fisher, A.V.; Ross, D.W.; Hyslop, J.J.; Richardson, R.I.; Simm, G.; Roehe, R. Assessing beef carcass tissue weights using computed tomography spirals of primal cuts. *Meat Sci.* **2010**, *84*, 30–38. [CrossRef] [PubMed]
46. May, S.G.; Mies, W.L.; Edwards, J.W.; Williams, F.L.; Wise, J.W.; Morgan, J.B.; Savell, J.W.; Cross, H.R. Beef carcass composition of slaughter cattle differing in frame size, muscle score, and external fatness. *J. Anim. Sci.* **1992**, *70*, 2431–2445. [CrossRef]

MDPI

Article

Effect of Feeding Barley, Corn, and a Barley/Corn Blend on Beef Composition and End-Product Palatability

Wilson Barragán-Hernández [1,2,3], Michael E. R. Dugan [1], Jennifer L. Aalhus [1], Gregory Penner [4], Payam Vahmani [5], Óscar López-Campos [1], Manuel Juárez [1], José Segura [1], Liliana Mahecha-Ledesma [2] and Nuria Prieto [1,*]

1 Lacombe Research and Development Centre, Agriculture and Agri-Food Canada, Lacombe, AB T4L 1W1, Canada; wbarraganh@agrosavia.co (W.B.-H.); mike.dugan@canada.ca (M.E.R.D.); jennifer.aalhus@outlook.com (J.L.A.); oscar.lopezcampos@canada.ca (Ó.L.-C.); manuel.juarez@canada.ca (M.J.); jose.seguraplaza@canada.ca (J.S.)
2 Faculty of Agricultural Sciences, University of Antioquia, Medellín 1226, Colombia; Liliana.mahecha@udea.edu.co
3 Corporación Colombiana de Investigación Agropecuaria (AGROSAVIA), El Nus Research Centre, San Roque 053037, Colombia
4 College of Agriculture and Bioresources, University of Saskatchewan, Saskatoon, SK S7N 5A8, Canada; greg.penner@usask.ca
5 Department of Animal Science, University of California, 2201 Meyer Hall, Davis, CA 95616, USA; pvahmani@ucdavis.edu
* Correspondence: nuria.prietobenavides@canada.ca

Citation: Barragán-Hernández, W.; Dugan, M.E.R.; Aalhus, J.L.; Penner, G.; Vahmani, P.; López-Campos, Ó.; Juárez, M.; Segura, J.; Mahecha-Ledesma, L.; Prieto, N. Effect of Feeding Barley, Corn, and a Barley/Corn Blend on Beef Composition and End-Product Palatability. *Foods* **2021**, *10*, 977. https://doi.org/10.3390/foods10050977

Academic Editors: Huerta-Leidenz Nelson and Markus F. Miller

Received: 20 March 2021
Accepted: 27 April 2021
Published: 29 April 2021

Abstract: This study evaluated the relationship among palatability attributes, volatile compounds, and fatty acid (FA) profiles in meat from barley, corn, and blended (50:50, barley and corn) grain-fed steers. Multiple correspondence analysis with three dimensions (Dim) explained 62.2% of the total variability among samples. The Dim 1 and 2 (53.3%) separated pure from blended grain-fed beef samples. Blended grain beef was linked to a number of volatiles including (E,E)-2,4-decadienal, hexanal, 1-octen-3-ol, and 2,3-octanedione. In addition, blended grain-fed beef was linked to fat-like and rancid flavors, stale-cardboard, metallic, cruciferous, and fat-like aroma descriptors, and negative categories for flavor intensity (FI), off-flavor, and tenderness. A possible combination of linoleic and linolenic acids in the blended diet, lower rumen pH, and incomplete biohydrogenation of blended grain-fed polyunsaturates could have increased ($p \leq 0.05$) long-chain n-6 fatty acids (LCFA) in blended grain-fed beef, leading to more accumulation of FA oxidation products in the blended than in barley and corn grain-fed meat samples. The Dim 3 (8.9%) allowed corn separation from barley grain beef. Barley grain-fed beef was mainly linked to alkanes and beef positive FI, whereas corn grain-fed beef was associated with pyrazines, in addition to aldehydes related to n-6 LCFA oxidation.

Keywords: barley; corn; blend; eating quality; volatile compounds; fatty acids; beef

1. Introduction

Beef is a valuable food for human nutrition, offering rich contents of available protein, fat, vitamins, and minerals. A European study found that consumption between 75 and 211 g/d of meat contributed to the intake of protein and saturated, monounsaturated, and polyunsaturated fats in a range of 29 to 41%, 19 to 24%, 23 to 28%, and 11 to 20%, respectively, as well as contributing with a variety of vitamins and minerals such as B12 vitamin (29–37%) and zinc (27–37%) [1]. However, beyond health interest, beef consumption is strongly influenced by overall consumer liking, with flavor explaining between 38% and 48% of the variability [2–4].

Beef flavor and other sensory attributes are influenced by ante and post-mortem factors related to genetics, feeding systems, ageing/storage conditions, and cooking methods [5–7]. Volatile compounds responsible for beef flavor originate from water-soluble substances and lipid precursors resultant from the Maillard reaction and thermal lipid oxidation [8]. For beef, volatiles include alkanes, aldehydes, ketones, alcohols, furans, esters, and pyrazines as commonly reported, among other volatiles [9]. These compounds alone or together can positively or negatively stimulate the complex system of consumer senses located in the tongue, mouth, and nasal cavity to develop an opinion regarding its acceptability [8].

The study of beef volatiles is a tool that can be used to interpret complex flavors and anticipate consumer satisfaction, as volatile organic compounds have been associated with consumer perception and satisfaction [10]. For instance, reviews by Mottram [11] and Calkins and Hodgen [12] indicate pyrazines, furans, and alkanes can be related to pleasant and roasted/grilled flavor in beef, whereas some aldehydes and alcohols can contribute to some off-flavors. In this context, knowing the effect of dietary manipulation on volatile profiles could help anticipate consumer beef acceptance [3].

On the Canadian prairies, finishing cattle on barley grain-based diets is the norm [13]. However, corn grain finishing is growing due to development of new varieties adapted to low-luminosity and low-temperature conditions [14]. Additionally, barley and corn blended diets have been proposed to support least-cost ration formulation, and take advantage of greater rates of starch bypass from the rumen, which could improve energetic efficiency and marbling fat deposition [15,16].

Canadian and Japanese consumers have shown a preference for barley over corn grain-fed beef [17]. O'Quinn et al. [3] reported differences in volatile profiles from barley or corn grain-fed beef. However, Jeremiah et al. [18] and McEwen et al. [19] reported no differences in flavor between corn and barley grain-fed beef. Hence, there is some controversy over effects of grain type fed on flavor and their association with volatile profiles, and no data are available for comparing barley, corn, and blended grain-fed beef. Additionally, discrepancies in the fatty acid composition from barley and corn grain-fed beef have also been reported among studies [3,17,20], fatty acids being key precursors of volatile compounds [21].

The objective of the present study was to evaluate the effect of feeding barley, corn, and a barley/corn blend on descriptive sensory attributes, volatile compounds, and flavor and fatty acid profiles from beef, following a qualitative approach.

2. Materials and Methods

2.1. Animals, Diets, and Collection of Samples

A complete description of the animals, diets, and butchering process used herein is available in Johnson et al. [13]. The research protocol for this study was preapproved by the University of Saskatchewan Animal Research Ethics Board (protocol 20100021), according to the guidelines of the Canadian Council on Animal Care (Ottawa, ON, Canada). In short, 288 commercial crossbred steers (464 ± 1.7 kg) were randomly assigned to 24 pens (12 steers/pen) at the University of Saskatchewan. Four pens were randomly assigned to six treatments following a factorial (2×3) design of silage source (corn or barley, 8% dry matter—DM) and grain source: barley (86% DM), corn (85% DM), or blend (50:50 barley and corn, 85% DM). Silage was balanced across grain treatments, and all diets included the same minerals and vitamins and were isoproteic. Following 89 days on feed, steers (623 ± 86 kg) were taken to a federally inspected abattoir (Cargill Meat Solution, High River, AB, Canada) and slaughtered according to Canadian Council on Animal Care principles and guidelines [22].

For this study, a total of 85 steers were randomly selected (4 steers/pen) considering only the grain-fed source (27 for corn, 29 for barley, and 29 for blend treatment). After slaughter, the *longissimus thoracis* (LT) between the 6th and 12th ribs (bone-in ribeye) from each carcass was collected, transported in a refrigerated vehicle (2–4 °C) to the Lacombe Research and Development Centre (Lacombe, AB, Canada), and aged in a cooler at 2 °C, 0.5 m·s^{-1} of wind speed, and 80% of relative humidity for 15 days. Following ageing, four 25 mm steaks (LT between the 8th and 12th ribs) were taken from each bone-in ribeye, trimmed of all subcutaneous and seam fat, and assigned respectively to fatty acids, volatile compounds, flavor profile, and descriptive sensory analyses.

2.2. Sensory Analyses

Steaks designated for descriptive sensory and flavor profile analyses were cooked after being thawed for 24 h at 4 °C. A 10 cm spear point Type T thermocouple probe (Wika Instruments, Edmonton, AB, Canada) was inserted into the center of the steak and connected to a Hewlett Packard HP34970A Data Logger (Hewlett Packard Co., Boise, ID, USA) to monitor the internal temperature of the steaks while cooking. On a Garland grill (Model ED30B, Condon Barr Food Equipment Ltd., Edmonton, AB, Canada) that was preheated to 210 °C, steaks were grilled to an internal temperature of 35.5 °C, flipped, and were taken off when they reached 71 °C. After removal from the grill, steaks were cooled for 3 min, and then each steak was subsampled by cutting 1.3 × 1.3 × 1.3 cm cubes, avoiding areas with high levels of connective tissue or fat. Prior to sensory analysis, the temperature of the steak cubes was equilibrated by putting samples in covered glass containers in a circulating water bath (68 °C). Samples were presented to a trained expert, nine-member meat evaluation panel in a balanced design with sample assignment determined using Compusense 5 Software, version 4.6 (Compusense Inc., Guelph, ON, Canada).

The panelists rated the following attributes from steak samples for descriptive attribute sensory analyses: initial and overall tenderness, initial and sustained juiciness, beef flavor and off-flavor intensity, amount of connective tissue, and residual mouth coating. nine-point descriptive scales were used to assign the scores: 9 = extremely tender, extremely juicy, extremely intense beef flavor, extremely bland off-flavor, no connective tissue, and no residual mouth coating; 1 = extremely tough, extremely dry, extremely bland beef flavor, extremely intense off-flavor, extremely abundant connective tissue, and extremely abundant residual mouth coating.

The AMSA flavor lexicon was used for flavor profile analysis [20]. Samples were evaluated using a 15 cm line scale with standard reference points for detected tastes (sweet, sour, bitter, salty, and umami), aromas, and flavors (brown-roasted, beef identity, cruciferous, oily, grainy, bloody-serumy, corn, liver-like, sour-dairy, green-hay, burnt, barnyard, buttery, metallic, stale-cardboard, other, and unidentified) (0 = none; 15 = extremely intense).

Paid panelists, who had served as trained experts for an average of 6 years, were recruited, screened, and trained [23] to exclusively evaluate meat samples. The guidelines set forth by the AMSA were used to evaluate and monitor panelists' performance [20]. For each session, the number of steaks evaluated included 6 for descriptive sensory analyses and 3 for flavor profile analyses. Four sessions were conducted per day in total: 2 in the morning and 2 in the afternoon, with a 20 min break between sessions in the morning and afternoon. All panel evaluations were performed in partitioned booths that were well-ventilated and illuminated by 180 lux green lighting. Unsalted soda crackers and distilled water were supplied to cleanse the palate of residual flavor notes between samples [24].

2.3. Volatile Compounds

Analysis of beef volatile compounds was performed as outlined in Ruan et al. [9]. Briefly, steaks were grilled to a final temperature of 71 °C as described above. Steaks were then ground for 15 s at 10 × 1000 rpm with Grindomix GM200 (Rest GmbH, Haan, Germany) and subsampled in triplicate (1 g each) for stir bar sorptive extraction. This was coupled with thermal desorption–gas chromatography–mass spectrometry analyses (SBSE-TD-GC-MS). Each subsample was placed into a 10 mL sample vial with 8 mL of extraction solution (75% saturated NaCl with 25% MeOH, *v*/*v*). A commercial sorptive stir bar (TwisterTM GERSTEL GmbH & Co.KG, Mülheiman der Ruhr, Germany) was added to each vial to agitate the meat slurry for 120 min at 35 °C × 1000 rpm on a Gerstel Twister® stir plate (Gerstel GmbH & Co. KG, Mülheiman der Ruhr, Germany). Stir bars were then thermally desorbed by programming the TDS 2 from 40 °C (held for 1 min) to 200 °C (held for 5 min) at 60 °C/min. The desorbed compounds were cryofocused in the CIS 4 at −120 °C. Following desorption, the CIS 4 was programmed from 40 to 275 °C (held for 5 min) at 12 °C/s to inject the trapped compounds onto the analytical column. The separations were executed on an HP-5 MS fused-silica capillary column (30 m × 0.25 mm

I.D., 0.25 μm film thickness, Agilent Technologies, Santa Clara, CA, United States). Oven temperature was programmed from 50 °C (held for 1 min) to 100 °C (held for 2 min) at 10 °C/min, then to 280 °C (held for 1 min) at 30 °C/min. The carrier gas was helium with a flow rate of 1.2 mL/min. Volatile compounds were tentatively identified using a mass spectral library in NIST 08 (NIST 08 version 2.0) requiring a match factor over 85 with the retention index for all the volatile compounds. The base peak (m/z) of each volatile was then standardized to the internal reference peak (Nonanal).

2.4. Fatty Acid Analysis

Intramuscular lipid extraction and fatty acid analysis were conducted as outlined in Vahmani et al. [25]. In short, chloroform−methanol (2:1, v/v) was used to extract subsamples from steaks between the 11th and 12th ribs. Acid (5% methanolic HCl) and base (0.5 N sodium methoxide) were then used to methylate an aliquot of lipids from each tissue. Fatty acid methyl esters (FAME) were then analyzed using a CP-3800 gas chromatograph equipped with a 100 m CP-Sil 88 fused capillary column (Varian Inc., Mississauga, ON, Canada) with *c*10-17:1 methyl ester (Nu-Check Prep Inc., Elysian, MN, USA) as the internal standard. To quantify the FAME, chromatographic peak area and internal standard-based calculations were employed.

2.5. Statistical Analyses

Principal component analysis (PCA) was applied to all the volatile compounds tentatively identified in the samples from this study to select the most important compounds from the dataset. The volatile compounds were selected based on their loadings in the PCA and then binarized with a threshold of 0.4 times internal reference peak: compounds > 0.4 times internal reference peak were considered as present (Volatile_y), and anything else as absent (Volatile_n). The descriptive sensory attributes and flavor profile datasets were categorized as well. Descriptive sensory attributes with an average rating ≥6 were categorized as positive (Sensory attribute_Positive), 5 = neither positive or negative (Sensory attribute_Neither), and <5 as negative (Sensory attribute_Negative). A flavor descriptor was present when the average of panelists' rates was higher than 1 (aroma/taste/flavor descriptor—A/T/F y), and anything else was considered absence (aroma/taste/flavor descriptor—A/T/F n).

The descriptive sensory attributes, flavor profile, and volatile compounds from barley, corn, and blended grain-fed beef were submitted to multiple correspondence analysis (MCA), using a FactoMiner package [26] in the software R-Project (version 3.6.1., 2019, Team Core R, Vienna, Austria). Additionally, to improve insight about fatty acid and volatile compound relationships, a canonical correlation analysis (CCA) was performed using a CCA package in the software R-Project (version 3.6.1., 2019). The results were presented using biplot, structure correlation, and weights in representative dimensions of CCA [27]. Moreover, fatty acid analyses and sensory attributes were conducted using the MIXED procedure of SAS (Version 9.2 Institute Inc., Cary, NC, USA). In fatty acid analysis, grain was used as a fixed effect, and pen as the random effect. For sensory analysis, grain was used as a fixed effect, and pen, trained panelist and session for sensory analysis, as the random effect. In rejection of the null hypothesis, the least-square means difference was conducted by the PDIFF statement with alpha = 0.05. Preliminary analyses showed no effect of silage or silage grain interaction either on sensory attributes or on fatty acid and volatile profiles, and, hence, only the effect of grain type was evaluated in this study.

3. Results and Discussion

In this study, a total of 162 volatile compounds were tentatively identified in meat samples from barley, corn, and blended grain-fed steers. After applying a PCA with all the volatiles, five principal components (PC variance > 1) were retained, explaining 84.5% of the total variance. In these PCs, the volatile compounds with a loading higher than 0.2 in absolute value were selected, resulting in a total of 22 tentatively identified volatile compounds considered for this study (Table 1). The rest of the volatile compounds not selected are presented in the Supplementary Table S1. Among volatiles selected, the most significant groups were alkanes (31.8%), aldehydes (18.1%), ketones (13.6%), alcohols (9%) derived from lipids, and pyrazines (9%) originated from water-soluble compounds resulting from lipid thermal and Maillard reactions [28].

Table 1. Mean of standardized base peak of selected volatile compounds from barley, corn, and blended grain-fed beef samples.

Volatile Compounds	Barley	Blended	Corn
(E,E)-2,4-Decadienal	1.837978	1.80720552	1.55708039
(Z)-7-Hexadecenal	5.30218346	4.48371819	5.38537755
1-Octadecanol	0.96836213	0.86885953	0.99385898
1-Octen-3-Ol	0.5153148	0.44414982	0.42838899
1,2-Epoxyoctadecane	0.54637487	0.33669352	0.55077623
2-Isopentyl-3,6-Dimethylpyrazine	0.29760656	0.40774939	0.36947751
2-Pentadecanone	0.49167099	0.44082738	0.49833478
2-Undecanone	0.24967149	1.22870434	0.40275022
2,3-Dimethyl-5-Isopentylpyrazine	0.3877261	0.19724228	0.28495414
2,3-Octanedione	0.59298402	0.51187674	0.52339562
Heptadecane	0.74794743	0.56457323	0.55966006
Heptane	0.14459152	0.12024081	0.06092762
Hexadecanal	0.85396753	0.53034371	0.87460321
Hexadecane	0.38918876	0.37397273	0.35616684
Hexanal	0.55325571	0.39576164	0.53619463
Hexane	0.34493859	0.19225675	0.17337753
Methyl Oleate	2.29272729	1.82602211	1.73166007
n-Hexadecanoic Acid	0.07806632	0.07472788	0.02894254
Nonadecane	0.13397839	0.229666	0.06135028
Octadecane	1.08206796	0.92397752	0.89970811
Oleic Acid	0.57770987	0.56190167	0.39991017
Undecane	0.05524978	0.23207027	0.39906141

A full description of the descriptive sensory and flavor profile (aromas, tastes, and flavors) attributes from barley, corn, and blended grain-fed beef are presented in the Supplementary Table S2. The MCA performed among sensory attributes, flavor profile, and volatile compounds from barley, corn, and blended grain-fed beef achieved 62.2% of the total variability in three dimensions (Figure 1). This total variability was reached by considering the variables showing a high contribution to the construction of the dimensions. Hence, only those variables explaining a high variability are presented in Figure 1.

The first and second dimensions (Dim 1 and Dim 2) of the MCA explained 36.6% and 16.7% of the variability, respectively (Figure 1A). Both Dim 1 and 2 contributed to discriminate pure grains (barley or corn) against the blended grain-fed beef based on descriptive sensory attributes, flavor profile, and volatile compounds. The barley and corn grain-fed categories were located in the lower-left quadrant, whereas the blended grain-fed beef was observed in the upper-right quadrant. Specifically, the positive axis of Dim 1 was associated with the blended grain-fed category and lipid products such as aldehydes [(E, E)-2,4-decadienal and hexanal] and alcohols (1-octen-3-ol). These volatile compounds contribute to beef flavor characteristic [29]. However, in their highest concentrations, these volatiles are associated with adverse lipid oxidation odors such as rancidity, fishy, and grassy [8,30,31]. Additionally, the blended grain-fed category was also associated by Dim 1 with 2,3-octanedione that has an oxidized fat and warmer over flavor derived from lipid oxidation [32]. Indeed, the positive space of the Dim 1 was also characterized by the presence of stale-cardboard, fat-like, metallic, and cruciferous aromas, fat-like and rancidity flavor descriptors, and the negative categories for off-flavor (from slightly to extremely intense off-flavor) and beef flavor intensity (from slightly to extremely bland beef flavor). The association among aroma and flavor descriptors related to oxidation, low beef flavor intensity, and the above-mentioned volatile compounds was previously described by Larick and Turner [30] and Kerth and Miller [8] in beef. Burnett et al. [21] and Therkildsen et al. [33] described a negative association between polyunsaturated fatty acids (PUFA) and beef flavor intensity and overall palatability due to fatty acid oxidation. In this study, the CCA between fatty acids and volatile compounds showed an association among the C18:2 n-6 ratio (r = −0.432), n-6 FA group (r = −0.397), PUFAs (r = −0.385), C22:6 n-3 (r = −0.380), PUFA/saturated fatty acids (SFA) ratio (r = −0.334), hexanal (r = −0.422) 2,3-octanedione (r = 0.3462), and 1-octen-3-ol (r = 0.2866) in the negative space of the Dim 1 (Roy's statistics test $p < 0.001$). These results agree with Calkins and Sullivan [34] and Larick and Turner [30], who reported hexanal and 1-octen-ol are related to lipid oxidation from PUFAs.

The meat from blended grain-fed cattle was also characterized by negative categories for initial tenderness, overall tenderness, and connective tissue (tougher meat and with an abundant connective tissue), with overall tenderness and connective tissue being located in the positive space of both Dim 1 and 2, and initial tenderness in the positive and negative space of Dim 1 and 2, respectively. This agrees with Legako et al. [35], who reported a negative correlation between aldehydes and consumer palatability scores for overall liking, tenderness, and overall tenderness, and also desirable flavor descriptors such as beef identity, bloody-serumy, brown-roasted, and umami. However, in contrast to our results, these authors also reported a negative correlation between aldehydes and fat-like flavor descriptor. Starowicz and Zieliński [36] suggested that the Maillard reaction could be related to tenderness improvement via alteration of protein cross-linking based on the amino acids involved. This hypothesis was previously corroborated by Sun et al. [37], who used Maillard reaction modificated proteins from mechanically deboned chicken prepared at 90 °C to fabricate Cantonese sausages with less hardness and chewiness texture. The present study may corroborate this theory due to the association between the negative categories for initial and overall tenderness and connective tissue and the absence of volatiles from the Maillard reaction in the positive space of the Dim 1.

Unlike blended grain, the negative space of Dim 1 and Dim 2 (lower-left quadrant) was associated with both barley and corn grain-fed beef and associated with the presence of (1) alkanes (hexadecane, heptadecane, and octadecane) and ketones (2-pentadecanone), which have been related to pleasant flavors such as meaty and brown-roasted [38]; (2) oleic acid positively related to desirable beef palatability; (3) the positive category for off-flavor intensity (from slightly bland to no off-flavor), sweet taste, and corn aroma. Barley and corn-fed beef were also linked to hexadecanal, which has been related to fatty odor, however, this aldehyde has a small contribution to meat flavor [31]. Additionally, some pyrazine compounds (2,3-dimethyl-5-isopentylpyrazine and 2-isopentyl-3,6-dimethylpyrazine) and alkanes (nonadecane, hexane, and heptane) linked to meaty and roasted flavor [11,12,28], ketones (2-undecanone), a positive category for beef flavor intensity (from slightly to extremely intense beef flavor), and green-hay aroma were associated to pure grain-fed meat by the negative axis of the Dim 1. On the other hand, the positive space of Dim 1 of the CCA (Figure 2) showed an association between fatty acids such as C18:1t15 (r = 0.494), C19:1 (r = 0.362), cMUFAs (r = 0.344), and C18:1c13 (r = 0.343), and volatile compounds mentioned above, such as octadecane (r = 0.626), heptadecane (r = 0.512), 2,3-dimethyl-5-isopentylpyrazine (r = 0.496), and 2-pentadecanone (r = 0.451). These results agree with those of Mottram [11], who reported that alkanes come from the oxidation of long-chain fatty acids and, together with pyrazines and other volatile compounds, contribute to the pleasant beef flavor [8]. Likewise, O'Quinn et al. [3] and Hwang and Joo [39] reported an association between MUFAs and beef flavor desirability and overall palatability in beef. Overall, the volatile profiles associated with barley and corn grain-fed beef in this study are in agreement with that found by O'Quinn et al. [3]. Moreover, both barley and corn grain-fed meat were associated with positive initial and overall tenderness and connective tissue categories, although to a lesser extent due to their proximity to the origin of both Dim 1 and 2.

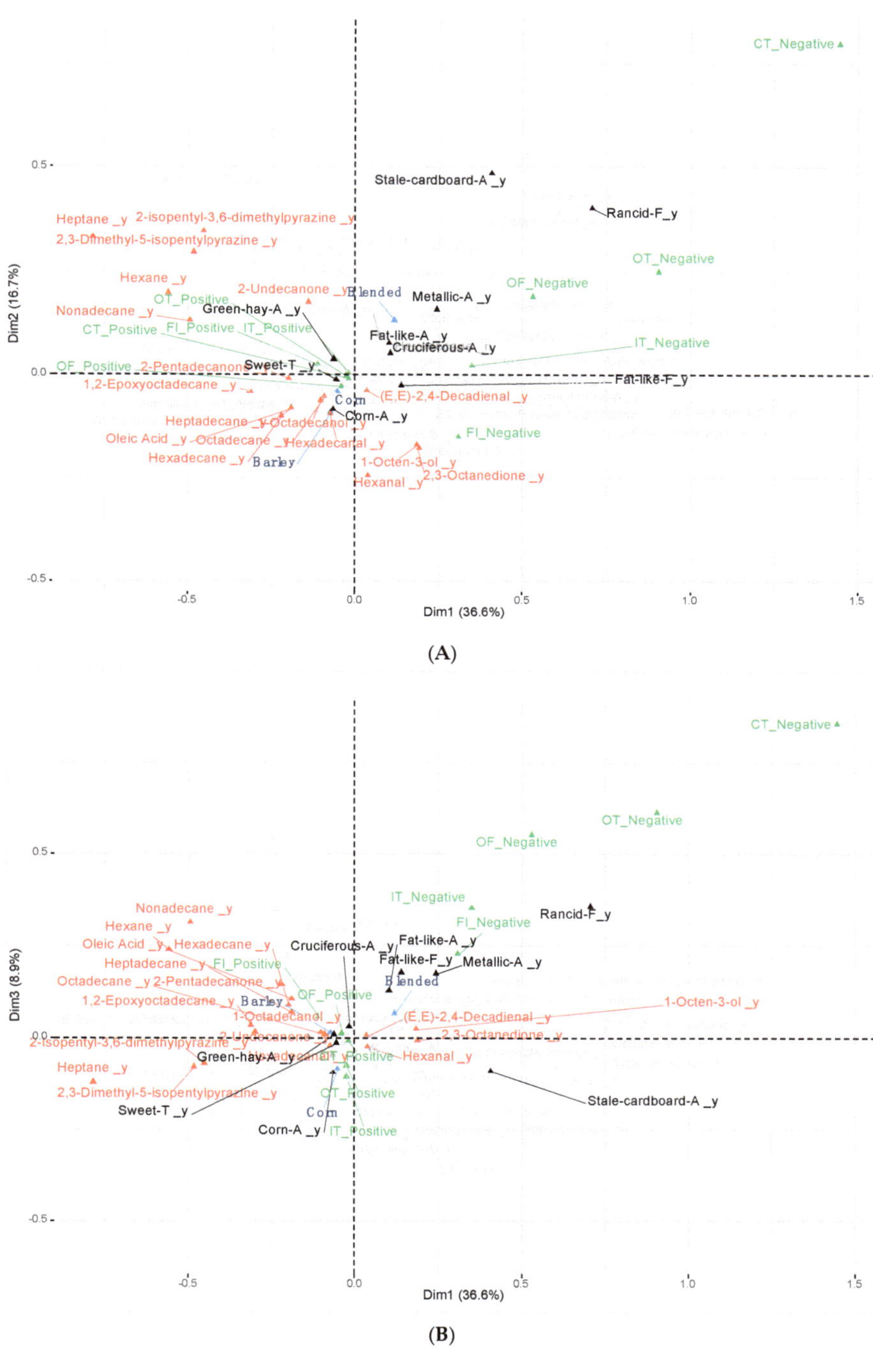

Figure 1. Multiple correspondence analysis among descriptive sensory attributes, flavor profile, and volatile compounds from barley, corn, and blended grain-fed beef. Blue: grain diet; red: volatile compounds; black: aroma/taste/flavor descriptors; green: descriptive sensory attributes. (**A**) Dimensions 1 and 2. (**B**) Dimensions 1 and 3.

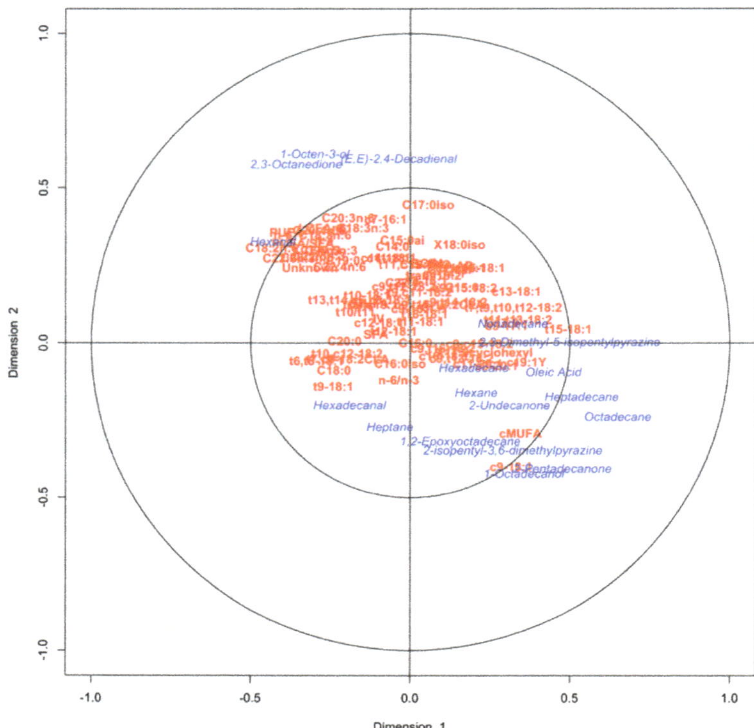

Figure 2. Canonical correlation between fatty acids (red) and volatile compounds (blue) from barley, corn, and blended grain-fed beef.

The reason why meat from blended grain-fed cattle was associated with aroma and flavor descriptors related to lipid oxidation, negative categories for descriptive sensory attributes, and volatile compounds related to PUFA oxidation could be related to differences in the total fatty acids (mg/g of LT) and the fatty acid composition (% of total fatty acids) (Table 2). Many of the LT fatty acids for the blended grain-fed beef are between or close to the values for either barley- or corn-fed beef (n-3, atypical dienes, cisMUFA, transMUFA, and SFA), and would thus not explain why feeding blended grain would enhance oxidation. There was some indication that feeding the blend might enhance ∑ PUFAs, ∑ n-6 fatty acids, and specifically, 18:2n-6 in LT fatty acids, as their percentages were numerically greater than in barley- or corn-fed meat, but differences did not reach significance. Significant differences were, however, found for 18:2n-6 elongation and desaturation products (i.e., 20:3n-6, $p < 0.05$; 20:4n-6, $p = 0.05$), which would support greater oxidation potential for blended grain-fed LT as susceptibility to oxidation increases geometrically as the number of double bonds increases. In addition, long-chain PUFAs are preferentially incorporated into cell membrane phospholipids, which is where initiation of fatty acid oxidation is thought to occur [40]. The reason for elevated n-6 fatty acids when feeding the blend are not related to the dietary supply, but instead likely relate to effects on rumen pH. Lower rumen pH was found in blended compared with barley and corn grain-fed steers in the present study, as previously reported by Johnson et al. [41]. Under these conditions, lipolysis of dietary lipids could have been inhibited [42], reducing PUFA biohydrogenation by rumen bacteria, and allowing for greater bypass of n-6 fatty acids.

Table 2. Fatty acid composition [a] in longissimus thoracis samples from barley, corn, and blended grain-fed cattle.

	Barley	Blend	Corn	*p*-Value [1]
∑ Total Fatty Acids (mg/g of tissue)	40.06 ± 2.08	40.91 ± 2.01	46.34 ± 2.13	0.08
Fatty Acids (% of total fatty acids)				
∑ PUFAs [b]	4.83 ± 2.03	5.28 ± 0.22	4.72 ± 0.24	0.18
∑ n-3	0.69 ± 0.04 ^{A}	0.65 ± 0.04 ^{A}	0.52 ± 0.04 ^{B}	0.01
C18:3n-3	0.26 ± 0.01 ^{A}	0.24 ± 0.01 ^{A}	0.20 ± 0.01 ^{B}	0.01
C20:5n-3	0.11 ± 0.01 ^{A}	0.11 ± 0.01 ^{A}	0.07 ± 0.01 ^{B}	0.02
C22:5n-3	0.28 ± 0.02 ^{A}	0.28 ± 0.02 ^{A}	0.22 ± 0.02 ^{B}	0.02
C22:6n-3	0.04 ± 0.00	0.03 ± 0.00	0.03 ± 0.00	0.16
∑ n-6 [c]	4.15 ± 0.20	4.63 ± 0.19	4.20 ± 0.20	0.16
C18:2n-6	2.81 ± 0.13	3.21 ± 0.13	3.03 ± 0.14	0.10
C20:3n-6	0.25 ± 0.01 ^{A,B}	0.26 ± 0.01 ^{A}	0.21 ± 0.01 ^{B}	0.04
C20:4n-6	0.91 ± 0.06 ^{A,B}	0.96 ± 0.05 ^{A}	0.77 ± 0.06 ^{B}	0.05
∑ Atypical Dienes [d]	0.45 ± 0.01 ^{A}	0.40 ± 0.01 ^{B}	0.38 ± 0.01 ^{B}	<0.0001
*t*11,*c*15-18:2	0.10 ± 0.01 ^{A}	0.06 ± 0.01 ^{B}	0.05 ± 0.01 ^{B}	0.00
∑ Conjugated Linoleic Acids [e]	0.29 ± 0.01	0.30 ± 0.01	0.30 ± 0.01	0.55
*c*9,*t*11-18:2	0.18 ± 0.01	0.18 ± 0.01	0.19 ± 0.01	0.54
∑ *cis* MUFAs [f]	44.50 ± 0.50	44.90 ± 0.48	45.10 ± 0.50	0.63
*c*9-16:1	3.31 ± 0.10	3.26 ± 0.10	3.04 ± 0.10	0.09
*c*9-18:1	36.33 ± 0.43	37.21 ± 0.42	37.80 ± 0.44	0.07
∑ *trans* MUFAs [g]	2.79 ± 0.13	2.68 ± 0.13	2.67 ± 0.13	0.77
*t*10-18:1	1.31 ± 0.10	1.13 ± 0.09	1.07 ± 0.10	0.20
*t*11-18:1	0.51 ± 0.02	0.53 ± 0.02	0.57 ± 0.02	0.15
∑ SFA [h]	45.62 ± 0.47	45.10 ± 0.45	45.64 ± 0.48	0.62
C16:0	26.68 ± 0.30	26.36 ± 0.28	26.18 ± 0.30	0.50
C18:0	13.9 ± 0.26 ^{B}	14.0 ± 0.25 ^{B}	14.8 ± 0.27 ^{A}	0.02
n-6/n-3	6.22 ± 0.21 ^{C}	7.26 ± 0.19 ^{B}	8.32 ± 0.21 ^{A}	<0.0001

[1] Different uppercase letter ^{A,B,C} in the same row means significant difference according to Tukey's test ($p < 0.05$). [a] Least-square means ± standard error of mean. [b] Σ PUFAs = sum of polyunsaturated fatty acids (∑ n-6 + ∑ n-3). [c] Σ n-6 = C18:2n-6; C18:3n-6; C20:2n-6; C20:3n-6; C20:4n-6; C22:4n-6. [d] Σ Atypical dienes = *c*9,*t*14-18:2; *c*9,*t*13-18:2; 17:0-cyclohexyl; *c*9,*t*15-18:2; *c*9,*t*12-18:2; *t*11,*c*15-18:2; *c*9,*t*16-18:2; *c*9,*c*15-18:2. [e] Σ Conjugate linoleic acid = *t*7,*c*9-18:2 (CLA); *c*9,*t*11-18:2 (CLA); *t*9,*c*11-18:2; *t*10,*c*12-18:2; *t*11,*t*13-18:2; *t*7,*t*9-t10,*t*12-18:2. [f] Σ *c*MUFAs = sum of monounsaturated cis fatty acids (*c*9-14:1; *c*7-16:1; *c*9-16:1; *c*11-16:1; *c*13-16:1; *c*9-17:1; *c*9-18:1; *c*11-18:1; *c*12-18:1; *c*13-18:1; *c*14-18:1; *c*15-18:1; *c*16-18:1; *c*-19:1-Y; *c*9-20:1; *c*11-20:1). [g] Σ *trans* MUFAs = sum of monounsaturated trans fatty acids (*t*6-*t*8-18:1; *t*9-18:1; *t*10-18:1; *t*11-18:1; *t*12-18:1; *t*13-*t*14/*c*6-*c*8-18:1; *t*15-18:1; *t*16-18:1). [h] Σ Sum SFA = sum of saturated fatty acid (C14:0; C15:0; C16:0; C17:0; C18:0; C19:0; C20:0; C22:0).

The differences in descriptive sensory attributes and flavor profiles between meat samples from blended and pure (barley or corn) grain-fed cattle found in this study are in contrast to those reported by Miller et al. [43], who found no differences in the eating quality between meat samples from those feeding regimes. These discrepancies between studies could be due to the lack of effect of grain type or blend on beef fatty acid composition, which may relate to lower levels of grain in their diets and differences in grain processing, and limited statistical power due to low number of experimental units (n = 6). In the present experiment, both corn and barley were dry-rolled, whereas Miller et a. [43] steam-rolled corn and crimped barley, which may have impacted rumen fermentation rates and pH.

When the whole dataset of sensory attributes, flavor profile, and volatile compounds were represented on an XY plane according to the Dim 1 and 3 (Figure 1B), Dim 3 explained 8.9% of variability among these variables for meat samples from barley, corn, and blended grain-fed cattle. Similar to that observed in the bi-dimensional plane described by Dim 1 and 2, the blended grain-fed beef was located in the upper-right quadrant and was associated with undesirable aromas, flavors, and sensory attributes, and volatiles originating from lipid oxidation. However, the Dim 3 allowed the quadrant separation of meat samples from each pure grain-finished treatment. The upper-left quadrant of MCA linked barley grain-fed beef to a combination of alkanes (nonadecane, hexadecane, hexane, heptadecane, and octadecane), epoxide (1,2-epoxyoctadecane), ketones (2-pentadecanone and 2-undecanone), and oleic acid, which were previously described by Calkins and Hodgen [12] and Mottram [11] as a combination of volatiles that produce a pleasant meat and roasted flavor. Indeed, positive categories for beef flavor intensity (from slightly intense to extremely intense) and off-flavor (from slightly bland to none) as well as green-hay and cruciferous aromas presented in the same quadrant.

The negative space for Dim 3 related the meat samples from corn-fed cattle to volatile compounds such as pyrazines (2,3-dimethyl-5-isopentylpyrazine and 2-isopentyl-3,6-dimethylpyrazine), alkane (heptane), and ketone (2,3-octanedione), as well as to positive sweet taste and corn aroma. However, corn grain-fed beef was also associated with hexadecanal and hexanal. As previously mentioned, these aldehydes come from the oxidation of n-6 fatty acids [30,35], and, in this study, corn had higher ($p < 0.05$, Table 2) n-6/n-3 ratio than barley grain-fed beef samples. Similar results were reported by Vahmani et al. [20] in subcutaneous fat samples from the same animals. In contrast, Brassard et al. [44] reported lower n-6/n-3 ratio in meat from barley and corn concentrate-fed goats, probably due to hay ad libitum access and different concentration of linolenic intake. As previously mentioned, hexadecanal has been described as a fatty odor contributor with low participation in flavor development [31]. Hexanal is a common volatile found when feeding grain-based diets [45], related to grassy, leafy, and green flavor descriptors, and with low odor threshold [8,28,38]. However, in higher concentrations, hexanal has been previously described as an indicator of meat flavor deterioration [46]. The association of meat from corn-fed steers with hexanal found in this study is in agreement the with the findings of O'Quinn et al. [3], who found higher concentrations of hexanal in meat from corn-fed compared with barley-fed steers. In contrast to the positive association of barley grain-fed beef with beef flavor intensity, no flavor characteristics were related to corn grain-fed beef in this study. This, in part, disagrees with the findings of McEwen et al. [19], who did not report differences in beef flavor intensity between meat from corn-fed steers and barley grain-fed steers, probably due to the different cattle breed (Angus) and lower barley participation in diet composition (70% DM). Nevertheless, Jeremiah et al. [18] found no differences in beef flavor intensity from crossbred cattle fed barley and corn in a concentration similar to that used in this study. Differences in the taste panel (semitrained panel vs. trained panel) and more sophisticated statistical analyses used in the present study could have also contributed to the differences between studies.

Corn grain-fed beef was also associated (negative space of Dim 1 and 3) with positive categories for initial and overall tenderness (from slightly to extremely tender) and connective tissue (from slight to none). These results agree with the findings of Wismer et al. [17], who reported higher initial and overall tenderness in beef from corn- compared with barley-fed cattle. In contrast, several studies have found no differences in descriptive sensory attributes, such as tenderness and amount of connective tissue, evaluated by a trained panel in beef from barley- and corn-fed cattle [18,19,47] despite using cattle breed, grain concentration, and storage time and thawing/cooking conditions of meat samples similar to those used in the current study. Nevertheless, recent advancements in statistical methods could have, in part, contributed to the discrepancies among studies.

4. Conclusions

Use of integral analysis of MCA was able to separate barley and corn from blended grain-fed beef based on aroma/flavor profile, descriptive sensory attributes, and volatile compounds. The fatty acid profile of the meat samples suggested an influence of barley and corn in blended grain-fed beef on n-6 LCFA deposition, which could have increased their oxidation potential. This effect was supported by some aldehydes and alcohols from PUFA oxidation, undesirable aromas and flavor descriptors, and negative categories for descriptive sensory traits associated with the blended grain-fed beef samples. Barley and corn grain-fed beef were also differentiated, although with lower explained variance in the MCA; barley grain-fed beef was associated with volatile compounds originating a pleasant beef flavor, whereas corn grain-fed beef was linked to some volatile compounds from lipid oxidation. However, corn grain-fed beef was more associated with positive categories for tenderness. Nevertheless, apart from the positive association of beef flavor intensity with barley grain-fed beef, no flavor descriptors were associated with either barley or corn grain-fed beef, which may suggest that the different volatiles associated with barley and corn grain-fed beef in this study did not translate into differences in meat flavor detected by the trained panelists. Hence, a further quantitative approach to understand the volatile thresholds and their influence on meat palatability would be warranted in order to maximize the potential of volatile compounds to anticipate consumer satisfaction. In addition, even though feeding blended grain-fed diets may at times be economically feasible, the interaction between different fed grains leading to oxidative instability may be an unanticipated outcome, and thus deserves further attention.

Supplementary Materials: The following are available online at https://www.mdpi.com/article/10.3390/foods10050977/s1, Table S1: Mean of standardized base peak of non-selected volatile compounds from barley, corn and blended grain-fed beef samples. Table S2: Mean and standard deviation of sensory descriptive and flavour profile attributes from barley, corn and blended grain-fed beef samples.

Author Contributions: Conceptualization, N.P., W.B.-H., and J.L.A.; methodology, N.P., J.S., Ó.L.-C., M.J., G.P., P.V., M.E.R.D., and J.L.A.; formal analysis, N.P. and W.B.-H.; resources, N.P., J.L.A., and G.P.; data curation, M.E.R.D.; writing—original draft preparation, N.P., W.B.-H., and M.E.R.D.; writing—review and editing, N.P., J.S., Ó.L.-C., M.J., G.P., P.V., M.E.R.D., L.M.-L., and J.L.A.; visualization, N.P. and W.B.-H.; supervision, N.P. and J.L.A.; project administration, J.L.A.; funding acquisition, J.L.A. and G.P. All authors have read and agreed to the published version of the manuscript.

Funding: The production study was funded by the Saskatchewan Barley Development Commission (Saskatoon, SK, Canada), DuPont Pioneer, the Saskatchewan Cattleman's Association, and the Saskatchewan Ministry of Agriculture and the Canada-Saskatchewan Growing Forward 2.

Institutional Review Board Statement: The study was conducted according to the guidelines of the Declaration of Helsinki, and approved by the Institutional Ethics Committee of University of Saskatchewan (protocol 20100021).

Informed Consent Statement: Informed consent was obtained from all subjects involved in the study.

Acknowledgments: G. Penner thanks Jordan Johnson, Britney Sutherland, and Teresa Binetruy for their assistance in live animal husbandry. AAFC-Lacombe authors thank Chuck Pimm and the meat-processing staff, as well as Ivy Larsen and the meat technical team for their assistance in meat processing, laboratory analyses, and statistical analyses of the data. All authors thank Ryan Clisdell, Cargill Meat Solutions, for tracking the cattle lot and in-plant rib-eye collection.

Conflicts of Interest: The authors declare no conflict of interest.

References

1. Cocking, C.; Walton, J.; Kehoe, L.; Cashman, K.D.; Flynn, A. The role of meat in the European diet: Current state of knowledge on dietary recommendations, intakes and contribution to energy and nutrient intakes and status. *Nutr. Res. Rev.* **2020**, *33*, 181–189. [CrossRef]
2. Drey, L.; Legako, J.; Brooks, J.; Miller, M.; O'quinn, T. The contribution of tenderness, juiciness, and flavor to overall consumer beef eating experience. *Meat Muscle Biol.* **2017**, *1*, 13. [CrossRef]
3. O'Quinn, T.G.; Woerner, D.R.; Engle, T.E.; Chapman, P.L.; Legako, J.F.; Brooks, J.C.; Belk, K.E.; Tatum, J.D. Identifying consumer preferences for specific beef flavor characteristics in relation to cattle production and postmortem processing parameters. *Meat Sci.* **2016**, *112*, 90–102. [CrossRef] [PubMed]
4. Liu, J.; Ellies-Oury, M.P.; Chriki, S.; Legrand, I.; Pogorzelski, G.; Wierzbicki, J.; Farmer, L.; Troy, D.; Polkinghorne, R.; Hocquette, J.F. Contributions of tenderness, juiciness and flavor liking to overall liking of beef in Europe. *Meat Sci.* **2020**, *168*, 108190. [CrossRef] [PubMed]
5. Khan, M.I.; Jo, C.; Tariq, M.R. Meat flavor precursors and factors influencing flavor precursors-A systematic review. *Meat Sci.* **2015**, *110*, 278–284. [CrossRef]

6. Marrone, R.; Salzano, A.; Di Francia, A.; Vollano, L.; Di Matteo, R.; Balestrieri, A.; Anastasio, A.; Barone, C.M.A. Effects of feeding and maturation system on qualitative characteristics of buffalo meat (Bubalus bubalis). *Animals* **2020**, *10*, 899. [CrossRef] [PubMed]
7. Smaldone, G.; Marrone, R.; Vollano, L.; Peruzy, M.F.; Barone, C.M.A.; Ambrosio, R.L.; Anastasio, A. Microbiological, rheological and physical-chemical characteristics of bovine meat subjected to a prolonged ageing period. *Ital. J. Food Saf.* **2019**, *8*, 131–136. [CrossRef]
8. Kerth, C.R.; Miller, R.K. Beef flavor: A review from chemistry to consumer. *J. Sci. Food Agric.* **2015**, *95*, 2783–2798. [CrossRef]
9. Ruan, E.D.; Aalhus, J.L.; Juárez, M.; Sabik, H. Analysis of volatile and flavor compounds in grilled lean beef by stir bar sorptive extraction and thermal desorption—Gas chromatography mass spectrometry. *Food Anal. Methods* **2015**, *8*, 363–370. [CrossRef]
10. Frank, D.; Oytam, Y.; Hughes, J. Sensory perceptions and new consumer attitudes to Meat. In *New Aspects of Meat Quality*; Elsevier Ltd.: Amsterdam, The Netherlands, 2017; pp. 667–698, ISBN 9780081005934.
11. Mottram, D.S. Flavour formation in meat and meat products: A review. *Food Chem.* **1998**, *62*, 415–424. [CrossRef]
12. Calkins, C.R.; Hodgen, J.M. A fresh look at meat flavor. *Meat Sci.* **2007**, *77*, 63–80. [CrossRef]
13. Johnson, J.A.; Sutherland, B.D.; Mckinnon, J.J.; Mcallister, T.A.; Penner, G.B. Use of barley or corn silage when fed with barley, corn, or a blend of barley and corn on growth performance, nutrient utilization, and carcass characteristics of finishing beef cattle. *Transl. Anim. Sci.* **2019**, *4*, 129–140. [CrossRef] [PubMed]
14. Lardner, H.A.; Pearce, L.; Damiran, D. Evaluation of low heat unit corn hybrids compared to barley for forage yield and quality on the Canadian prairies. *Sustain. Agric. Res.* **2017**, *6*, 90. [CrossRef]
15. Humer, E.; Zebeli, Q. Grains in ruminant feeding and potentials to enhance their nutritive and health value by chemical processing. *Anim. Feed Sci. Technol.* **2017**, *226*, 133–151. [CrossRef]
16. Nikkhah, A. Barley grain for ruminants: A global treasure or tragedy. *J. Anim. Sci. Biotechnol.* **2012**, *3*, 1–9. [CrossRef]
17. Wismer, W.V.; Okine, E.K.; Stein, A.; Seibel, M.R.; Goonewardene, L.A. Physical and sensory characterization and consumer preference of corn and barley-fed beef. *Meat Sci.* **2008**, *80*, 857–863. [CrossRef]
18. Jeremiah, L.E.; Beauchemin, K.A.; Jones, S.D.M.; Gibson, L.L.; Rode, L.M. The influence of dietary cereal grain source and feed enzymes on the cooking properties and palatability attributes of beef. *Can. J. Anim. Sci.* **1998**, *78*, 271–275. [CrossRef]
19. McEwen, P.L.; Mandell, I.B.; Brien, G.; Campbell, C.P. Effects of grain source, silage level, and slaughter weight endpoint on growth performance, carcass characteristics, and meat quality in Angus and Charolais steers. *Can. J. Anim. Sci.* **2007**, *87*, 167–180. [CrossRef]
20. Vahmani, P.; Johnson, J.A.; Sutherland, B.D.; Penner, G.B.; Prieto, N.; Aalhus, J.L.; Juárez, M.; Lopez-Campos, O.; Dugan, M.E.R. Changes in the fatty acid composition of steer subcutaneous fat, including biohydrogenation products, are minimal when finished on combinations of corn and barley grains and silages. *Can. J. Anim. Sci.* **2020**. [CrossRef]
21. Burnett, D.D.; Legako, J.F.; Phelps, K.J.; Gonzalez, J.M. Biology, strategies, and fresh meat consequences of manipulating the fatty acid composition of meat. *J. Anim. Sci.* **2020**, *98*, 98. [CrossRef]
22. CCAC. *Guidelines on: The Care and Use of Farm Animals in Research, Teaching, and Testing*; Canadian Council on Animal Care: Ottawa, ON, Canada, 2009; ISBN 091908737X.
23. AMSA. *Research Guidelines for Cookery, Sensory Evaluation, and Instrumental Tenderness Measurements of Meat*; American Meat Science Association, Ed.; AMSA: Champaign, IL, USA, 2016; ISBN 8005172672.
24. Larmond, E. *Laboratory Methods for Sensory Evaluation of Foods*; Canada Department of Agriculture: Ottowa, ON, Canada, 1977; ISBN 0662012712.
25. Vahmani, P.; Rolland, D.C.; McAllister, T.A.; Block, H.C.; Proctor, S.D.; Guan, L.L.; Prieto, N.; López-Campos, Ó.; Aalhus, J.L.; Dugan, M.E.R. Effects of feeding steers extruded flaxseed on its own before hay or mixed with hay on animal performance, carcass quality, and meat and hamburger fatty acid composition. *Meat Sci.* **2017**, *131*, 9–17. [CrossRef] [PubMed]
26. Lê, S.; Josse, J.; Husson, F. FactoMineR: An R package for multivariate analysis. *J. Stat. Softw.* **2008**, *25*, 1–18. [CrossRef]
27. ter Braak, C.J.F. Interpreting canonical correlation analysis through biplots of structure correlations and weights. *Psychometrika* **1990**, *55*, 519–531. [CrossRef]
28. Legako, J.F.; Brooks, J.C.; O'Quinn, T.G.; Hagan, T.D.J.; Polkinghorne, R.; Farmer, L.J.; Miller, M.F. Consumer palatability scores and volatile beef flavor compounds of five USDA quality grades and four muscles. *Meat Sci.* **2015**, *100*, 291–300. [CrossRef]
29. Song, S.; Zhang, X.; Hayat, K.; Liu, P.; Jia, C.; Xia, S.; Xiao, Z.; Tian, H.; Niu, Y. Formation of the beef flavour precursors and their correlation with chemical parameters during the controlled thermal oxidation of tallow. *Food Chem.* **2011**, *124*, 203–209. [CrossRef]
30. Larick, D.K.; Turner, B.E. Headspace volatiles and sensory characteristics of ground beef from forage- and grain-Fed heifers. *J. Food Sci.* **1990**, *55*, 649–654. [CrossRef]
31. Tao, N.P.; Wu, R.; Zhou, P.G.; Gu, S.Q.; Wu, W. Characterization of odor-active compounds in cooked meat of farmed obscure puffer (*Takifugu obscurus*) using gas chromatography-mass spectrometry-olfactometry. *J. Food Drug Anal.* **2014**, *22*, 431–438. [CrossRef]
32. Stetzer, A.J.; Cadwallader, K.; Singh, T.K.; Mckeith, F.K.; Brewer, M.S. Effect of enhancement and ageing on flavor and volatile compounds in various beef muscles. *Meat Sci.* **2008**, *79*, 13–19. [CrossRef]
33. Therkildsen, M.; Spleth, P.; Lange, E.-M.; Hedelund, P.I. The flavor of high-quality beef—A review. *Acta Agric. Scand. Sect. A Anim. Sci.* **2017**, *67*, 85–95. [CrossRef]

34. Calkins, B.C.R.; Ph, D.; Sullivan, G. Ranking of Beef Muscles for Tenderness. 2006. Available online: https://www.beefcentral.com/wp-content/uploads/2014/05/ranking-of-beef-muscles-for-tenderness.pdf (accessed on 22 March 2021).
35. Legako, J.F.; Dinh, T.T.N.; Miller, M.F.; Adhikari, K.; Brooks, J.C. Consumer palatability scores, sensory descriptive attributes, and volatile compounds of grilled beef steaks from three USDA Quality Grades. *Meat Sci.* **2016**, *112*, 77–85. [CrossRef]
36. Starowicz, M.; Zieliński, H. How Maillard reaction influences sensorial properties (color, flavor and texture) of food products? *Food Rev. Int.* **2019**, *35*, 707–725. [CrossRef]
37. Sun, W.; Zhao, M.; Cui, C.; Zhao, Q.; Yang, B. Effect of Maillard reaction products derived from the hydrolysate of mechanically deboned chicken residue on the antioxidant, textural and sensory properties of Cantonese sausages. *Meat Sci.* **2010**, *86*, 276–282. [CrossRef] [PubMed]
38. Migita, K.; Iiduka, T.; Tsukamoto, K.; Sugiura, S.; Tanaka, G.; Sakamaki, G.; Yamamoto, Y.; Takeshige, Y.; Miyazawa, T.; Kojima, A.; et al. Retort beef aroma that gives preferable properties to canned beef products and its aroma components. *Anim. Sci. J.* **2017**, *88*, 2050–2056. [CrossRef]
39. Hwang, Y.-H.; Joo, S.-T. Fatty acid profiles, meat quality, and sensory palatability of grain-fed and grass-fed beef from Han-woo, American, and Australian crossbred cattle. *Korean J. Food Sci. Anim.* **2017**, *37*, 1225–8563. [CrossRef]
40. Wood, J.D.; Enser, M.; Fisher, A.V.; Nute, G.R.; Sheard, P.R.; Richardson, R.I.; Hughes, S.I.; Whittington, F.M. Fat deposition, fatty acid composition and meat quality: A review. *Meat Sci.* **2008**, *78*, 343–358. [CrossRef]
41. Johnson, J.A.; Sutherland, B.D.; McKinnon, J.J.; McAllister, T.A.; Penner, G.B. Effect of feeding barley or corn silage with dry-rolled barley, corn, or a blend of barley and corn grain on rumen fermentation, total tract digestibility, and nitrogen balance for finishing beef heifers. *J. Anim. Sci.* **2020**, *98*, 1–11. [CrossRef]
42. Enjalbert, F.; Combes, S.; Zened, A.; Meynadier, A. Rumen microbiota and dietary fat: A mutual shaping. *J. Appl. Microbiol.* **2017**, *123*, 782–797. [CrossRef]
43. Miller, R.K.; Rockwell, L.C.; Lunt, D.K.; Carstens, G.E. Determination of the flavor attributes of cooked beef from cross-bred Angus steers fed corn-or barley-based diets. *Meat Sci.* **1996**, *44*, 235–243. [CrossRef]
44. Brassard, M.-E.; Chouinard, Y.; Gervais, R.; Pouliot, E.; Gariépy, C.; Cinq-Mars, D. Effects of level of barley and corn in concentrate-fed Boer kids on growth performance, meat quality and muscle fatty acid composition. *Can. J. Anim. Sci.* **2017**. [CrossRef]
45. Elmore, J.S.; Mottram, D.S.; Enser, M.; Wood, J.D. Effect of the polyunsaturated fatty acid composition of beef muscle on the profile of aroma volatiles. *J. Agric. Food Chem.* **1999**, *47*, 1619–1625. [CrossRef]
46. Shahidi, F.; Pegg, R.B. Hexanal as an indicator of meat flavor deterioration. *J. Food Lipids* **1994**, *1*, 177–186. [CrossRef]
47. Nelson, M.L.; Busboom, J.R.; Ross, C.F.; O'Fallon, J. V Effects of supplemental fat on growth performance and quality of beef from steers fed corn finishing diets. *J. Anim. Sci.* **2008**, *86*, 936–948. [CrossRef] [PubMed]

Article

Quality of Chicken Fat by-Products: Lipid Profile and Colour Properties

Lina María Peña-Saldarriaga [1], Juana Fernández-López [2] and José Angel Pérez-Alvarez [2,*]

[1] Research & Development Department, Bios Group, Cra 48 No. 274 Sur-89 Envigado, 055422 Antioquia, Colombia; lina.pena@grupobios.co

[2] IPOA Research Group, Agro-Food Technology Department, Higher Polytechnic School of Orihuela, Miguel Hernández University, Orihuela, 03312 Alicante, Spain; j.fernandez@umh.es

* Correspondence: ja.perez@umh.es; Tel.: +34-966749784

Received: 7 July 2020; Accepted: 1 August 2020; Published: 3 August 2020

Abstract: The growth in the production and consumption of chicken meat and related products is responsible for the formation a large number of by-products. Among these, abdominal and gizzard fat is usually considered as waste and thus is discarded, creating an environmental problem. This work aims to characterize chicken fat by-products, evaluating their lipid profile and colour properties for their potential use as fat sources in meat products in substitution of traditionally used fats. In addition, the role of farm location, keeping the feeding and other farmer routines fixed, in the lipid profile was also evaluated. "Parrilleros" Colombian chickens from three farms located in various geographical zones of the Antioquia region were selected. After slaughtering, abdominal and gizzard fat was obtained. Lipid profile was evaluated by gas chromatography and the CIELAB colour properties were assessed. The production results and the lipid profile of chicken fat by-products (abdominal and gizzard fat) was similar in the three farms studied, which is important for their potential application as fat source in the formulation of meat products. The predominant fatty acids were oleic, palmitic and linoleic acids, showing a higher amount of unsaturated fatty acids than the fat sources traditionally used for this purpose. Valorization of chicken by-products by the use of abdominal and gizzard fat as fat source in chicken meat products formulation could be a feasible alternative contributing to the poultry sector sustainability.

Keywords: chicken fat by-products; unsaturated fatty acids; colour properties; lipid profile

1. Introduction

Poultry farming has been the main impetus for the sustained and steady economic development of Colombian agriculture in recent years, and is considered a determining variable in the growth of the Gross Domestic Product of the agricultural sector in the country. The growth in the poultry industry in Colombia is mainly due to increased domestic consumption. A decade ago the per capita consumption of poultry meat in Colombia was about 23 kg of chicken meat per year, while today it is 35.5 kg [1]. The industry has developed to such an extent that poultry products are now the most important source of animal protein in Colombia (contributing 50%), a trend that underlines the importance of this industry in the country and its constant growth. Not only in Colombia is the poultry industry important, but it also plays a relevant role in feeding much of the rest of the world. According to the Organization for Economic Cooperation and Development (OECD) and the Food and Agricultural Organization (FAO), the worldwide per capita consumption of chicken meat in the last decade has increased by 15%, growth which has outstripped that registered for beef and pork. The main consumers are the United States and Brazil, whose annual consumption exceeds 40 kg per capita [2]. Such an increase in the consumption of chicken meat is mainly due to the perception by health-conscious consumers that chicken meat is a low-fat source of healthy nutrition, rich in unsaturated fat and a high in protein [3]. In addition, chicken meat is increasingly used in the development of new chicken-based convenience products (chicken bologna, chicken nuggets, chicken hotdogs, chicken wings), which have been successfully marketed for consumption at home and also in the growing fast-food industry [4].

However, the rapid growth of poultry production has led to the massive generation of food-processing by-products likes bones, viscera, abdominal fat, feet, head, blood and feathers. If these by-products were regarded as having greater nutritional value, their use would contribute to the development of a sustainable food industry while increasing the value of this sector [5]. Until now, these by-products have only been sold as animal feed and to pet food processors [6–8] and, recently, for the production of biodiesel [9]. However, there are no references about the possible use of some of these by-products as raw materials for use in human food processing. For example, it may be possible that the abdominal and gizzard fat that remains inside the poultry carcass, where it represents approximately 2–2.5% of the total weight of the slaughtered chicken [10], could be used as fat source for the production of chicken sausages or other meat products, especially taking into account its characteristic content of unsaturated fatty acids. Until now, this abdominal and gizzard fat has been discarded by small producers, together with the viscera, feathers and blood, thus creating and environmental problem.

The production of high quantities of by-products by the poultry industry and the potential of abdominal and gizzard fat as a healthy fat source in different applications, about which little information is available, led to the development of this study. The main objective was to determinate the fatty acid profile and colour properties of poultry fat by-products (in this case, abdominal and gizzard fat) and to assess whether these properties remain stable and whether they depend on the farm conditions (feeding and geographical location).

2. Materials and Methods

2.1. Experimental Design: Animals and Diets

One-day-old "Parrilleros" Colombian chickens from three commercial farms (La Nirvana, La Goleta and Villa Rita) located in various geographical zones of the Antioquia region in Colombia (Barbosa, Yolombó and Caldas, respectively), characterized by their different climatic conditions, were reared on litter floors (wood shavings) in open-sided housing conditions with feed and water provided *ad libitum*. The average number of birds reared and the average density (bird/m^2) in each farm were: Villa Rita, 78,432 and 12.9, respectively; La Nirvana, 314,200 and 13.0, respectively; and La Goleta 1,132,002 and 12.4, respectively. In all the farms the photoperiod was 12 h (±30 min) (12L/12D). In each farm, 75 birds were selected for the experiment (kept in pens on litter) and were divided into 3 replications with 25 birds per group. Each bird had a padlock badge for identification during measurements. All the chickens were initially fed the same balanced diet: a "pre-starter" diet until they reached 150 g in weight and a "starter" diet until 900 g (approx. 16 days); this was followed by a "finisher" diet based on standard formulations used in different fattening periods, until slaughter at 45 days of age. Nine different "finisher" diets were assessed (Table 1) depending on the availability of raw materials and prices in attempt to minimize costs for the companies, while maintaining the same nutritional levels. The finisher experimental diets and water were offered *ad libitum*.

Productivity parameters (final body weight (FBW), daily body weight gain (BWG), feed intake (FI), and feed conversion ratio (FCR)) were monitored and recorded for the entire flock (75 birds) per farm. At 45 days, thirteen chickens per farm, with an FBW close to the mean of the whole group were slaughtered in an abattoir (previous electrical stunning) licensed by The National Institute for the Surveillance of Drugs and Foods (Colombia) and abdominal and gizzard fat was obtained (Figure 1). The fat samples were refrigerated and sent to the Food Science and Technology Institute laboratory to assess their fatty acid content and colour.

Table 1. Experimental finisher diets.

Ingredients (%)	Finisher Diets								
	1	2	3	4	5	6	7	8	9
Sorghum	10.00	10.00	10.00	10.00	10.00	10.00	10.00	10.00	10.00
Soy oil	2.95	2.92	3.24	3.20	4.00	4.00	4.00	4.00	3.50
Yellow corn	10.00	10.00	10.00	10.00	10.00	10.00	10.00	10.00	10.00
White corn	43.68	43.79	44.45	44.30	45.00	44.95	45.03	34.84	32.84
Corn gluten	4.16	4.26	4.38	4.85	4.72	4.09	4.11	-	-
Wheat	-	-	-	-	-	-	-	10.00	10.00
Bone flour	3.18	3.00	3.06	2.87	2.89	3.23	3.19	4.50	3.45
Soybean meal	7.61	7.61	11.17	11.01	14.94	14.68	14.77	11.77	11.51
Full fat soybean	12.50	12.50	10.00	10.00	4.49	5.16	5.03	10.00	14.99
Sunflower meal	2.00	2.00	-	-	-	-	-	-	-
Nutrients (% of diet)									
Protein	18.56	18.51	18.54	18.57	18.58	18.70	18.59	18.46	18.45
Lipids	8.46	8.97	8.41	8.97	8.30	8.24	8.16	8.28	8.25
Fibre	3.90	3.88	3.90	3.89	3.63	3.62	3.57	3.58	3.58
Minerals	3.32	3.37	3.26	3.26	3.26	3.20	3.22	3.34	3.33
Fatty acids (% total fat)									
C14:0 (Myristic acid)	0.19	0.19	0.18	0.17	0.15	0.16	0.16	0.22	0.20
C16:0 (Palmitic acid)	11.51	11.48	11.45	11.41	11.27	11.35	11.34	11.70	11.54
C16:1 (Palmitoleic acid)	0.86	0.84	0.91	0.89	1.04	1.05	1.05	1.06	0.90
C18:0 (Stearic acid)	4.04	4.00	4.01	3.97	3.97	4.04	4.03	4.32	4.15
C18:1 (Oleic acid)	24.90	24.86	24.93	24.88	24.97	25.03	25.03	24.85	24.36
C18:2 (Linoleic acid)	52.29	52.42	52.34	52.48	52.47	52.22	52.24	51.36	52.12
C18:3 (Linolenic acid)	5.62	5.62	5.57	5.57	5.43	5.46	5.45	5.81	6.09
C > 19	0.59	0.59	0.63	0.62	0.71	0.70	0.70	0.69	0.64

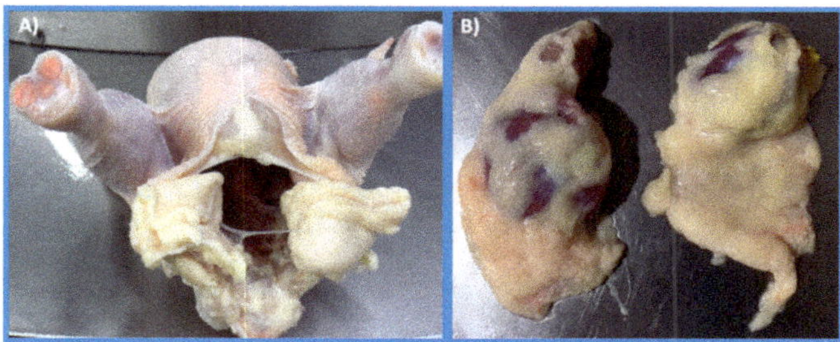

Figure 1. Chicken fat by-products: (**A**) Abdominal fat, (**B**) gizzard fat.

2.2. Chemical analysis

Samples of fat (200 g; 65% abdominal fat and 35% gizzard fat, the normal fat proportions of the carcass) were dried and extracted following the Soxhlet procedure and using diethyl ether as the extraction solvent [11]. The methyl esters from fatty acids (FAME) were prepared using BF_3 in methanol and stored at −80 °C until chromatographic analysis.

The FAME were analysed using a gas chromatograph (GC-2014 Gas Chromatograph, Shimadzu, Chiyoda-ku, Tokyo, Japan) equipped with a flame ionization detector, a split/splitless injector, and a fused silica capillary column containing polyethylene glycol as stationary phase (db-wax, 60 m × 0.25 mm, J&W Scientific, Santa Clara, CA, USA). The injector temperature was set to 230 °C. The initial column temperature was 80 °C for 2 min at a rate of 3 °C per minute, was raised to 180 °C at 30 °C per minute and was kept at this temperature for 30 min. After this time, the temperature was increased to 200 °C at a rate of 3 °C per minute and remained at this temperature for 108 min. The fatty acids were quantified using C11:0 methyl ester as internal standard. Identification of fatty acids was performed by comparison of the retention times with those of known fatty acids and the results expressed as percentage of the area of each fatty acid over the total area of fatty acids (%).

2.3. Colour Properties

The CIELAB space was chosen for colour determination following American Meat Science recommendations [12]. The following colour coordinates were determined: lightness (L*), redness (a*, ±red-green), and yellowness (b*, ± yellow-blue). The chroma saturation index [$C^* = (a^{*2} + b^{*2})^{1/2}$] and the hue angle ($h^* = \tan^{-1} b^*/a^*$) were also estimated. The reflectance spectra between 400 and 700 nm were also obtained at every 20 nm. These colour coordinates were determined by a SP62 spectrophotometer X-RITE (X-RITE, Grand Rapids, MI, USA). Measurements were made using D65 illuminant, 64 mm area and a 10° observer angle. These colour measurements were made in 43 samples of chicken fat by-products in their original solid form and also after heating at 78 °C for 3 min and re-solidifying at room temperature (re-solidified fat), simulating the thermal treatment applied for processing cooked meat products.

2.4. Statistical Analysis

The calculation of production results considered the entire flock, i.e., 75 birds in each farm. To determine the sampling of chicken fat, considering the chicken live weight and the farm of origin with different geographical location, confidence interval for one proportion—confidence interval Ross Lenth's Piface- was used. The number of samples to be analysed for the lipid profile according to the statistical analysis performed on the sampling of fat in the plant was thirteen, for a variance of 6.6 obtained from the sum of the two types of chicken fat by-products. Colour data are reported as average ± standard deviation. The data were analysed statistically using IBM SPSS Statistics for Windows, version 23 (IBM Corp., Armonk, NY, USA).

3. Results

3.1. Chickens' Performance

The production results (final body weight, average daily weight gain, feed intake and feed conversion ratio) were similar in the three farms (Table 2) and there were no statistically significant differences that depended on the feed used ($p > 0.05$).

Table 2. Productivity parameters of "Parrilleros" chickens in the three farms under study (n = 75 per farm).

Item [1]/Farm	La Goleta	Villa Rita	La Nirvana
FBW (kg)	2.31 ± 0.02	2.37 ± 0.05	2.32 ± 0.04
BWG (g/day)	57.4 ± 1.23	57.3 ± 1.20	57.8 ± 1.68
FI (kg)	3.66 ± 0.07	3.80 ± 0.09	3.78 ± 0.08
FCR (kg/kg)	1.58 ± 0.03	1.59 ± 0.05	1.63 ± 0.04

n = number of birds (whole flock); [1] Each value represents the mean of 3 replicates (25 birds per pen). No significant differences ($p < 0.0$) were found between farms. FBW: final body weight; BWG: average of daily body weight gain; FI: feed intake; FCR: feed conversion ratio.

3.2. Chemical Analysis

Importantly, there were no significant differences ($p > 0.05$) in the total fat content between the 8 different finisher diets used for chicken feed (Table 1). These diets were elaborated considering the composition and content of the ingredients used in each formula. In all the diets, linoleic (52.22%), oleic (24.87%) and palmitic (11.45%) fatty acids were identified as the predominant fatty acids.

The weight of fat by-products per chicken carcass was approximately 40 g, of which 65% corresponded to abdominal fat and 35% to gizzard fat. The total yield for lipid extraction obtained in chicken fat by-products was 75%.

Fatty acid profiles (% of total lipids) of chicken fat by-products from the 3 farms used in this study are shown in Table 3. No differences were found ($p > 0.05$) between the lipid profiles of chicken fat by-products from the 3 farms under study.

Table 3. Lipid profile (% of total lipids) of chicken fat by-products from the three farms under study.

Fatty Acid	Common Name	Farms			Variation Coefficient (%)
		Villa Rita	La Goleta	La Nirvana	
C14:0	Myristic acid	0.52	0.50	0.50	3.9
C16:0	Palmitic acid	24.18	23.63	23.81	4.0
C16:1	Palmitoleic acid	5.01	4.83	5.16	3.6
C18:0	Stearic acid	5.69	6.00	5.92	5.4
C18:1w9	Oleic acid	36.15	36.83	35.31	6.0
C18:2w6	Linoleic acid	22.55	22.22	23.75	0.6
C18:3w3	Linolenic acid	1.46	1.49	1.64	1.0
$\sum$ SFA		30.4	30.1	30.2	3.6
$\sum$ MUFA		41.2	41.7	40.5	5.4
$\sum$ PUFA		24.0	23.7	25.4	0.5
$\sum$ PUFA/$\sum$ SFA		0.8	0.8	0.8	

The predominant fatty acids in chicken fat by-products were oleic (C18:1), palmitic (C16:0) and linoleic (C18:2) acids (Table 3), which reflects the lipid profile of the diets (Table 1). The chicken fat by-products showed a higher unsaturated fat content (65.5%) than of saturated fat (30.3%), which also reflects the values of the diets (Table 1).

3.3. Colour Properties

The colour parameters of chicken fat by-products are shown in Table 4. Solid fat had statiscally higher ($p < 0.05$) L* and a* values than the melted and re-solidified fat. By contrast, the b* coordinate, saturation index and hue values were higher ($p < 0.05$) when chicken fat by-products were previously melted.

Table 4. Colour parameters [(L*) lightness, (a*) redness, (b*) yellowness, (C*) chroma or saturation index, (h*) hue] of chicken fat by-products (solid fat and melted and re-solidified fat).

Chicken Fat by-Product	L*	a*	b*	C*	h*
Solid	71.52 ± 2.22a	3.44 ± 0.09 ^{a}	24.65 ± 1.56 ^{b}	24.89 ± 1.47 ^{b}	82.06 ± 1.25 ^{b}
Melted and re-solidified	40.26 ± 1.03b	0.96 ± 0.03 ^{b}	65.89 ± 1.02 ^{a}	65.9 ± 2.14 ^{a}	89.17 ± 1.22 ^{a}

^{a,b}: different letters indicate significant differences ($p < 0.05$). $n = 39$.

Figure 2 presents the reflectance spectra (400–700 nm) obtained for the solid fat and the melted and re-solidified fat. As it can be seen, the shape of the spectra for both types of fat is completely different. At all the wavelengths studied, solid fat showed higher ($p < 0.05$) reflectance percentages than the other fat. Melted and re-solidified fat did not show any reflectance from 400 to 480 nm (mainly corresponding to violet and blue), while from 480 to 540 nm (green) the reflectance values showed the higher increase (approx.13%), these reflectance values remaining constants until the end of the spectrum (corresponding to yellow, orange and red).

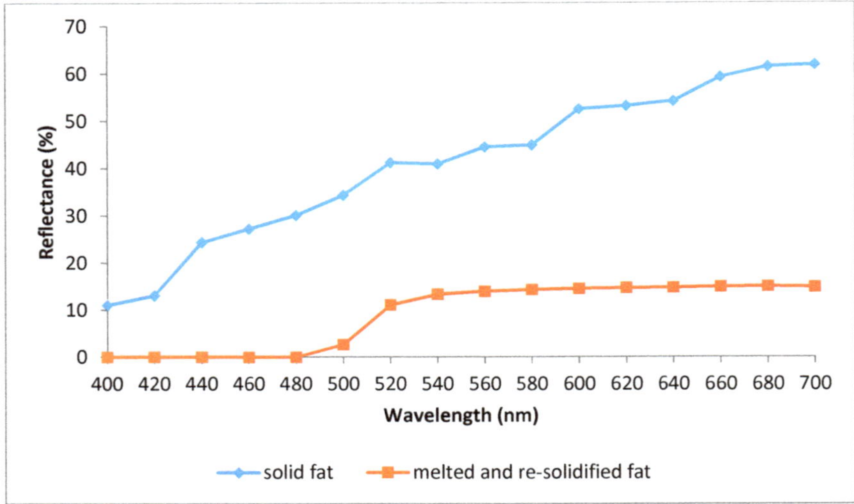

Figure 2. Reflectance spectra (400–700 nm) of the chicken fat by-products (solid fat and, melted and re-solidified fat).

4. Discussion

4.1. Chickens' Performance

Our study found no significant effect of farm or finisher diet on the productivity parameters of chickens, and all the values obtained agree with the normal productive parameters reported for chickens reared in Antioquia (Colombia) [13].

4.2. Chemical Analysis

As explained above, the fat deposits of a chicken carcass come mainly from the diet, so that the lipid profile in these tissues reflect the lipid profile of the diet [14]. The interactions that take place between the nutrients that compose the diet and the synthesis and activity of lipogenic enzymes are responsible for a wide range of possibilities regarding lipid deposition in adipose tissue. Moreover, the biological activity of some fatty acids stimulates or inhibits specific lipogenic genes encoding enzymes [15].

The yields obtained for the lipid extraction in chicken fat by-products are markedly higher than the levels reported for chicken skin (<30%) [16,17], the usual chicken fat source in the meat industry.

The fact that the lipid profile of chicken fat by-products from the 3 farms under study did not show differences is probably due to the stability of the feed used in each farm. Since the feeding and other farms' routines were the same in all three farms, so that the only difference was their respective geographical location and climatic conditions, it seem safe to conclude that neither factor was important enough to modify this composition. This is very important because if chicken fat by-products are to be used as fatty ingredients in the meat industry, the greater the homogeneity in their composition, the easier it will be to formulate meat products.

The lipid profile of fat by-products was within the range reported in the literature for chicken skin fat (Table 3) [17–19] which was to be expected because the lipid profile of different chicken carcass parts (skin, adipose tissue and meat) does not any show statistical differences [20].

Of other sources of animal fat commonly used in meat products (Table 5), chicken fat by-products have the highest amount of unsaturated fatty acids (UFA, 65.5%) and bovine tallow the lowest (44–50%). It must be noted that the chicken fat by-products analysed in this study contained a higher proportion of polyunsaturated fatty acids (PUFA, approx. 40% of total UFA) than pork or beef fat (has less than 20%).

Unsaturated fatty acids include essential fatty acids that play beneficial roles in human health. Oleic acid may help decrease the circulating concentration of low density lipoprotein (LDL) cholesterol in humans and is considered a "healthy" fat [21]. High oleic acid values are desirable for their hypocholesterolemic action, and have the added advantage of not lowering high density lipoprotein (HDL) cholesterol ("good cholesterol"), and protecting against coronary heart diseases [22,23]. The essential fatty acids include the w3 and w6 families, which are not biologically synthetized by humans, but which are necessary for biological processes and therefore should be included in the human diet [23].

Table 5. Lipid profile (%) of traditional fat sources in the meat industry, according to the literature, and of chicken fat by-products analyzed in this work.

Fatty Acid	Common Name	Beef Tallow (1)	Pork Lard (2)	Poultry Skin (3)	Chicken Fat by-Products
C14:0	Myristic acid	1–1.5	1–1.5	–	0.51
C16:0	Palmitic acid	24–28	24–28	20–24	23.87
C16:1	Palmitoleic acid	2–3	2–3	5–9	5.00
C18:0	Stearic acid	20–24	13–14	4–6	5.87
C18:1w9	Oleic acid	40–43	43–47	33–44	36.10
C18:2w6	Linoleic acid	2–4	8–11	18–20	22.84
C18:3w3	Linolenic acid	<1	<1	1–2	1.53
$\sum$ SFA		46–55	38–43.5	25–31.5	30.23
$\sum$ MUFA		42–46	45–50	38–53	41.13
$\sum$ PUFA		2–4	8–11	19–22	24.37
$\sum$ PUFA/$\sum$ SFA		<0.1	0.3	0.8	0.8

(1) Mottram et al. [24]; Alm [25]. (2) Mottram et al. [24]; Ospina-E et al. [22]; Alm [25]. (3) Sheu & Chen [20]; Feddern et al. [17]; Alm [25].

By contrast, the highest saturated fatty acid (SFA) levels are found in beef tallow (46–55%) and the lowest in poultry fat by-products (30.2%). Taking into consideration that the high consumption of saturated fatty acids has been associated with increased levels of serum cholesterol and LDL, both risk factors for cardiovascular diseases [26,27], using chicken fat by-products as fatty raw material in the meat industry could be considered advantageous. However, some studies suggest that the role of saturated fat in heart diseases is complex because of the heterogeneous biological effects of different saturated fatty acids and the diversity of food sources [27,28], so that not all SFAs should be considered hypercholesterolemic. These findings suggest that the specific matrix of different foods, including other fatty acids, nutrients, and bioactives, may biologically modify the effect of saturated fat in cardiovascular diseases.

According to French et al. [29] the most undesirable fatty acid is myristic acid, which only represents 1.3% in chicken fat by-products (Table 3), 3% in beef tallow and 3.5% in pork backfat (Table 5). Several authors have reported that palmitic acid has a low hypercholesterolemic effect and stearic acid has no effect because it becomes oleic acid in the body [30] and so does not influence blood cholesterol levels.

These results suggest that chicken fat can be used as fatty ingredient in formulating sausages, for example, as a partial or total substitute of traditional solid fat sources with their higher SFA concentrations, or be used together with chicken skin, thus increasing the amount of useful fat that can be obtained from poultry [31]. In addition, the high levels of UFA in chicken fat by-products could allow them to be used as frying oil as well as mixed with other solid fats to increase their plasticity.

4.3. Colour Properties

The colour of foods is the first characteristic that makes an impression on consumers and is one of the most intuitive factors influencing consumer purchase decisions [32,33]. Contrary to what might be expected, pure fats and oils are colourless. The characteristic colours usually associated with some of them are imparted by foreign substances that are lipid-soluble and have been absorbed by these lipids. In the case of the fat from carcasses, the colour basically depends on the feed that the live animal received [34]. In the case of chickens, when maize (rich in carotenes and xanthophylls) is included in their diet, the fatty deposits take on a yellow colour. Another factor influencing fat colour is the concentration of haemoglobin retained in the capillaries of the adipose tissue and also the connective tissue that is included [35]. According to this author, mature adipose cells or adipocytes can easily reach a diameter of micron size and are almost filled by a single large droplet of triglyceride. Thus, the nucleus and cytoplasm of an adipose cell are restricted to a thin layer under the plasma membrane, which accounts for the low water content of fat. Mature adipose cells with very little cytoplasm contain few organelles. The large triglyceride droplet that fills most of the cell is not directly bounded by a membrane, but is restrained by a cytoskeletal meshwork of 10-nm filaments, which is most conspicuous in the adipose cells of poultry.

From a technological point of view, fat fulfils several functions in meat product processing (e.g., appearance, taste and textural properties) although, in the case of colour, its principal role is in the brightness of the resulting meat products. The colour coordinate values (L*, a* and b*) of the analysed chicken fat by-products are into the range reported by Sirri et al. [36] for chicken skin. These authors measured the colour coordinates in the skin of different parts of the chicken carcass (breast, thigh and shank) and reported the following values: 65.8-81.7 for lightness, −3.75–7.52 for redness, and 7.45–39.12 for yellowness. These data point to high variability in skin colour, especially in the case of b*, even taking into account that the total xanthophyll content of the feeds used was homogeneous (from 12 to 15 mg/kg of feed) in the different flocks studied. This suggests that, in addition to pigment concentrations, other factors could play an important role in determining the final skin colour of poultry.

The observed reduction in lightness and the increase in yellowness due to melting (Table 4) could be due to the reduction in moisture and the consequent increase in the concentration of yellow pigment (carotenes). Based on the differences in the reflectance spectra obtained for solid fat by-products, and melted and re-solidified fat (Figure 2), it is clear that the heat applied to melt the fat caused severe changes in its ultrastructure, which were not reversed when the fat re-solidified.

5. Conclusions

The lipid profile of chicken fat by-products from the three farms was similar (with low coefficients of variation), despite factors associated with their different geographical locations (as long as the birds were fed a similar diet) which is very important finding for their potential application as a fat source in the formulation of meat products. The predominant fatty acids in chicken fat by-products were oleic, palmitic and linoleic acids, showing higher amount of unsaturated fatty acids than recorded for traditional fat sources used to make meat products. As regards the colour properties, chicken fat by-products had colour coordinate values that were in the range of those of chicken skin, which is the usual fat source in the meat industry. However, melting and re-solidification caused severe changes in the reflectance spectrum. In view of these results, chicken fat by-products could be used as fat ingredient in sausage formulations to partially or totally substitute traditionally used solid fat sources with their higher saturated fatty acid concentrations.

Author Contributions: Conceptualization, J.F.-L. and L.M.-P.S.; methodology, L.M.-P.S. and J.A.P.-A.; validation, L.M.-P.S. and J.A.P.-A.; investigation, J.F.-L. and J.A.P.-A.; resources, L.M.-P.S.; writing—original draft preparation, L.M.-P.S.; writing—review and editing, J.F.-L. and J.A.P.-A. All authors have read and agreed to the published version of the manuscript.

Funding: This research received no external funding.

Acknowledgments: IPOA researchers are members of the HealthyMeat network, funded by CYTED (ref. 119RT0568).

Conflicts of Interest: The authors declare no conflict of interest.

Ethics statement: This research and the rearing and slaughter of chickens were carried out in accordance with the National Committee Guidelines (Colombia) on animal testing and care.

References

1. FENAVI. Consumo per Capita Mundial de Carne de Pollo. Federación Nacional de Avicultores de Colombia, 2019. Available online: https://fenavi.org/estadisticas/consumo-per-capita-mundo-pollo/ (accessed on 12 February 2020).
2. FIRA. Carne de Pollo. Fidecomisos Instituidos en Relación con la Agricultura-Panorama Alimentario, Mexico. 2019. Available online: https://s3.amazonaws.com/inforural.com.mx/wp-content/uploads/2019/09/29173801/Panorama-Agroalimentario-Carne-de-pollo-2019.pdf (accessed on 15 March 2020).
3. Mozdziak, P. Species of meat animals: Poultry. In *Encyclopedia of Meat Sciences*; Dikeman, M., Devine, C., Eds.; Academic Press: Oxford, UK, 2014; pp. 369–373.
4. Choi, Y.S.; Han, D.J.; Choi, J.H.; Hwang, K.E.; Song, D.H.; Kim, H.W.; Kim, Y.B.; Kim, C.J. Effect of chicken skin on the quality characteristics of semi-dried restructured jerky. *Poult. Sci.* **2016**, *95*, 1198–1204. [CrossRef] [PubMed]
5. Herrera, M. Taking Advantage of by-Products or Waste in the Poultry Industry for the Production of Animal Meal. Virtual Pro, Bogotá, Colombia. 2008. Available online: https://www.virtualpro.co/biblioteca/aprovechamiento-de-los-subproductos-o-residuos-en-la-industria-avicola-para-la-produccion-de-harinas-de-origen-animal (accessed on 15 January 2020).
6. El Boushy, A.R.Y.; Van der Poel, A.F.B. *Poultry Feed from Waste*; Springer: Boston, MA, USA, 1994.
7. Barbut, S. Waste treatment and by-products. In *the Science of Poultry and Meat Processing*; Chapter 18; Barbut, S., Ed.; Guelph, ON, Canada, 2015; pp. 1–27. Available online: https://atrium.lib.uoguelph.ca/xmlui/bitstream/handle/10214/9300/Barbut_SciPoultryAndMeatProcessing_18ByProductsAndWaste_v01.pdf?sequence=18&isAllowed=y (accessed on 3 August 2020).
8. Vikman, M.; Siipola, V.; Kanerva, H.; Slizyte, R.; Wikberg, H. Poultry by-products as potential source of nutrients. *Adv. Recycl. Waste Mang.* **2017**, *2*, 1–5.
9. Abid, M.; Touzani, M. First step of waste chicken skin valorization by production of biodiesel in Morocco. *J. Mater Environ. Sci.* **2017**, *8*, 2372–2380.
10. Chiu, M.C.; Gioielli, L.A.; Sotero-Solis, V. Fraccionamiento de la grasa abdominal de pollo. *Grasas y Aceites* **2002**, *53*, 298–303.
11. AOAC. *Official Methods of Analysis of AOAC International*, 18th ed.; Association of Official Analytical Chemistry: Gaithersburg, MD, USA, 2005.
12. AMSA. *Meat Color Measurement Guidelines*; American Meat Science Association: Champaign, IL, USA, 2012.
13. Díez-Arias, D. Pronutrients and productive parameters in broilers. *Veterinaria Digital*. 12 September 2018. Available online: https://www.veterinariadigital.com/en/post_blog/pronutrients-and-productive-parameters-in-broilers/ (accessed on 15 December 2019).
14. Sirri, F.; Tallarico, N.; Meluzzi, A.; Franchini, A. Fatty acid composition and productive traits of broiler fed diets containing conjugated linoleic acid. *Poult. Sci.* **2003**, *82*, 1356–1361. [CrossRef]
15. Jump, D.B. Dietary polyunsaturated fatty acids and regulation of gene transcription. *Curr. Opin. Lipidol.* **2002**, *13*, 155–165. [CrossRef]
16. De Souza, S.A.B.; Matsushita, M.; de Souza, N.E. Protein, lipids and cholesterol in roasted chicken. *Acta Sci.* **1997**, *19*, 1069–1073.
17. Feddern, V.; Kupski, L.; Cipolatti, E.P.; Giacobbo, G.; Mendes, G.L.; Badiale-Furlong, E.; de Souza-Soares, L.A. Physico-chemical composition, fractionated glycerides and fatty acid profile of chicken skin fat. *Eur. J. Lipid. Sci. Technol.* **2010**, *112*, 1277–1284. [CrossRef]
18. Chiu, M.C.; Gioielli, L.A.; Grimaldi, R. Structured lipids from chicken fat, its stearin and medium chain triacylglycerol blends. I-Fatty acid and triacylglycerol compositions. *Quim. Nova* **2008**, *31*, 232–237. [CrossRef]
19. Lee, K.T.; Foglia, T.A. Synthesis, purification, and characterization of structured lipids produced from chicken fat. *J. Am. Oil Chem. Soc.* **2000**, *77*, 1027–1034. [CrossRef]
20. Sheu, K.S.; Chen, T.C. Yield and quality characteristics of edible broiler skin fat as obtained from five rendering methods. *J. Food Eng.* **2002**, *55*, 263–269. [CrossRef]
21. Kwon, H.N.; Choi, C.B. Comparison of lipid content and monounsaturated fatty acid composition of beef by country of origin and marbling score. *J. Korean Soc. Food Sci. Nutr.* **2015**, *44*, 1806–1812. [CrossRef]

22. Ospina-E, J.C.; Cruz-S, A.; Pérez-Alvarez, J.A.; Fernández-López, J. Development of combinations of chemically modified vegetable oils as pork backfat substitutes in sausages formulation. *Meat Sci.* **2010**, *84*, 491–497. [CrossRef] [PubMed]
23. Da Silva Martins, T.; Antunes de Lemos, M.V.; Freitas Mueller, L.; Baldi, F.; Rodrigues de Amorim, T.; Matias Ferrinho, A.; Muñoz, J.A.; de Souza Fuzikawa, I.H.; Vespe de Moura, G.; Gemelli, J.L.; et al. Fat deposition, fatty acid composition, and its relationship with meat quality and Human Health. In *Meat Science and Nutrition*; Arshad, M.S., Ed.; Intech Open Ltd.: London, UK, 2018; pp. 17–37.
24. Mottram, H.R.; Crossman, Z.M.; Evershed, R.P. Regiospecific characterization of the triacylglycerols in animal fats using high performance liquid chromatography-atmospheric pressure chemical ionization mass spectrometry. *Analyst* **2001**, *126*, 1018–1024. [CrossRef]
25. Alm, M. *Edible Oil Processing*; AOCS Lipid Library; The American Oil Chemists' Society: Urbana, IL, USA, 2019.
26. FAO. Fats and fatty acids in human nutrition. Report of an expert consultation. *FAO Food Nutr. Pap.* **2010**, *91*, 1–166.
27. Liu, A.G.; Ford, N.A.; Hu, F.B.; Zelman, K.M.; Mozaffarian, D.; Kris-Etherton, P.M. A healthy approach to dietary fats: Understanding the science and taking action to reduce consumer confusion. *Nutr. J.* **2017**, *16*, 53–68. [CrossRef]
28. Mozaffarian, D.; Micha, R.; Wallace, S. Effects on coronary heart disease of increasing polyunsaturated fat in place of saturated fat: A systematic review and metaanalysis of randomized controlled trials. *PLoS Med.* **2010**, *7*, e1000252. [CrossRef]
29. French, M.A.; Sundram, K.; Clandinin, M.T. Cholesterolaemic effect of palmitic acid in relation to other dietary fatty acids. *Asia Pac. J. Clin. Nutr.* **2002**, *11*, 401–407. [CrossRef]
30. Sinclair, A.J. Dietary fat and cardiovascular disease: The significance of recent developments for the food industry. *Food Aust.* **1993**, *45*, 226–232.
31. Peña-Saldarriaga, L.; Pérez-Alvarez, J.A.; Fernández-López, J. Quality properties of chicken emulsion-type sausages formulated with chicken fatty byproducts. *Foods* **2020**, *9*, 507. [CrossRef]
32. Pérez-Alvarez, J.A.; Fernández-López, J. Chemical and biochemical aspects of color in muscle foods. In *Handbook of Meat, Poultry & Seafood Quality*; Nollet, L.M.L., Ed.; Blackwell Publishing Professional: Ames, IA, USA, 2007; pp. 25–44.
33. Font-I-Furnols, M.; Guerrero, L. Consumer preference, behavior and perception about meat and meat products: An overview. *Meat Sci.* **2014**, *98*, 361–371. [CrossRef] [PubMed]
34. Irie, M. Optical evaluation of factors affecting appearance of bovine fat. *Meat Sci* **2001**, *57*, 19–22. [CrossRef]
35. Swatland, H.J. Reversible pH effect on pork paleness in a model system. *J. Food Sci.* **1995**, *60*, 988–991. [CrossRef]
36. Sirri, F.; Petracci, M.; Bianchi, M.; Meluzzi, A. Survey of skin pigmentation of yellow-skinned broiler chickens. *Poult. Sci.* **2010**, *89*, 1556–1561. [CrossRef] [PubMed]

Article

Multivariate Relationships among Carcass Traits and Proximate Composition, Lipid Profile, and Mineral Content of *Longissimus lumborum* of Grass-Fed Male Cattle Produced under Tropical Conditions

Lilia Arenas de Moreno [1], Nancy Jerez-Timaure [2], Nelson Huerta-Leidenz [3,*], María Giuffrida-Mendoza [4], Eugenio Mendoza-Vera [5] and Soján Uzcátegui-Bracho [6]

[1] Facultad de Agronomía, Instituto de Investigaciones Agronómicas, Universidad del Zulia, Box 15205, Maracaibo, Zulia 4001, Venezuela; lilia.arenas@gmail.com
[2] Instituto de Ciencia Animal, Facultad de Ciencias Veterinarias, Universidad Austral de Chile, Valdivia 5090000, Chile; nancy.jerez@uach.cl
[3] Department of Animal and Food Sciences, Texas Tech University, Box 42141, Lubbock, TX 79409-2141, USA
[4] Facultad de Medicina, Universidad del Zulia, Box 15131, Maracaibo, Zulia 4001, Venezuela; mariagvm@gmail.com
[5] Facultad de Ingeniería, Universidad del Zulia, Box 15131, Maracaibo, Zulia 4001, Venezuela; mendozaeb64@gmail.com
[6] Facultad de Ciencias Veterinarias, Universidad del Zulia, Box 15131, Maracaibo, Zulia 4001, Venezuela; sojanuzcategui@gmail.com
* Correspondence: nelson.huerta@ttu.edu; Tel.: +1-956-250-4337

Citation: Arenas de Moreno, L.; Jerez-Timaure, N.; Huerta-Leidenz, N.; Giuffrida-Mendoza, M.; Mendoza-Vera, E.; Uzcátegui-Bracho, S. Multivariate Relationships among Carcass Traits and Proximate Composition, Lipid Profile, and Mineral Content of *Longissimus lumborum* of Grass-Fed Male Cattle Produced under Tropical Conditions. *Foods* **2021**, *10*, 1364. https://doi.org/10.3390/foods10061364

Academic Editor: Thierry Astruc

Received: 6 May 2021
Accepted: 10 June 2021
Published: 12 June 2021

Abstract: Hierarchical cluster (HCA) and canonical correlation (CCA) analyses were employed to explore the multivariate relationships among chemical components (proximate, mineral and lipidic components) of lean beef *longissimus dorsii lumborum* (LDL) and selected carcass traits of cattle fattened on pasture under tropical conditions (bulls, $n = 60$; steers, $n = 60$; from 2.5 to 4.0 years of age, estimated by dentition). The variables backfat thickness (BFT), Ca, Mn, Cu, C14:0, C15:0, and C20:0 showed the highest coefficients of variation. Three clusters were defined by the HCA. Out of all carcass traits, only BFT differed significantly ($p < 0.001$) among clusters. Clusters significantly ($p < 0.001$) differed for total lipids (TLIPIDS), moisture, dry matter (DM), fatty acid composition, cholesterol content, and mineral composition (except for Fe). The variables that define the canonical variate "CARCASS" were BFT and degree of marbling (MARBLING). TLIPIDS was the main variable for the "PROXIMATE" canonical variate, while C16:0 and C18:1*c* had the most relevant contribution to the "LIPIDS" canonical variate. BFT and MARBLING were highly cross-correlated with TLIPIDS which, in turn, was significantly affected by the IM lipid content. Carcass traits were poorly correlated with mineral content. These findings allow for the possibility to develop selection criteria based on BFT and/or marbling to sort carcasses, from grass-fed cattle fattened under tropical conditions, with differing nutritional values. Further analyses are needed to study the effects of sex condition on the associations among carcass traits and lipidic components.

Keywords: *longissimus dorsii lumborum*; multivariate analyses; proximate composition; fatty acid profile; mineral content; carcass traits; tropical beef cattle

1. Introduction

Beef is known as one of the main sources of protein with high biological value, bioavailable minerals (Fe, Zn and P), vitamins of the B-complex (B1, B2, B3 B6 and B12) and other nutritional components (D, E, and β-carotenes) [1–5]. It is also a nutritional source of monounsaturated (MUFA) and essential polyunsaturated (PUFA) fatty acids (omega 3 and omega 6) with dietary and functional properties, and therapeutic effects [6–8]. The main benefits of beef consumption for nutrition and health are closely related to its unique chemical composition [2–4].

It is well known that intrinsic factors like species, breed, gender, age, and the structure of the type of muscle [3,9–15], and extrinsic factors such as animal nutrition and pre-slaughter conditions [16–18] are largely responsible for the variation found in carcass traits, beef sensory attributes, and nutrient composition. Nutrient composition of grass-fed beef has been a subject of study worldwide [15,19–23] and although there is a general perception that its consumption brings health benefits to consumers, there is no consensus on this matter [24]. Numerous studies on the effect of castration reviewed by Huerta and Ríos [25] have demonstrated that carcasses of castrated males (steers) accumulate more fat than their non-castrated counterparts (bulls); however, the influence of castration on the nutrient composition of lean, grass-fed beef (i.e., fatty acids, cholesterol, and minerals) has been less studied in the tropics, particularly in cattle with *Bos indicus* influence. A couple of reports in Brazil [26,27] indicate that the intramuscular fat (IMF) of bulls contains more PUFA and exhibit a higher PUFA/SFA ratio than steers. These findings are explained by the larger muscle mass and leaner beef of bulls, and therefore, a more abundant content of membrane phospholipids of muscle cells [28]. The comparison of lean meats from grass-fed bulls vs. steers in cholesterol or mineral content has not indicated significant differences [3].

For decades, the meat industry and scientists have used carcass characteristics to predict palatability-related attributes and/or consumer acceptability. Indeed, most of the carcass quality grading systems rely on the relationships between individual (or combined) carcass traits and sensory attributes of meat [25]. Key characteristics that describe the beef carcass include carcass weight, physiological maturity (often using dentition or ossification as a proxy for age), sex, fat cover and colour, and conformation. Depending upon the country, marbling and lean colour and/or texture have often been added as quality traits to refine the carcass evaluation technique [29,30].

To our knowledge, there is limited information regarding the nutritional quality/value of meat specifically focused on a possible relationship between the anatomical or other physical characteristics of the intact beef carcass and its meat nutrient composition. This information gap needs to be addressed/closed particularly for beef produced under tropical, grass feeding conditions given the alleged health claims linked to its consumption [31]. We propose that, to determine any relationship, all of these traits must be simultaneously considered by using a multivariate analysis approach. Jeong et al. [32] investigated the relationships between the content of IMF, the fatty acid composition, and characteristics of the muscle fibre in the longissimus thoracis of pork. These researchers employed the principal components analysis (PCA) and hierarchical cluster analysis (HCA), an appropriate example of the applicability of this type of statistical approach. Similarly, Patel et al. [33] used multivariate analyses to explore the relationship among animal and carcass characteristics, beef (*longissimus thoracis*) quality traits, and lean meat mineral composition (20 elements). In this case, the researchers employed a combination of univariate (simple correlation) and multivariate (factorial analysis) techniques that allowed them to compare the relationship between minerals, not only individually but also in a factorial fashion (five factors) with the animal/carcass performance and the beef quality traits. This study [33] only included the carcass weight as one of the three performance characteristics. Both previous investigations [32,33] indicate the need for studying complex relationships employing a multivariate approach, that may include a large number of variables. In this case the Canonical Correlation analysis (CCA), offers a promising multivariate method to complement other techniques. CCA has been widely used in agricultural science [34–37] to explore the interrelation between multiple variables, relationships that could be symmetric, that is, without a dependency relationship among them, or asymmetric, when one of the sets is dependent and the other is independent.

The underlying principle of CCA is to investigate the relationship between the variables by developing several independent canonical functions that maximize the correlation between the linear composites known as canonical variates [38]. The CCA represents the bivariate correlation between the two canonical variates in a canonical function. The canonical correlation coefficient measures the strength of association between the variable sets under concern. This technique can assist in the analysis of several traits; furthermore, it may indicate the most relevant factors to the set of variables under study [39–41].

Knowing the degree of association of the multivariate relationships between the nutrient composition and the quality traits of dressed beef may allow identifying predictors of the meat nutrient composition that can be assessed on the hanging carcass and, eventually, the possibility to develop selection criteria for sorting carcasses with different nutritional values.

This study aimed to explore the multivariate relationships among chemical components (proximate, mineral and lipid components) of lean beef *longissimus dorsii lumborum* (LDL) and selected carcass traits of cattle fattened on pasture under tropical conditions.

2. Materials and Methods

2.1. Characteristics of the Sample

Carcass traits and nutrient composition data from a randomly selected group of 120 slaughtered cattle (60 bulls and 60 steers; 2.5 to 4.0 years of age, estimated by dentition) were collected for this observational study. This sample was representative of slaughter male cattle derived from the prevailing production systems in the Venezuelan tropics where livestock is mostly fattened on pasture with little or no supplementation [42]. Out of this group, 9 animals were mixed-breed dairy (predominantly Holstein, Brown Swiss, or dual-purpose cattle without a defined breed predominance) x Zebu breeds; and 110 were mixed-breed cattle with a phenotypic predominance of Zebu breeds.

2.2. Harvesting, Carcass Classification and Sample Collection

The animals were harvested at a commercial packing house following the procedures of the Venezuelan Standards of Bioethics and Biosecurity for Research with Living Organism [43], and the Venezuelan Standard for *Postmortem* Inspection of Cattle [44]. After being weighed, carcasses were chilled at 2–4 °C. After 48 h *postmortem*, the chilled carcasses were subjected to evaluation. Skeletal and lean maturity (SM and LM, respectively) scores and subcutaneous backfat thickness (BFT) were determined following USDA guidelines [45]. The subcutaneous fat cover (CFINISH) was evaluated using a four-level scale: 1 = Uniform; 2 = Uneven; 3 = In patches; 4 = Devoid [46]. The degree of marbling (MARBLING) was evaluated according to Decreto Presidencial N° 181, using a descriptive scale: 1 = practically devoid, 2 = traces, 3 = slight and 4 = small amount [47].

After evaluation, chilled carcasses were cut out following conventional butchering procedures according to regulation 792-82 of the Venezuelan Commission for Industrial Standard [48], trimmed to 6.4 mm fat cover, and fabricated into commercial cuts. Muscle samples (2.5 cm thick) from the most anterior (cranial) part of the LDL muscle were excised, individually vacuum packaged, identified by animal number, frozen at −30 °C and stored at −20 °C until the final preparation for the proximate analyses. Samples were partially thawed at 4 °C (to avoid fluids losses), trimmed of visible adipose and connective tissue, and homogenized in a Black & Decker™ food processor. Each homogenized sample was subdivided into smaller portions (subsamples) which were packaged in 50 g-zip-lock bags (4–5 bags) and identified by animal number. Bags containing homogenized subsamples were assigned to each type of chemical analysis (proximate, mineral, or lipid profile analysis) and immediately processed accordingly. The remaining bags were preserved at −20 °C as spare samples in the event that a confirmatory analysis was needed. A flowchart (Figure S1, supplementary material) illustrates sample handling for chemical analyses. All the samples were analyzed in duplicate [49].

2.3. Proximate Composition Analysis

Duplicate samples were analyzed for crude protein (CP) content following the Kjeldahl procedure; moisture (WATER) and dry matter (DM) were estimated by weight loss at

105 °C for 24 h, and ash at 550 °C during 6 h [50]. Total lipids (TLIPIDS) content was determined by extracting with a 2:1 chloroform:methanol mixture according to the method of Folch et al. [51] with some modifications as described by Slover & Lanza [52].

2.4. Mineral Analysis

Duplicates of 10.0 g of ground meat were calcined in a furnace at 550 °C for 6 h. Sample handling and mineral analyses were conducted according to the methodology described by Giuffrida-Mendoza et al. [1]. Mineral content was expressed in mg.100 g^{-1} of fresh tissue.

2.5. Lipid Profile Analysis

Cholesterol content of each steak sample was determined in triplicate, according to the procedure described by Rhee et al. [53].

Fatty acids (FA) were determined by gas chromatography as described by Slover and Lanza [54]. A duplicate of an aliquot of the lipid extract, corresponding to 25 mg of the total lipids of each sample, mixed with the internal standard (Margaric acid, C17:0 methyl ester) was saponified and esterified with BF_3/CH_3OH [55] to yield fatty acid methyl esters (FAME). FAME were analyzed following the procedure described by Uzcátegui-Bracho et al. [49].

2.6. Data Analysis

The data analysis was performed using the IBM SPSS 23 statistical software [56]. The original, historical data consisted of 120 samples, being reduced to 109 after carrying out preliminary analyses. Univariate analyses were used to evaluate descriptive statistics, kurtosis, skewness, and detection of outliers. Multivariate analyses allowed to detect and treat the possible atypical values and to verify conformity with the basic assumptions of randomness, multivariate normality, and homoscedasticity of the variance. For exploring if any noise was caused for the inclusion of 9 observations (mixed dairy x zebu breed types) the statistical analyses were run again with 100 subjects phenotypically classified as predominantly Zebu crossbreds. The statistical output of this exploratory analysis showed the canonical correlations between the selected carcass traits and the three groups of chemical variables (proximate components, lipid profile, mineral components) were like those found in the previous run with 109 subjects, thus proving that the inclusion of these mixed dairy x Zebu cattle did not cause significant changes in the results. In fact, its inclusion introduced more variability to the sample, which enriched the results.

Two hierarchical cluster analyses (HCA) were performed. The first HCA was applied to explore the presence of any pattern or relationship between the 32 variables under study (except for the categorical variables CFINISH and MARBLING), using the linkage (between groups) method. To measure the degree of association between variables, Pearson's correlation coefficient was applied with the measurement transformed into absolute values. The second HCA was applied to group all the samples using Ward's method with the squared Euclidean distance measure and considering the sex condition to describe how the variables are presented within each cluster.

To validate the clusters obtained, an ANOVA with two main factors (sex condition and cluster) was applied on each variable. The results from the two HCA were represented by dendrograms. To analyze the relationship among the subgroups of the variables proximate, mineral, and lipid components with respect to the subgroups of carcass traits, a canonical correlation analysis (CCA) was carried out. Wilk's Lambda and Bartlett tests were used to determine the significance of canonical correlations.

The acronyms of the variables studied in this research and their definitions are shown in Table 1.

Table 1. Acronyms of the variables studied and their definitions.

Abbreviation	Definition
SM	Skeletal maturity
LM	Lean maturity
CW	Carcass weight
BFT	Back fat thickness
CFINISH	Carcass finish
MARBLING	Degree of marbling
TLIPIDS	Total lipids content
DM	Dry matter content
CP	Crude protein content
CHOLEST	Cholesterol content
C14:0	Myristic acid
C14:1	Myristoleic acid
C15:0	Pentadecilic acid
C16:0	Palmitic acid
C16:1	Palmitoleic acid
C18:0	Stearic acid
C18:1*c*	Oleic acid
C18:1*t*	Elaidic acid
C18:2	Linoleic acid
C18:3	α-Linolenic acid
C20:0	Arachidic acid
C20:4ω6	Arachidonic acid
C22:6ω3	Docosahexaenoic acid
SFA	Sum of saturated fatty acids
UFA	Sum of unsaturated fatty acids
MUFA	Sum of monounsaturated fatty acids
PUFA	Sum of polyunsaturated fatty acids
HCA	Hierarchical cluster analysis
CCA	Canonical correlation analysis
"CARCASS"	Canonical variate of the variable of carcass traits
"PROXIMATE"	Canonical variate of the variables of proximate composition
"LIPIDS"	Canonical variate of the variables of lipidic composition
"MINERAL"	Canonical variate of the variables of mineral content
SEXC	Sex condition

3. Results

3.1. Descriptive Statistics for Carcass Traits, Proximate Composition, Mineral Content, and Fatty Acid Composition of Beef

The descriptive statistics of the experimental data are presented in Tables 2 and 3. The variables BFT, Ca, Mn, Cu, C14:0, C15:0, and C20:0 showed the highest coefficients of variation. In general, this sample of grass-fed beef carcasses had relative low values of BFT (0.1–1.2 cm) and MARBLING levels. Frequency and percentage distribution of MARBLING levels (i.e., Practically devoid, Traces and Slight amounts) in the filtered data (N = 109) were 48 (44%), 24 (22%) and 37 (33.9%), respectively (values are not presented in tabular form). Among the proximate components, TLIPIDS from separable lean only presented the greatest variation (between 0.93 and 6.67 g.100 g^{-1}). Out of the 30 fatty acids under study, the most abundant were C16:0 (0.028–1.288), C18:0 (0.053–0.705), C18:1c (0.27–1.749) and C18:1t (0.117–0.981). Overall, MUFA constituted 57.65% of the total; PUFA represented less than 5% of the total and the rest (37.35%) corresponded to SFA.

Table 2. Descriptive statistics for carcass traits proximal, and proximate and mineral contents in beef *longissimus lumborum* muscle.

Variable		Mean	SD	Minimum	Maximum	CV
Carcass traits	SM [1]	212.8	37.96	150.00	350.00	0.179
	LM [1]	193.1	23.46	150.00	260.00	0.121
	CW, kg	279.5	34.41	207.00	380.00	0.123
	BFT, cm	0.41	0.28	0.10	1.20	0.692
	MARBLING [2]	1.90	0.88	1	3	0.464
	CFINISH [3]	2.11	0.69	1	3	0.325
Proximate g.100 g^{-1}	DM	26.02	1.29	23.19	29.64	0.05
	Moisture	73.99	1.32	70.36	77.39	0.02
	Ash	1.05	0.15	0.70	1.43	0.14
	CP	20.79	1.53	16.90	24.00	0.07
	TLIPIDS	2.79	1.09	0.93	6.67	0.39
Mineral content mg.100 g^{-1}	Ca	2.83	1.58	1.00	8.27	0.560
	Mg	21.73	3.05	14.34	29.27	0.140
	P	210.05	34.68	100.13	322.53	0.165
	Na	82.69	19.97	41.03	119.00	0.242
	K	241.73	59.56	119.78	395.87	0.246
	Fe	1.87	0.49	0.44	3.76	0.265
	Zn	4.14	0.78	2.79	6.60	0.189
	Cu	0.086	0.04	0.024	0.19	0.457
	Mn	0.026	0.014	0.008	0.08	0.533

[1] Carcasses within the 100–199 maturity range score represent the youngest group (100 is equal to A00 and 199 is equal to A99) ; 200–299: represent carcasses with intermediate, more advanced maturity (200 is equal to B00 and 299 is equal to B99). [2] 1 = Practically devoid, 2 = Traces, 3 = Slight, 4 = Small; [3] 1 = Uniform; 2 = Uneven; 3 = In patches; 4 = Devoid. Description of acronyms is presented in Table 1.

Table 3. Descriptive statistics for cholesterol content and fatty acid profile in beef *longissimus lumborum* muscle.

Variable		Mean	SD	Minimum	Maximum	CV
Lipid profile mg.100 g^{-1}	CHOLEST	64.96	13.86	30.16	97.34	0.213
	C14:0	0.068	0.033	0.018	0.156	0.481
	C14:1	0.038	0.022	0.002	0.110	0.596
	C15:0	0.079	0.044	0.004	0.223	0.560
	C16:0	0.534	0.245	0.028	1.288	0.460
	C16:1	0.092	0.042	0.023	0.249	0.456
	C18:0	0.285	0.125	0.053	0.705	0.440
	C18:1*c*	0.876	0.320	0.274	1.749	0.365
	C18:1*t*	0.489	0.202	0.117	0.981	0.413
	C18:2	0.076	0.034	0.010	0.163	0.443
	C18:3	0.006	0.003	0.001	0.016	0.453
	C20:0	0.005	0.003	0.001	0.015	0.554
	C20:4ω6	0.013	0.006	0.003	0.034	0.447
	C22: 6ω3	0.025	0.011	0.002	0.062	0.442
	SFA	0.973	0.429	0.275	2.432	0.442
	UFA	1.632	0.563	0.544	3.032	0.345
	MUFA	1.502	0.525	0.488	2.827	0.349
	PUFA	0.121	0.044	0.036	0.234	0.362
	Cis	1.156	0.423	0.368	2.637	0.366
	Trans	0.491	0.202	0.117	0.981	0.411
	UFA/SFA	1.794	0.552	0.773	4.583	0.308
	MUFA/SFA	1.646	0.488	0.673	4.075	0.298
	PUFA/SFA	0.136	0.057	0.051	0.508	0.418
	Cis/Trans	2.528	0.790	1.229	5.709	0.312

Description of acronyms is presented in Table 1.

3.2. Characterization of the Carcass Traits and Chemical Components of Beef Longissimus Lumborum Muscle by HCA

The first HCA allowed to determine how variables were grouped by degree of similarity as calculated by the squared Euclidean distance with a similarity index ranging from 0 (higher similarity) to 25 (lower similarity); a close distance between variables indicating high correlation. This first HCA also allowed to provide a simple representation of the total data composed of 32 variables and to explore how the variables correlated to each other. Figure 1 shows that most of the variables in this dataset, tended to cluster in the same subgroup (carcass traits, proximate composition, mineral content or lipidic composition). The variable CHOLEST was grouped with LM. The variables Fe and Cu were clustered with ash and CP, and the variables TLIPIDS and C16:1 were proximate with the second smallest distance (Figure 1).

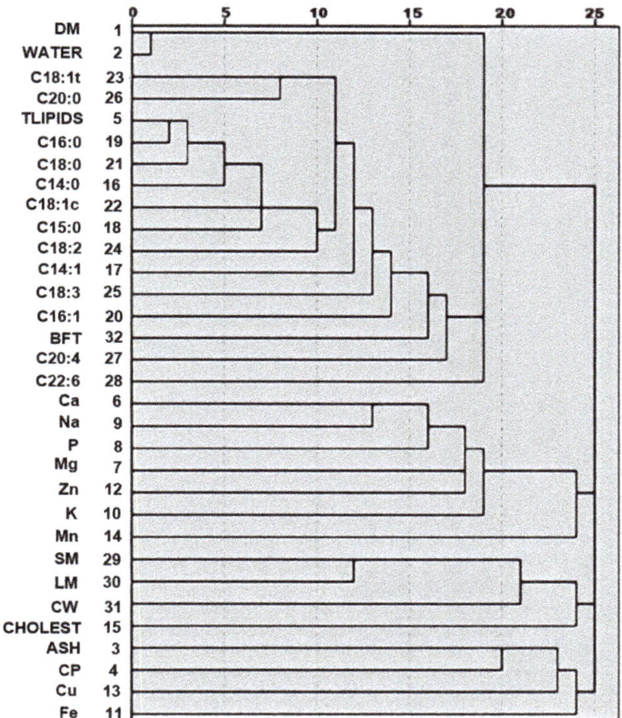

Figure 1. Dendrogram for variables from the hierarchical cluster analysis. Description of acronyms is presented in Table 1.

The HCA by samples resulted in a dendrogram with three clusters, sufficiently distant to expect relatively different values among the groups (Figure S2, Supplementary Material). ANOVA and multiple range tests (at 5% of significance) were applied to validate and describe the clusters. Results of these analyses are presented in Tables 4 and 5. Among the carcass traits, only BFT differs significantly ($p < 0.001$) among clusters (Table 4). In the proximate composition data, TLIPIDS, moisture, and DM resulted different among clusters. With the only exception of Fe, the mineral composition also differed ($p < 0.0001$). Clusters differed ($p < 0.0001$) in fatty acid composition and CHOLEST (Table 5; $p < 0.05$).

Table 4. Comparison of carcass traits, proximate composition, and mineral content of beef *longissimus lumborum* muscle between sex conditions and the three clusters.

Variable	Cluster 1		Cluster 2		Cluster 3		*p* Value	
	Steer (n = 25)	Bull (n = 3)	Steer (n = 7)	Bull (n = 26)	Steer (n = 22)	Bull (n = 26)	Cluster	SEXC
				Carcass Traits				
SM [1]	213.60	193.33	228.57	201.92	211.82	220.38	0.581	0.527
LM [1]	198.40	173.33	195.71	190.00	195.94	191.56	0.588	0.108
BFT [2]	0.724	0.333	0.357	0.226	0.510	0.215	<0.0001	<0.0001
CW [3]	289.48	301.33	247.14	276.42	280.00	278.96	0.066	0.863
				Proximate composition (g.100 g^{-1})				
DM	26.78	25.95	25.27	24.97	26.55	26.10	<0.0001	<0.0001
Moisture	73.20	74.05	74.69	75.07	73.47	73.91	<0.0001	<0.0001
Ash	1.04	1.03	1.05	1.08	1.02	1.06	0.669	0.276
CP	20.17	20.50	21.17	21.09	20.83	20.97	0.052	<0.0001
TLIPIDS	4.16	3.59	2.24	1.81	2.83	2.45	<0.0001	<0.0001
				Mineral content (mg.100 g^{-1})				
Ca	1.88	2.09	2.58	1.98	4.16	3.59	<0.0001	0.618
Mg	20.26	22.58	18.29	20.47	23.08	24.09	<0.0001	0.049
P	198.27	226.87	201.72	189.04	223.45	231.33	<0.0001	0.075
Na	88.32	96.42	87.70	93.54	72.87	71.79	<0.0001	0.703
K	214.84	215.53	208.57	213.45	268.79	284.93	<0.0001	0.322
Fe	1.98	1.93	1.78	1.83	1.87	1.83	0.445	0.446
Zn	4.04	3.69	3.78	3.75	4.49	4.49	<0.0001	0.541
Cu	0.067	0.064	0.093	0.086	0.109	0.082	0.010	0.535
Mn	0.021	0.024	0.017	0.022	0.026	0.036	<0.0001	0.015

Description of acronyms is presented in Table 1. [1] Carcasses within the 100–199 maturity range score represent the youngest group (100 is equal to A00 and 199 is equal to A99) ; 200–299: represent carcasses with intermediate, more advanced maturity (200 is equal to B00 and 299 is equal to B99). [2] expressed in cm. [3] expressed in kg.

Table 5. Comparison of lipid profile of beef *longissimus lumborum* muscle between sex conditions within the clusters.

Variable [1]	Cluster 1		Cluster 2		Cluster 3		*p* Value	
	Steer (n = 25)	Bull (n = 3)	Steer (n = 7)	Bull (n = 26)	Steer (n = 22)	Bull (n = 26)	Cluster	SEXC
				Lipid profile (mg.100 g^{-1})				
CHOLEST	65.99	64.88	62.45	63.27	61.244	69.49	<0.0001	0.311
C14:0	0.103	0.092	0.051	0.043	0.068	0.059	<0.0001	<0.0001
C14:1	0.060	0.045	0.025	0.026	0.036	0.030	<0.0001	<0.0001
C15:0	0.131	0.104	0.064	0.046	0.078	0.060	<0.0001	<0.0001
C16:0	0.8280	0.715	0.434	0.311	0.564	0.456	<0.0001	<0.0001
C16:1	0.131	0.117	0.071	0.085	0.081	0.075	<0.0001	0.002
C18:0	0.425	0.402	0.218	0.191	0.283	0.248	<0.0001	<0.0001
C18:1c	1.216	1.209	0.724	0.611	0.917	0.781	<0.0001	<0.0001
C18:1t	0.713	0.601	0.363	0.311	0.493	0.464	<0.0001	<0.0001
C18:2	0.108	0.097	0.056	0.051	0.071	0.075	<0.0001	0.001
α-C18:3	0.008	0.007	0.005	0.004	0.006	0.005	<0.0001	<0.0001
C20:0	0.007	0.005	0.002	0.003	0.005	0.005	<0.0001	0.010
C20:6ω6	0.016	0.011	0.011	0.010	0.013	0.013	<0.0001	0.023
C22:6ω3	0.032	0.024	0.025	0.020	0.025	0.021	<0.0001	<0.0001

[1] Description of acronyms is presented in Table 1.

Cluster 1 is mainly represented by steers with the highest values in BFT (0.724 cm) and TLIPIDS (4.16 g.100 g^{-1}); therefore, this subgroup also exhibited the highest values in most of the fatty acids evaluated ($p < 0.0001$). Cluster 2 is mostly composed of bulls with the lowest BFT (0.226 cm) and total lipids (1.81 mg.100 g^{-1}; $p < 0.05$). On the other hand, cluster 3 was more balanced, comprised of 56.5% of bulls and 53.5% of steers, with similar values in BFT and TLIPIDS; this cluster represented the group of samples with similarities in fat-related traits and differences in mineral content. In general, cluster 3 showed the highest values in Ca, Mg, P, K, Zn, Cu and Mn ($p < 0.0001$). Figures S3–S6 (Supplementary Materials) illustrated the projection means of variables by each cluster.

3.3. Relationship among Subgroups of Variables by CCA

The first canonical correlation between U (representing the proximate composition traits) and V (representing the subgroup of carcass traits) was significant ($p < 0.0001$). Canonical redundancy analysis revealed that the first canonical correlation represents 75.96% of the explained variance, which indicates that there is a high degree of association between carcass traits and the proximate composition.

Standardized canonical coefficients or canonical weights of original variables represent their relative contribution to the corresponding canonical variates U (named as "PROXIMATE") and V (named as "CARCASS"), respectively. Standardized canonical coefficient or canonical weights of TLIPIDS had the greatest contribution to the canonical variate "PROXIMATE" (Table 6). The correlation coefficient (−0.993) also indicated that this variable makes an important contribution to the constitution of this canonical variate.

Table 6. Standardized canonical coefficient and canonical correlation coefficient between the original variables and its canonical variate "PROXIMATE".

Original Variables	Standardized Canonical Coefficient	Canonical Correlation Coefficient
DM	0.230	−0.502
Moisture	0.283	0.510
Ash	0.088	0.071
CP	0.034	0.173
TLIPIDS	−0.966	−0.993

Description of acronyms is presented in Table 1.

Table 7 shows the standardized coefficients and correlations coefficients between the original carcass traits and the canonical variate "CARCASS". The variables MARBLING and BFT were the largest contributors to the formation of this canonical variate ($r = -0.836$ and $r = -0.855$, respectively). On the other hand, the variable CFINISH had a moderate contribution to the canonical variate "CARCASS" ($r = 0.751$). The very low standardized coefficients of SM, LM, and CW indicated an irrelevant contribution of these variables to the canonical variate "CARCASS".

Table 7. Standardized canonical coefficient (canonical weights), and canonical correlation coefficient between the original variables and the canonical variate "CARCASS".

Original Variables	Standardized Canonical Coefficient	Canonical Correlation Coefficient
SM	0.143	−0.023
LM	−0.037	−0.099
CW	0.027	−0.196
CFINISH	0.297	0.751
BFT	−0.424	−0.855
MARBLING	−0.501	−0.836

Description of acronyms is presented in Table 1.

Canonical cross correlation describes the correlation between variables and the opposite canonical variate. The variable TLIPIDS showed the highest canonical cross-loading with the canonical variate "CARCASS" (Table 8). On the other hand, the variables MARBLING, BFT, and CFINISH showed an important canonical cross correlation with the canonical variate "PROXIMATE". This result indicates a strong and significant linear correlation between these four variables.

Table 8. Cross correlations (canonical cross loadings) between the variables and the opposite canonical variate.

Original Variables	Canonical Variate "CARCASS"	Original Variables	Canonical Variate "PROXIMATE"
DM	−0.316	SM	−0.014
Moisture	0.320	LM	−0.062
Ash	0.045	CW	−0.123
CP	0.109	CFINISH	0.472
TLIPIDS	−0.624	BFT	−0.538
		MARBLING	−0.526

Description of acronyms is presented in Table 1.

The variance of the canonical variate "PROXIMATE" associated with the variables of its own group represented 30.70% of the total data variation, which could be attributed to the high loading value of the variable TLIPIDS. The cross variance between "PROXIMATE" and "CARCASS" only accounted for 12.10% of the total variance; a low value that is also associated with the contribution of the TLIPIDS variable. The contribution of the canonical variate "CARCASS" was 34.10% of the total variance. The canonical correlation between PROXIMATE and CARCASS subgroups of variables aligns with the results obtained from the linear correlation between the variables related to fatness (Table 9). The correlation coefficients are all significant and moderate to strong, with values very close to those reached by the components of the canonical variates.

Table 9. Pearson correlation coefficients among carcass fatness-related variables.

Variables	TLIPIDS	CFINISH	BFT	MARBLING
TLIPIDS	1	−0.471 **	0.5532 **	0.519 **
CFINISH	−0.471 **	1	−0.580 **	−0.432 **
BFT	0.532 **	−0.580 **	1	0.549 **
MARBLING	0.519 **	−0.432 **	0.549 **	1

** Significant correlation at $p < 0.01$ (bilateral). Number of observations = 109. Description of acronyms is presented in Table 1.

The CCA between the subgroups of variables related to lipid composition traits and the subgroup of carcass traits revealed six canonical correlations. The first canonical correlation was highly significant ($p = 0.002$), showing a not very strong correlation coefficient ($r = 0.629$), and represented 59.17% of the explained variability. Table 10 shows the standardized canonical and correlation coefficients between the original variables and their canonical variate "LIPIDS".

The C16:0 (loading value = −0.796; $r = -0.912$); and C18:1c (loading value = −0.566; $r = -0.878$) showed the highest correlation coefficients with the canonical variety "LIPIDS". This indicates that Palmitic and Oleic acids were the fatty acids that mostly contributed to the conformation of the canonical variate "LIPIDS". The variable cholesterol content presented a low canonical weight and, therefore, low correlation with its canonical variety.

The main fatty acids (C16:0 and C18:1c) with the highest standardized canonical coefficient also exhibited the highest correlations with the carcass trait variables (Table 11). Other fatty acids, like C14:0, C20:0 and C18:1t presented moderate correlations, but with lesser impact given their low standardized coefficients. On the other hand, carcass traits: CFINISH, BFT, and MARBLING presented a not very strong cross correlation with the canonical variate "LIPIDS" (Table 11).

The variance of the canonical variate "LIPIDS" accounted for 38.8% of the total data variation. The cross variance between "LIPIDS" and "CARCASS" only represented 19.10% of the total variance. This low value is potentially related to a low correlation between these two subgroups of variables, and/or the smaller number of CARCASS variables as compared to the lipidic components.

Six possible canonical correlations were obtained between mineral content and carcass traits; however, none of them were statistically significant. Based on the available data and the statistical technique applied, there is insufficient evidence to demonstrate the existence of any relationship between mineral components and carcass traits.

Table 10. Standardized canonical coefficient (canonical weights), and canonical correlation coefficient between the original variables and their canonical variate "LIPIDS".

Original Variable	Standardized Coefficient	Correlation Coefficient
CHOLEST	0.082	0.064
C14:0	0.125	−0.738
C14:1	0.065	−0.517
C15:0	−0.030	−0.742
C16:0	−0.796	−0.912
C16:1	0.127	−0.501
C18:0	−0.032	−0.798
C18:1*c*	−0.566	−0.878
C18:1*t*	0.144	−0.679
C18.2	0.257	−0.549
α-C18:3	−0.073	−0.542
C20:0	−0.072	−0.474
C20:4ω6	−0.226	−0.423
C22:6ω3	0.047	−0.344

Description of acronyms is presented in Table 1.

Table 11. Cross correlations (canonical cross loadings) between the original variables and "CARCASS" and "LIPIDS" canonical variates.

OriginalVariables	"CARCASS"	Variables	"LIPIDS"
CHOLEST	0.045	SM	0.026
C14:0	−0.518	LM	−0.036
C14:1	−0.363	CW	−0.177
C15:0	−0.521	CFINISH	0.603
C16:0	−0.640	BFT	−0.567
C16:1	−0.352	MARBLING	−0.534
C18:0	−0.560		
C18:1c	−0.617		
C18:1t	−0.477		
C18:2	−0.385		
α-C18:3	−0.380		
C20:0	−0.332		
C20:4ω6	−0.297		
C22:6ω3	−0.241		

Description of acronyms is presented in Table 1.

4. Discussion

The carcasses evaluated in this study are representative of South American tropical cattle fattened on pasture, which presents more variation in degrees of fatness, carcass finish, and conformation traits than their counterparts that are subjected to more standardized, intense feeding protocols. The results obtained in this study also concur with carcass characteristics [18,57,58] and nutrient values [59–62] reported in previous studies conducted in Venezuela with samples of *Longissimus lumborum* taken from crossbred cattle varying in age, sex condition, and diet. Also, these values are similar to those reported for the fresh *Longissimus* muscle derived from Bangladeshi beef (zebu type) finished on pasture [63] and other types of tropical cattle [15]. The mineral content found in this study is within the ranges reported for raw meat from tropical cattle subjected to different management practices [3,15,64,65].

The relatively low Ca content in these beef samples is potentially related to the quality of vegetation consumed/grazed by these animals. Pastures and forages constitute the main food sources for fattening cattle in Venezuela. The nutritional value of pastures depends on the amount of nutrients present in these plant species, which are absorbed from the soil; consequently, it will be the characteristics of the soil that defines the development of the plant with respect to the concentration and availability of the mineral elements present.

According to Araujo [66], tropical grasses can hardly supply all the minerals in amounts adequate to the needs of the animals. The factors that affect the mineral content in forages, in addition to the soil, are: the forage species, the age of the pasture, the yield, the management of the pastures, and climate [66]. Low fertility and high acidity stand out among the most important limitations in quality of most soils in Venezuela. Research carried out in Venezuela to study to the state of mineral nutrition in livestock systems [67,68], showed that calcium levels in pastures were generally poor and this deficiency is reflected in the cattle's animal tissues. A review of the nutritional value of beef produced under tropical conditions [15], also confirms the low concentrations of this mineral.

The relationship between meat sensory traits and physicochemical characteristics has been studied using multivariate analysis techniques in beef [41,69]; however, no available information was found about the relationships of carcass traits with the chemical composition of the derived meat.

The first HCA allowed a simple representation of the total data with 32 variables and to explore how the variables correlated. The second HCA grouped all the observations by similarities in three clusters with a significant variation in BFT, TLIPIDS, mineral and fatty acid composition. It has been demonstrated that sex greatly influences protein and fat deposition in cattle and defines distinct differences in body composition [27,70,71]; nevertheless, our observations suggest that in grass-fed tropical cattle, the expected differences in body and lipid composition between sexes are not as noticeable. In fact, our results suggest that BFT, TPLIPIDS, and fatty acid composition represented the main variables that defined the clusters. Cluster 3 contained a similar proportion of bulls and steers with the lowest values in BFT and total lipids.

It is noteworthy to highlight that this was not a controlled experiment designed to compare sex conditions. Therefore, genetics, management (slaughter weight and age), nutrition and other confounding factors [72] could affect this type of non-controlled comparison. The inclusion of bulls (intact males) and steers (castrates) in the study was just to provide a balanced random sample of these two sex conditions of slaughter male cattle in the country. Since the fat content is the most variable proximate component in beef, we can hypothesize that the low variability of the sample in levels of marbling (and as a result of IMF), is responsible for this outcome. The low variability in IMF is likely due to two additive factors: genetics and plane of nutrition. Both sex conditions had a common genetic background (*Bos indicus*) and regardless of sex, it is known that *Bos indicus*-influenced cattle individuals, exhibits lower levels of marbling when compared to *Bos taurus* biological types. The second factor could be the low energy content of the grass-based diet which did not facilitate a greater expression of the inherent differences between the sex conditions in terms of lipid content. Finally, it should also be considered that only separable lean was used for chemical analyses. Otherwise, if the meat sample had not been devoid of the surrounding subcutaneous and intermuscular fat depots, the differences between the sex conditions would have been more noticeable because the steers would give loins with a more significant amount of fat per 100 g of fresh tissue than bulls. Needless to say, a relatively greater fat content would bring a concomitant reduction in the concentration of other proximate components.

There is a consensus that the use of a CCA as a multivariate approach is appropriate for evaluating the interrelations among meat quality and carcass traits [41,73]. For this study, CCA is suitable because it measures the magnitude of interrelations between sets of multiple variables [74]. The CCA allows for studying the interrelationship among groups of multiple independent variables and determines the magnitude of the relationships that may exist between subgroups.

We analyzed canonical correlations by paired groups of variables: proximate composition, lipids, and mineral contents with carcass traits. The correlation coefficients between the original variables and the canonical variates allow establishing the weight of each original variable in the conformation of the canonical variate. The main variables that define the "CARCASS" canonical variate were BFT and MARBLING, meaning that these two variables had the highest contribution to the corresponding canonical variate. TLIPIDS was the main contributing variable for the canonical variate "PROXIMATE", while C16:0 and C18:1*c* had the largest contribution for the "LIPIDS" canonical variate.

As expected, carcass fatness-related traits (BFT, MARBLING and CFINISH) exhibited the highest cross-correlations with TLIPIDS (Table 7), suggesting that MARBLING is not the only carcass trait that could affect the content of IMF. Canonical weights are important parameters for defining the contribution of original variables to the canonical variates. However, the understanding of canonical loadings and cross-loadings is critical because these values describe the correlation between original and canonical variates. Our findings represent the first evidence of a strong multivariate relationship of quantitative carcass traits with the chemical composition, particularly TLIPIDS, and the fatty acid composition of lean tissue.

In our study, the C16:0 and C18:1*c* components presented the highest correlations with carcass traits. The palmitic acid (C16:0) represents the main product of the *de novo* fatty acids synthesized from carbohydrates and volatile fatty acids of the diet, and it can be elongated to stearic acid (C18:0), and then to arachidic (C20:0) [75]. Also, the HCA revealed a high correlation between C16:1 and TLIPIDS (Figure 1).

According to the available data and the CCA applied, there is not enough evidence to assume that there is an association between mineral component variables and carcass traits. These results validate the findings of the HCA which indicate that carcass traits have a weak correlation with the mineral composition of the meat. The results obtained in this work are comparable with those of Duan et al. [11], who reported weak but significant correlations among beef mineral concentrations and carcass traits. According to Duan et al. [11], Mg concentration was positively correlated ($p < 0.05$) with all carcass traits but negatively correlated with hot carcass weight, while no significant correlation ($p > 0.05$) was detected between contents of Fe or Zn and carcass traits. Garmyn et al. [76] reported significant correlations between Fe, Zn and marbling levels. Castillo et al. [73] reported that the magnitude of the interrelations among protein, fat, and minerals are different between male (castrated and intact males) and female Saanen goats from 5 to 45 kgs live weight. In our study, individual mineral content did not correlate ($p > 0.05$) with any carcass trait. Age-related carcass traits (carcass weight, skeletal and lean maturity) were not correlated ($p > 0.05$) with the proximate, mineral or lipidic compositions.

BFT represents between 10 to 13% of total carcass fat tissue and it is dependent on genetics, nutrition, and finishing systems of ruminants; also, these may be influenced by sex and age, considering that the nutrient dynamics in the ruminant's body differs between sexes, and these differences become more evident with age [77]. A meta-analysis study by Al-Jammas et al. [78] reported that BFT and USDA yield grade were the variables most highly related to changes in the weight of adipose tissues in the carcass, suggesting that variations in USDA yield grade and BFT may properly explain the differences in meat chemical composition.

The process of IMF deposition depends on many factors such as sex condition, age, and nutrition [72,79,80]. MARBLING, was also significantly and strongly ($r > 0.5$) related to TLIPIDS, despite exhibiting a very low range of scores [between 1 (Practically devoid) and 3 (Slight); Table 2]. Most likely the reason why marbling was barely second to BFT in the r value (Table 9) was due to the aforementioned narrower range of marbling variation present in these carcasses derived from *Bos indicus*-influenced, grass-fed cattle. In fact, BFT data (Table 2) show a higher coefficient of variation (%) than that of marbling (69.2% vs. 46.4%).

Clearly, the use of a descriptive scale without subdividing each degree of marbling into 100 subunits like the USDA counterpart contributed to diminish its variation in our analysis. Differences in methods of assessment of MARBLING can affect its correlation with IMF%. For instance, Giaretta et al. [81] found that the IMF was more correlated with the percentage of marbling evaluated by the J-image analysis ($r = 0.62$) than when the USDA scores were used ($r = 0.56$) while Kruk et al. [82] had reported that the Meats Standards Australia (MSA) marbling ratings were poorly associated with IMF% when compared to other scoring systems. The correlations with IMF% ranged from 0.67–0.79 in this study [82]; however, the authors commented that in other Australian studies, the correlation coefficients with the marbling scores of the Australian Authority for the Uniform Specifications of Meat and Livestock (AUS-MEAT) were lower, ranging from 0.32 to 0.57. Another reason for the not very strong correlation detected between MARBLING and IMF in our study might be related to the very nature of these very lean meats (where "Practically devoid" and "Traces" levels of marbling comprised two thirds of the filtered data). Brackebusch et al. [83] reported a strong linear association between IMF (%) and marbling while Kornasla et al. [84] found that marbling percentage was not very strongly correlated with chemical IMF% ($r = 0.60$). In fact, Silva et al. [85] pointed out that the association between marbling and IMF is not very strong because part of the IMF is invisible and it also depends on the size and shape of the marbling specks. Furthermore, differences in methodologies of lipid extraction may explain the discrepancies among values reported for correlations between IMF% and marbling levels. According to Siebert et al. [86] when meat is low in fat, significantly more total lipids are extracted with polar solvent mixtures (e.g., chloroform:methanol) due to the phospholipid content of tissue membranes. Therefore, it can be speculated that the total lipid extract of these very lean meats having a much higher component of invisible membrane lipids (e.g., phospholipids and lipoproteins) makes it difficult to achieve the stronger correlations (with the marbling scores) found in highly-marbled carcasses.

Rhee et al. [53] reported the relationships of MARBLING (with eight levels from "Moderately Abundant" to "Practically Devoid") to CHOLEST of beef *longissimus* muscle, and showed that only raw steaks with "Practically Devoid" MARBLING level contained significantly less cholesterol (on wet basis) than did raw steaks with any of the other seven MARBLING scores. In the present study, CHOLEST showed a low loading value, indicating a weak association with carcass traits. Catillo et al. [87] reported a close relationship between pork carcass leanness as defined by the EUROP classification system, and the fatty acid composition of backfat. These authors [87] found that as the lean meat content of the carcass decreased from class E to class O, the backfat total content of SFA increased by more than 4 percent, while the total PUFA content decreased about the same percentage; however, we could not find similar reports in beef. Elucidation of the multivariate and quantitative relationship between BFT, MARBLING and fatty acid composition may be useful for a better understanding of the roles of fat deposition on the nutritional composition of beef produced under the conditions described herein.

5. Conclusions

Canonical correlation analysis is an optimized multivariate technique for evaluating the existence or non-existence of relationships between complex groups of variables. In this study it proves to be a powerful tool to study the relationship between the selected set of carcass traits and the proximate, lipid and mineral components, particularly when it is expected that certain degree of interaction exists among these three groups of chemical variables. This work demonstrates an important relationship between backfat thickness, marbling and the content of total lipids and fatty acids in beef LDL muscle from cattle fed on pastures under tropical conditions. Instead, carcass traits are poorly associated with beef mineral content. These findings allow for the possibility to develop selection criteria based on BFT and/or marbling to sort carcasses with differing nutritional values. For the experience gained in beef carcass grading in Venezuela, the evaluation of marbling levels requires more intense training and supervision of graders than the BFT measurement.

Moreover, marbling is seldom used to grade beef carcasses in the developing, tropical countries. Therefore, it is more practical to use BFT in future regression analyses to explain the variation in lipid composition of beef *longissimus* muscle.

Further analyses are needed to determine the potential influence of sex condition on the magnitude of the associations among carcass traits and beef fatty acid composition.

Supplementary Materials: The following are available online at https://www.mdpi.com/article/10.3390/foods10061364/s1, Figure S1: Graphical description of sample preparation and chemical analyses; Figure S2: Dendrogram for samples from the hierarchical cluster analysis identifying clusters. The samples were numbered randomly from 1 to 109; Figure S3: Projection of variables from the subgroup proximate composition by clusters; Figure S4: Projection of variables from the subgroup mineral content by clusters; Figure S5: Projection of variables from the subgroup lipid profile by clusters; Figure S6: Projection of variables from the subgroup carcass traits by clusters.

Author Contributions: N.H.-L. designed the investigation and field data that supported this research. L.A.d.M. was responsible for the selection of chemical procedures and training of personnel who conducted the chemical analyses and collected lab data that support this research. S.U.-B. conducted the chromatographic analysis of the samples. N.J.-T., E.M.-V. and L.A.d.M. performed statistical analyses, and tabulated results. L.A.d.M., N.J.-T., M.G.-M. and N.H.-L. interpreted, designed, and revised the structure of the manuscript. All authors searched and reviewed the literature, discussed the contents of the manuscript, and approved submission. All authors have read and agreed to the published version of the manuscript.

Funding: This research was funded by the Consejo de Desarrollo Científico y Humanístico de la Universidad del Zulia (CONDES-LUZ), AND Facultad de Agronomía, Universidad del Zulia, Venezuela.

Institutional Review Board Statement: Data collection was carried on in compliance with the criteria for animal care and welfare described in the Bioethics and Biosafety guide of the Fondo Nacional de Ciencia, Innovación y Tecnología (FONACIT) of Venezuela.

Data Availability Statement: Data are not available in public datasets, please contact the authors.

Acknowledgments: The authors are grateful to Matadero Industrial Centro Occidental (MINCO) for their valuable support during the live and carcass data collection, and donation of beef samples. We thank Martin O'Connors for providing further training in carcass evaluation according to the USDA standards. The authors also gratefully acknowledge the cooperation of Daniel Huerta-Sánchez (Department of Economics and Finance, Florida Gulf Coast University) in the review of technical English to improve the statistical discussion, and Javier Aracena in the editing of the supplementary figures.

Conflicts of Interest: The authors declare no conflict of interest.

References

1. Giuffrida-Mendoza, M.; Arenas de Moreno, L.; Uzcátegui-Bracho, S.; Rincón-Villalobos, G.; Huerta-Leidenz, N. Mineral content of *longissimus dorsi thoracis* from water buffalo and Zebu-influenced cattle at four comparative ages. *Meat Sci.* **2007**, *75*, 487–493. [CrossRef]
2. Binnie, M.A.; Barlow, K.; Johnson, V.; Harrison, C. Red meats: Time for a paradigm shift in dietary advice. *Meat Sci.* **2014**, *98*, 445–451. [CrossRef] [PubMed]
3. Giuffrida-Mendoza, M.; Arenas de Moreno, L.; Huerta-Leidenz, N. Composición nutritiva de la carne de ganado tropical venezolano. *An. Venez. Nutr.* **2014**, *27*, 167–176.
4. Williams, P. Nutritional composition of red meat. *Nutr. Diet.* **2007**, *64*, S113–S119. [CrossRef]
5. Klurfeld, M.D. What is the role of meat in a healthy diet? *Anim. Front.* **2018**, *8*, 5–10. [CrossRef]
6. Balić, A.; Vlaŝi, D.; Žužul, K.; Marinović, B.; Mokos, Z.B. Omega-3 Versus Omega-6 Polyunsaturated Fatty Acids in the prevention and treatment of inflammatory skin diseases. *Int. J. Mol. Sci.* **2020**, *21*, 741. [CrossRef] [PubMed]
7. Oppedisano, F.; Macri, R.; Gliozzi, M.; Musolino, V.; Carresi, C.; Maiuolo, J.; Bosco, F.; Nucera, S.; Zito, M.C.; Guarnieri, L.; et al. The anti-inflammatory and antioxidant properties of n-3 PUFAs: Their role in cardiovascular protection. *Biomedicines* **2020**, *8*, 306. [CrossRef] [PubMed]
8. Watanabe, Y.; Tatsuno, I. Omega-3 polyunsaturated fatty acids focusing on eicosapentaenoic acid and docosahexaenoic acid in the prevention of cardiovascular diseases: A review of the state-of-the-art. *Expert. Rev. Clin. Pharmacol.* **2021**, *14*, 79–93. [CrossRef] [PubMed]

9. Bureš, D.; Bartoň, L. Growth performance, carcass traits and meat quality of bulls and heifers slaughtered at different ages. *Czech J. Anim. Sci.* **2012**, *57*, 34–43. [CrossRef]
10. Cafferky, J.; Hamill, R.M.; Allen, P.; O'Doherty, J.V.; Cromie, A.; Torres, S. Effect of Breed, and gender on meat quality of *M. longissimus thoracis et lumborum* muscle from Crossbred Beef Bulls and Steers. *Foods* **2019**, *8*, 173. [CrossRef]
11. Duan, Q.; Tait, R.G., Jr.; Schneider, M.J.; Beitz, D.C.; Wheeler, T.L.; Shackelford, S.D.; Cundiff, L.V.; Reecy, J.M. Sire breed effect on beef longissimus mineral concentrations and their relationships with carcass and palatability traits. *Meat Sci.* **2015**, *106*, 25–30. [CrossRef] [PubMed]
12. Jung, E.-Y.; Hwang, Y.-H.; Joo, S.-T. The relationship between chemical compositions, meat quality, and palatability of the 10 primal cuts from Hanwoo steer. *Korean J. Food Sci. An.* **2016**, *36*, 145–151. [CrossRef]
13. Karakök, S.G.; Ozogul, Y.; Saler, M.; Ozogul, F. Proximate analysis. fatty acid profiles and mineral contents of meats: A comparative study. *J. Muscle Foods* **2010**, *21*, 210–223. [CrossRef]
14. Listrat, A.; Lebret, B.; Louveau, I.; Astruc, T.; Bonnet, M.; Lefaucheur, L.; Picard, B.; Bugeon, J. How muscle structure and composition influence meat and flesh quality. *Sci. World J.* **2016**, *3182746*, 1–14. [CrossRef] [PubMed]
15. Rubio-Lozano, M.S.R.; Ngapo, T.M.; Huerta-Leidenz, N. Tropical Beef: Is there an axiomatic basis to define the concept? *Foods* **2021**, *10*, 1025. [CrossRef] [PubMed]
16. Guerrero, A.; Valero, M.V.; Campo, M.M.; Sañudo, C. Some factors that affect ruminant meat quality: From the farm to the fork. Review. *Acta Sci. Anim. Sci.* **2013**, *35*, 335–347. [CrossRef]
17. Hwang, Y.-H.; Joo, S.-T. Fatty acid profiles, meat quality, and sensory palatability of grain-fed and grass-fed beef from Hanwoo, American, and Australian crossbred cattle. *Korean J. Food Sci. Anim. Resour.* **2017**, *37*, 153–161. [CrossRef]
18. Jerez-Timaure, N.; Huerta-Leidenz, N. Effects of breed type and supplementation during grazing on carcass traits and meat quality of bulls fattened on improved savannah. *Livest. Sci.* **2009**, *121*, 219–226. [CrossRef]
19. Comparin, M.A.S.; Morais, M.G.; Fernandes, H.J.; Coelho, R.G.; Coutinho, M.A.S.; Ribeiro, C.B.; Menezes, B.B.; Rocha, R.F.A.T. Chemical composition, and fatty acid profile of meat from heifers finished on pasture supplemented with feed additives. *Rev. Bras. Saúde Produção Anim.* **2015**, *16*, 606–616. [CrossRef]
20. Daley, C.A.; Abbott, A.; Doyle, P.S.; Nader, G.A.; Larson, S. A review of fatty acid profiles and antioxidant content in grass-fed and grain-fed beef. *Nutr. J.* **2010**, *9*, 1–12. [CrossRef]
21. Giuffrida-Mendoza, M. Beneficios de la alimentación a pastoreo en la calidad nutritiva de la carne del ganado doble propósito. In *Desarrollo Sostenible de la Ganadería de Doble Propósito*; González-Stagnaro, C., Soto Belloso, E., Eds.; Fundación Grupo de Investigadores de la Reproducción animal en la Región Zuliana: Maracaibo, Venezuela, 2008; pp. 852–863.
22. Uzcátegui-Bracho, S.; Rodas-González, A.; Hennig, K.; Arenas de Moreno, L.; Leal, M.; Vergara-López, J.; Jerez-Timaure, N. Composición proximal, mineral y contenido de colesterol del músculo *Longissimus dorsi* de novillos criollo limonero suplementados a pastoreo. *Rev. Cien.* **2008**, *18*, 589–594.
23. Leheska, J.M.; Thompson, L.D.; Howe, J.C.; Hentges, E.; Boyce, J.; Brooks, J.C.; Shriver, B.; Hoover, L.; Miller, M.F. Effects of conventional and grass feeding systems on the nutrient composition of beef. *J. Anim. Sci.* **2008**, *86*, 3575–3585. [CrossRef] [PubMed]
24. Provenza, F.D.; Scott, L.; Kronberg, S.L.; Gregorini, P. Is grassfed meat and dairy better for human and Environmental Health? *Front. Nutr.* **2019**, *6*, 1–12. [CrossRef] [PubMed]
25. Huerta-Leidenz, N.; Rios, G. Bovine castration at different stages of growth. II. Carcass characteristics. *Rev. Fac. Agron. LUZ* **1993**, *10*, 163–187.
26. Ruiz, M.R.; Matsushita, M.; Visentainer, J.V.; Hernandez, J.A.; Ribeiro, E.L.d.A.; Shimokomaki, M.; Reeves, J.J.; deSouza, N.E. Proximate chemical composition and fatty acid profiles of *Longissimus thoracis* from pasture fed LHRH immunocastrated, castrated and intact *Bos indicus* bulls. *S. Afr. J. Anim. Sci.* **2005**, *35*, 13–18.
27. Aricetti, J.A.; Rotta, P.P.; Prado, R.M.D.; Perotto, D.; Moletta, J.L.; Matsushita, M.; Prado, I.N.D. Carcass characteristics chemical composition and fatty acid profile of longissimus muscle of bulls and steers finished in a pasture system bulls and steers finished in pasture systems. *Asian Australas. J. Anim. Sci.* **2008**, *21*, 1441–1448. [CrossRef]
28. Eichhorn, J.M.; Coleman, L.J.; Wakayama, E.J.; Blomquist, G.J.; Bailey, C.M.; Jenkins, T.G. Effects of breed type and restricted versus ad libitum feeding on fatty acid composition and cholesterol content of muscle and adipose tissue from mature bovine females. *J. Anim. Sci.* **1986**, *63*, 781–794. [CrossRef] [PubMed]
29. Polkinghorne, R.J.; Thompson, J.M. Meat standards and grading. A world view. *Meat Sci.* **2010**, *86*, 227–235. [CrossRef]
30. Rodas-González, A.; Huerta-Leidenz, N.; Jerez-Timaure, N. Benchmarking Venezuelan quality grades for grass-fed cattle carcasses. *Meat Muscle Biol.* **2017**, *1*, 71–80. [CrossRef]
31. Butler, G.; Ali, A.M.; Oladokun, S.; Wang, J.; Davis, D. Forage-fed cattle point the way forward for beef? *Future Foods* **2021**, *3*, 100012. [CrossRef]
32. Jeong, J.-Y.; Jeong, T.-C.; Yang, H.-S.; Kim, G.-D. Multivariate analysis of muscle fiber characteristics, intramuscular fat content and fatty acid composition in porcine *longissimus thoracis* muscle. *Livest. Sci.* **2017**, *202*, 13–20. [CrossRef]
33. Patel, N.; Bergamaschi, M.; Magro, L.; Petrini, A.; Bittante, G. Relationships of a detailed mineral profile of meat with animal performance and beef quality. *Animals* **2019**, *9*, 1073. [CrossRef]
34. Chen, D. Analysis of input and output of China's agriculture based on canonical correlation. *Asian J. Agric. Res.* **2011**, *3*, 9–15.

35. Kim, T.W.; Kim, C.W.; Noh, C.W.; Kim, S.W.; Kim., I.S. Identification of association between supply of pork and production of meat products in Korea by canonical correlation analysis. *Korean J. Food Sci. Anim. Resour.* **2018**, *38*, 794–805. [CrossRef] [PubMed]
36. Saba, J.; Tavana, S.; Qorbanian, Z.; Shadan, E.; Shekari, F.; Jabbari, F. Canonical Correlation Analysis to Determine the Best Traits for Indirect Improvement of Wheat Grain Yield under Terminal Drought Stress. *JAST* **2018**, *20*, 1037–1048. Available online: http://jast.modares.ac.ir/article-23-19920-en.html (accessed on 11 June 2021).
37. Tukimat, N.N.A.; Harun, S.; Tadza, M.Y.M. The potential of canonical correlation analysis in multivariable screening of climate model. *IOP Conf. Ser. Earth Environ. Sci.* **2019**, *365*, 012025. [CrossRef]
38. Hair, J.F.; Black, W.C.; Babin, B.J.; Anderson, R.E. *Multivariate Data Analysis. A Global Perspective*, 7th ed.; Pearson/Prentice Hall: Upper Saddle River, NJ, USA, 2009; ISBN 978-0138132637.
39. Diel, M.I.; Lúcio, A.D.; Lambrecht, D.M.; Pinheiro, M.V.M.; Sari, B.G.; Olivoto, T.; Valeriano, O.; de Melo, P.J.; de Lima Tartaglia, F.; Luis, A. Canonical correlations in agricultural research: Method of interpretation used leads to greater reliability of results. *IJERI* **2020**, *8*, 171–181. [CrossRef]
40. Wickramasinghe, N.D. Canonical correlation analysis: An introduction to a multivariate statistical analysis. *JCCPSL* **2019**, *25*, 37–38. [CrossRef]
41. Vargas, J.A.C.; Coutinho, J.E.S.; Gomes, D.I.; Alves, K.S.; Maciel, R.P. Multivariate relationship among pH, subcutaneous fat thickness, and color in bovine meat using canonical correlation analysis. *Rev. Colomb. Cienc. Pecu.* **2021**, *34*. [CrossRef]
42. Montero, A.; Huerta-Leidenz, N.; Rodas-González, A.; Arenas de Moreno, L. Deshuese y variación del rendimiento carnicero de canales bovinas en Venezuela: Descripción anatómica el proceso y nomenclatura de cortes equivalentes a los correspondientes norteamericanos. *Nacameh* **2014**, *8*, 1–22. Available online: http://nacameh.cbsuami.org (accessed on 22 February 2021). [CrossRef]
43. Ministerio del Poder Popular para Ciencia, Tecnología e Industrias Intermedias y Fondo Nacional de Ciencia, Tecnología e Innovación (MCT-FONACIT). *Código de Bioética y Bioseguridad*, 2nd ed.; Ministerio del Poder Popular para Ciencia, Tecnología e Industrias Intermedias y el Fondo Nacional de Ciencia, Tecnología e Innovación: Caracas, Venezuela, 2002; pp. 1–35. Available online: https://cupdf.com/download/bioetica-fonacit (accessed on 3 April 2021).
44. Comisión Venezolana de Normas Industriales. *Norma Venezolana 2072-83. Ganado Bovino. Inspección Postmortem*; FONDONORMA: Caracas, Venezuela, 1983; pp. 1–10. Available online: http://www.sencamer.gob.ve/sencamer/normas/2072-83.pdf (accessed on 25 February 2021).
45. United States Department of Agriculture (USDA). *Official United States Standards for Grades of Carcass Beef*; Agricultural Marketing Service: Washington, DC, USA, 2017. Available online: https://www.ams.usda.gov/grades-standards/carcass-beef-grades-and-standards (accessed on 11 June 2021).
46. Huerta Leidenz, N.; Alvarado, E.; Martínez, L.; Rincón, E. Conformación, acabado y características biométricas de la canal de diferentes clases de bovinos sacrificados en el Estado Zulia. *Rev. Fac. Agron. (LUZ)* **1979**, *5*, 522–536. Available online: https://produccioncientificaluz.org/index.php/agronomia/article/view/25841 (accessed on 25 January 2021).
47. *Decreto Presidencial No. 181: Gaceta Oficial de la República de Venezuela*; N° 35-486; Ministerio de Agricultura y Cria: Caracas, Venezuela, 1994.
48. Comisión Venezolana de Normas Industriales (COVENIN). *Norma Venezolana 792-82: Carne de Bovino. Definición e Identificación de las Piezas de una Canal*; FONDONORMA: Caracas, Venezuela, 1982; pp. 1–10. Available online: http://www.sencamer.gob.ve/sencamer/normas/792-82.pdf (accessed on 22 February 2021).
49. Uzcátegui-Bracho, S.; Huerta-Leidenz, N.; Arenas de Moreno, L.; Colina, G.; Jerez-Timaure, N. Contenido de humedad, lípidos totales y ácidos grasos del músculo *longissimus* crudo de bovinos en Venezuela. *Arch. Latinoamer. Nutr.* **1999**, *49*, 171–180.
50. Association of Official Analytical Chemists. *Official Methods of Analysis*, 15th ed.; Helrich, K., Ed.; Association of Official Analytical Chemists: Washington, DC, USA, 1990; Section 960.39; ISBN 978-093-558-442-4.
51. Folch, J.; Lees, M.; Stanley, G.H.S. A simple method for the isolation and purification of total lipid from animal tissues. *J. Biol. Chem.* **1957**, *226*, 497–509. [CrossRef]
52. Slover, H.T.; Lanza, E.; Thompson, R.H.; Davis, C.S.; Merola, G.V. Lipids in raw and cooked beef. *J. Food Compos. Anal.* **1987**, *1*, 26–37. [CrossRef]
53. Rhee, K.S.; Thayne, R.; Dutson, T.R.; Smith, G.C.; Hostetler, R.L.; Reiser, R. Cholesterol content of raw and cooked beef Longissimus muscles with different degrees of marbling. *J. Food Sci.* **1982**, *47*, 716–719. [CrossRef]
54. Slover, H.T.; Lanza, E. Quantitative Analysis of Food Fatty Acids by Capillary Gas Chromatography. *J. Am. Oil Chem. Soc.* **1979**, *56*, 933–943. [CrossRef]
55. Morrison, W.R.; Smith, L.M. Preparation of fatty acid methyl esters and dimethyl acetals from lipids with boron fluoride-methanol. *J. Lipid. Res.* **1964**, *5*, 600–608. [CrossRef]
56. *IBM SPSS Statistics 23*; IBM Corporation: Armonk, NY, USA, 2019.
57. Atencio-Valladares, O.; Huerta-Leidenz, N.; Jerez-Timaure, N. Predicción del rendimiento en cortes de carnicería de bovinos venezolanos. *Rev. Cien.* **2008**, *18*, 704–714. Available online: https://produccioncientificaluz.org/index.php/cientifica/article/view/15416 (accessed on 22 March 2021).
58. Huerta-Leidenz, N.; Rodriguez, R.; Vidal-Ojeda, A.; Vidal-Quintero, A.; Jerez-Timaure, N. Características cárnicas de búfalos de agua vs. vacunos acebuados. *Arch. Latinoam. Prod. Anim.* **1997**, *5* (Suppl. S1), 574–576.

59. Arenas de Moreno, L.; Vidal, A.; Huerta-Sánchez, D.; Navas, Y.; Uzcátegui-Bracho, S.; Huerta-Leidenz, N. Análisis comparativo proximal y de minerales entre carnes de iguana, res y pollo. *Arch. Latinoam. Nutr.* **2000**, *50*, 409–415.
60. Arenas de Moreno, L.; Ormos-Moreno, R.; Milli-París, S.; Huerta-Leidenz, N.; Uzcátegui-Bracho, S. Efecto de la dieta sobre la composición química de la carne de terneros. *Rev. Cien.* **2000**, *10*, 448–452.
61. Huerta-Montauti, D.; Villa, V.; Arenas de Moreno, L.; Rodas-González, A.; Giuffrida-Mendoza, M.; Huerta-Leidenz, N. Proximate and mineral composition of imported versus domestic beef cuts for restaurant use in Venezuela. *J. Muscle Foods* **2007**, *18*, 237–252. [CrossRef]
62. Uzcátegui-Bracho, S.; Giuffrida-Mendoza, M.; Arenas de Moreno, L.; Jerez-Timaure, N. contenido proximal, lípidos y colesterol de las carnes de res, cerdo y pollo obtenidas de expendios carniceros de la zona sur de Maracaibo. *RVTS* **2010**, *3*, 13–29.
63. Alam, M.K.; Rana, Z.H.; Akhtaruzzaman, M. Comparison of muscle and subcutaneous tissue fatty acid composition of Bangladeshi nondescript deshi bulls finished on pasture diet. *J. Chem.* **2017**, *2017*, 1–6. [CrossRef]
64. Huerta-Leidenz, N.; Arenas de Moreno, L.; Morón-Fuenmayor, O.; Uzcátegui-Bracho, S. Composición mineral del músculo *longissimus* crudo derivado de canales bovinas producidas y clasificadas en Venezuela. *Arch. Latinoam. Nutr.* **2003**, *53*, 96–101.
65. Arenas de Moreno, L.; Giuffrida-Mendoza, M.; Bulmes, L.; Uzcátegui-Bracho, S.; Huerta-Leidenz, N.; Jérez-Timaure, N. Efecto de la suplementación estratégica, régimen de implantes y condición sexual sobre la composición proximal y mineral de carne de bovinos cruda y cocida. *Rev. Cien.* **2008**, *18*, 65–72.
66. Araujo, O. La nutrición mineral del ganado vacuno. In *Desarrollo Sostenible de la Ganadería de Doble Propósito*; González-Stagnaro, C., Soto Belloso, E., Eds.; Fundación Grupo de Investigadores de la Reproducción animal en la Región Zuliana: Maracaibo, Venezuela, 2008; pp. 463–475.
67. Depablos, L.; Godoy, S.; Claudio, F.; Chicco, C.F.; Ordoñez, J. Nutrición mineral en sistemas ganaderos de las sabanas centrales de Venezuela. *Zootec. Trop.* **2009**, *27*, 25–37.
68. López, M.; Godoy, S.; Alfaro, C.; Chicco, C.F. Evaluación de la nutrición mineral en sabanas bien drenadas al Sur del estado Monagas, Venezuela. *Rev. Cien.* **2008**, *18*, 197–206.
69. Combes, S.; González, I.; Déjean, S.; Baccini, A.; Jehl, N.; Juin, H.; Cauquil, L.; Gabinaud, B.; Lebas, F.; Larzul, C. Relationships between sensory and physicochemical measurements in meat of rabbit from three different breeding systems using canonical correlation analysis. *Meat Sci.* **2008**, *80*, 835–841. [CrossRef]
70. Byers, F.M. Nutritional factors affecting growth of muscle and adipose tissue in ruminants. *Fed. Proc.* **1982**, *41*, 2562–2566.
71. Seideman, S.C.; Cross, H.R.; Oltjen, R.R.; Schanbacher, B.D. Utilisation of the intact male for red meat production: A review. *J. Anim. Sci.* **1982**, *55*, 826–840. [CrossRef]
72. Park, S.J.; Beak, S.H.; Jung, D.J.S.; Kim, S.Y.; Jeong, I.H.; Piao, M.Y.; Kang, H.J.; Fassah, D.M.; Na, S.W.; Yoo, S.P.; et al. Genetic, management, and nutritional factors affecting intramuscular fat deposition in beef cattle—A review. *Asian-Australas J. Anim. Sci.* **2018**, *31*, 1043–1061. [CrossRef]
73. Vargas, J.A.C.; Almeida, A.K.; Härter, C.J.; Souza, A.P.; Fernandes, M.H.M.D.R.; Resende, K.T.D.; Teixeira, I.A.M.D.A. Multivariate relationship among body protein, fat, and macrominerals of male and female Saanen goats using canonical correlation analysis. *Rev. Bras. Zootec.* **2018**, *47*, e20170289. [CrossRef]
74. Ventura, H.T.; Lopes, P.S.; Peloso, J.V.; Guimarães, S.E.F.; Carneiro, A.P.S.; Carneiro, P.L.S. A canonical correlation analysis of the association between carcass and ham traits in pigs used to produce dry-cured ham. *Genet. Mol. Biol.* **2011**, *34*, 451–455. [CrossRef] [PubMed]
75. Marinetti, G.V. Disorders of Fatty Acid Metabolism. In *Disorders of Lipids Metabolism*, 1st ed.; Plenum Press: New York, NY, USA, 1990; pp. 31–48.
76. Garmyn, A.J.; Hilton, G.G.; Mateescu, R.G.; Morgan, J.B.; Reecy, J.M.; Tait, R.G., Jr.; Beitz, D.C.; Duan, Q.; Schoonmaker, J.P.; Mayes, M.S.; et al. Nutrient components and beef palatability: Estimation of relationships between mineral concentration and fatty acid composition of Longissimus muscle and beef palatability traits. *J. Anim. Sci.* **2011**, *89*, 2849–2858. [CrossRef] [PubMed]
77. Rotta, P.P.; Prado, R.M.; Prado, I.N.; Valero, M.V.; Visentainer, J.V.; Silva, R.R. The effects of genetic groups, nutrition, finishing systems and gender of Brazilian cattle on carcass characteristics and beef composition and appearance: A review. *AJAS* **2009**, *22*, 1718–1734. [CrossRef]
78. Al-Jammas, M.; Agabriel, J.; Vernet, J.; Ortigues-Marty, I. Quantitative relationships between the tissue composition of bovine carcass and easily obtainable indicators. In Proceedings of the 61st International Congress of Meat Science and Technology (ICoMST), Clermont-Ferrand, France, 23–28 August 2015.
79. Monteiro, A.C.G.; Santos-Silva, J.; Bessa, R.J.B.; Navas, D.R.; Lemos, J.P.C. Fatty acid composition of intramuscular fat of bulls and steers. *Livest. Sci.* **2006**, *99*, 13–19. [CrossRef]
80. Moloney, A.P.; McGee, M. Factors Influencing the Growth of Meat Animals. In *Lawrie ´s Meat Science*, 8th ed.; Toldrá, F., Ed.; Woodhead Publishing: Cambridge, UK, 2017; pp. 19–47.
81. Giaretta, E.; Mordenti, A.L.; Canestrari, G.; Brogna, N.; Palmonari, A.; Formigoni, A. Assesment of muscle *Longissmus thoracis et lumborum* marbling by image analysis and relationships between meat quality parameters. *PLoS ONE* **2018**, *13*, 1–12. [CrossRef]
82. Kruk, Z.A.; Pitchford, W.S.; Siebert, B.D.; Deland, M.B.P.; Bottema, C.D.K. Factors affecting estimation of marbling in cattle and the relationship between marbling scores and intramuscular fat. *Anim. Prod. Aust.* **2002**, *24*, 129–132.
83. Brackebusch, S.; McKeith, F.; Carr, T.; McLaren, D. Relationship between longissimus composition and the composition of other major muscles of the beef carcass. *J. Anim. Sci.* **1991**, *69*, 631–640. [CrossRef]

84. Kornaska, M.; Kuchida, K.; Tarr, G.; Polkinghorne, R.J. Relationship between marbling measures across principal muscles. *Meat Sci.* **2017**, *123*, 67–78.
85. Silva, S.; Teixeira, A.; Font-i-Furnois, M. Intramuscular fat and marbling. In *A Handbook of Reference Methods for Meat Quality Assessment*, 1st ed.; Font-i-Furnois, M., Candek-Potokar, M., Maltin, C., Prevolnik Povse, M., Eds.; European Cooperation in Science and Technology (COST): Edinburgh, UK, 2015; Chapter 2; pp. 12–21.
86. Siebert, B.D.; Deland, M.P.; Pitchford, W.S. Breed differences in the fatty acid composition of subcutaneous and intramuscular lipid of early and late maturing, grain-finished cattle. *Aust. J. Agric. Res.* **1996**, *47*, 943–952. [CrossRef]
87. Catillo, G.; Zappaterra, M.; Lo Fiego, D.P.; Steri, R.; Davoli, R. Relationships between EUROP carcass grading and backfat fatty acid composition in Italian Large White heavy pigs. *Meat Sci.* **2021**, *171*, 108291. [CrossRef] [PubMed]

Article

Physicochemical and Sensory Assessments in Spain and United States of PGI-Certified *Ternera de Navarra* vs. *Certified Angus Beef*

María José Beriain [1,*], María T. Murillo-Arbizu [1], Kizkitza Insausti [1], Francisco C. Ibañez [1], Christine Leick Cord [2] and Tom R. Carr [3]

[1] Institute of Innovation & Sustainable Development in the Food Chain (IS-FOOD), Arrosadia Campus, Public University of Navarre (UPNA), Jerónimo de Ayanz Building, 31006 Pamplona, Spain; mariateresa.murillo@unavarra.es (M.T.M.-A.); kizkitza.insausti@unavarra.es (K.I.); pi@unavarra.es (F.C.I.)

[2] Nestle Purina PetCare Company, 1 Checkerboard Square, St. Louis, MO 63164, USA; christine.cord@purina.nestle.com

[3] Department of Animal Science, University of Illinois at Urbana-Champaign, Urbana, IL 61801, USA; trcarr1@illinois.edu

* Correspondence: mjberiain@unavarra.es; Tel.: +34-948-169136

Abstract: The physicochemical and sensory differences between the PGI-Certified *Ternera de Navarra* (*CTNA*) (Spanish origin) and *Certified Angus Beef* (*CAB*) (US origin) were assessed in Spain and the USA. To characterize the carcasses, the ribeye areas (REAs), and marbling levels were assessed in both testing places. Twenty striploins per certified beef program were used as study samples. For sensory analysis, the striploins were vacuum packaged and aged for 7 days at 4 °C and 85% RH in each corresponding laboratory. Thereafter, the samples were half cut and frozen. One of the halves was shipped to the other counterpart-testing place. The fat and moisture percentage content, Warner Bratzler Shear Force (WBSF), and total and soluble collagen were tested for all the samples. The *CAB* carcasses had smaller REAs ($p < 0.0001$) and exhibited higher marbling levels ($p < 0.0001$). The *CAB* striploins had a higher fat content ($p < 0.0001$) and required lower WBSF ($p < 0.05$) than the *CTNA* samples. Trained panelists rated the *CAB* samples as juicer ($p < 0.001$), more tender/less tough ($p < 0.0001$), and more flavorful ($p < 0.0001$) than the *CTNA* counterparts. This study shows that beef from both countries had medium-high tenderness, juiciness, and beef flavor scores and very low off-flavor scores. Relevant differences found between the ratings assigned by the Spanish and the US panelists suggest training differences, or difficulties encountered in using the appropriate terminology for defining each sensory attribute. Furthermore, the lack of product knowledge (i.e., consumption habits) may have been another reason for such differences, despite the blind sensory evaluation.

Keywords: *Pirenaica*; Protected Geographical Indication; *Ternera de Navarra*; *Certified Angus Beef*; country of origin; USDA standard; sensory profile

Citation: Beriain, M.J.; Murillo-Arbizu, M.T.; Insausti, K.; Ibañez, F.C.; Cord, C.L.; Carr, T.R. Physicochemical and Sensory Assessments in Spain and United States of PGI-Certified *Ternera de Navarra* vs. *Certified Angus Beef*. *Foods* **2021**, *10*, 1474. https://doi.org/10.3390/foods10071474

Academic Editors: Huerta-Leidenz Nelson and Markus F. Miller

Received: 5 May 2021
Accepted: 20 June 2021
Published: 25 June 2021

Publisher's Note: MDPI stays neutral with regard to jurisdictional claims in published maps and institutional affiliations.

1. Introduction

Globally, meat consumption demand is expected to rise in the coming years because of the population growth and rising incomes in developing countries [1]. According to the United Nations Food and Agriculture Organization, beef ranks third in the world's most consumed meats [2]. The understanding of meat quality factors and consumers preferences is crucial to bridge the gap between the quality approach objectives in the abattoirs, the product physicochemical approach, and the sensory assessment.

Food quality is a complex and multidimensional concept and consumers' quality expectations may not align with the definitions of food producers and academicians. Moreover, consumption habits and preferences may clearly influence the score assigned to meat coming from different origins. Hence, the country of origin and production practices can affect both meat quality and consumer acceptance [3,4]. As reported by Morales et al. [5], low- or trace-marbled beef was acceptable for many Chilean consumers, whereas consumers in Japan and Korea tend to prefer beef with abundant, evenly distributed marbling [6].

Marketing is about differentiation. Provenance and tradition serve to differentiate autochthonous, regional native products. In Europe, many food specialties are recognized by the name of the region, a term that involves the totality of natural factors (environmental conditions), and the activity of its inhabitants ("local know how") that altogether determine the quality of the product [7]. These authors point out that Protected Geographical Indications (PGI) offer a guarantee to consumers, defining the conditions, the procedures, and the extent of protection, and a safeguard to protect the names of regional products. In Spain, the certified PGI-Certified *Ternera de Navarra* (*CTNA*) is designated for the veal produced in the autonomous community and province of Navarre (Northern Spain) under specifications surveilled by a local Regulatory Council [8] and has been the subject of several characterization studies [9,10]. On the other hand, in the United States, beef or veal does not bear a PGI quality seal but is subjected to other marketing strategies that include Certified Beef Programs under specifications surveilled by the Agricultural Marketing Service of the US Department of Agriculture (AMS-USDA) [11]. Such is the case of *Certified Angus Beef* (*CAB*), a pioneering branding program with wide international visibility based on the Angus breed/breed type [11,12]. Table S1 (Supplementary Materials) depicts the main differences in minimum specifications for the respective certification of the *CAB* and CTNA programs.

The *Pireniaca* breed comprises 90% of the *CTNA* [13]. It is a meat-purpose breed that grazes in mountain areas, taking advantage of pastures mainly composed of native vegetation (graminoids stems, *Bachypodium pinnatum* leaves, *Festuca rubra* leaves, other graminoids, and forbs plus browse) [14]. Fattening is carried out in the same breeding farms, and feed diets are generally based on concentrate and straw. Typical young carcasses are obtained from young bullocks or heifers slaughtered at about 11–13 months of age.

Spanish cattlemen are currently looking to expand the local beef markets, including international markets. This strategy must consider the sensory evaluation and physiochemical traits of the final product to better predict the eating quality as it is perceived by the consumers in the target countries.

The United States' beef production system and management practices are very diverse, and ongoing regional assessments have been conducted in order to guide the development of representative production systems [8–10]. Male castration is commonly practiced; therefore, castrated males (steers) prevail. In the USA, beef quality grades are determined on combinations of carcass maturity and marbling levels [15], and typical young carcasses are derived from steers or heifers rather than bullocks. Steers and heifers are typically fed with grain-based diets and slaughtered at about 15–28 months of age.

The association of marbling degree with juiciness and the tenderness of the meat is well established. Frank et al. [16] showed positive associations between intramuscular fat and overall flavor as well as juiciness and tenderness for Australian beef. Fat makes the matrix soft and easier to chew, therefore influencing the meat's tenderness [17–19].

Initially, consumer's choices are decided by basic (visual) perceptions on meat at the sale point, such as marbling or color. Palatability features are perceived upon tasting and beef overall liking is primarily defined by texture and flavor [19]. In a consumer study performed in Chicago and San Francisco, domestic beef was rated as more acceptable than its Argentinean counterpart [20]. In this survey, US beef was scored with higher ratings for juiciness, tenderness, flavor, and overall acceptability. Sitz et al. [18] reported that consumers from Denver and Chicago preferred domestic beef steaks versus those coming from Canada and Australia and suggested that the more desirable sensory ratings for the domestic beef may be due to the familiar taste of this type of meat. On the contrary, in a consumer sensory acceptance study performed in Spain by Beriain et al. [9], USDA prime beef (*Longissimus dorsi*) scored better than Spanish beef, the former one had a higher fat content and exhibited more desirable ratings for tenderness, juiciness, and flavor.

The meat industry devotes a lot of efforts to avoid the high variability in beef sensory quality. For the most part, the inherent variability detected in beef eating quality depends on the muscle physical and chemical properties, but a high level of palatability differences still remains due to the extrinsic characteristics of beef [21]. Therefore, consumers' surveys are useful when assessing consumer's preferences or willingness to pay. However, they are probably not the best option to compare the sensory attributes of meats from different sources or origins that should be evaluated by experienced panelists. There are few papers comparing the evaluation of certified beef carried out by a panel of experts from different countries in terms of meat sensory characteristics.

The objectives of this study were (a) to assess the differences in the physicochemical traits of samples derived from two different certified beef programs; these are, the PGI-Certified *Ternera de Navarra* (*CTNA*) in Spain and the *Certified Angus Beef* (*CAB*) in the United States; and (b) to assess the effect of a certified beef program on beef/veal sensory attributes as evaluated by two, separately trained, descriptive panels in Spain and the USA.

2. Materials and Methods

In Spain, the experimental part was performed at the Public University of Navarre (UPNA), whereas the experimental part in the USA was performed at the Meat Science Laboratory, University of Illinois at Urbana-Champaign (UIUC).

2.1. Animals Handling and Sampling

This study was performed by testing a total of 40 meat samples derived from two different biological types of cattle, as follows: 20 striploins derived from *CTNA* (based on *Pirenaica*, a large-framed, slow-maturing, continental breed), and 20 *CAB* striploins imported from the USA, presumably derived from small-to-medium framed, earlier maturing Angus-influenced fed cattle.

The twenty yearling *CTNA* bullocks (507 $\pm$ 51 kg of BW and 366 $\pm$ 23 days of age) were born, raised, and harvested in Navarra. Animals were fed with commercial concentrate and straw. The handling experimental procedures of the *CTNA* cattle followed the European Directive 2010/63/EU, regulated by the Real Decreto 348/2000 in Spain. CNTA bullocks were slaughtered at a commercial abattoir in Pamplona (Navarre, Spain), in accordance with the Council Regulation for the protection of animals at slaughter [22]. Carcasses were chilled for 24 h in a conventional chamber at 2 °C (98% RH). The *Longisimus dorsi lumborum* (LDL) muscle was removed from each carcass from the first to the sixth lumbar vertebrae. Upon reception at the UPNA from the abattoir, the loins were vacuum packaged and aged for 7 days in the dark in controlled chambers at 85% RH and 4 °C, that had circulated air with one renovation every 24 h. Thereafter, the loins were half cut, and one portion was frozen at −20 °C and then shipped to UIUC. The other portion was cut into approximately 2.5-centimeter-thick striploins, vacuum packaged, and frozen at −20 °C until subsequent analysis. All samples were frozen by directly placing them in a freezer with a set temperature of −20 °C. Similarly, upon reception of the half part of the loins, the UIUC personnel cut the frozen loins into 2.5-centimeter-thick striploins using a band saw, the steaks were vacuum packaged, and kept frozen until the sensory analysis was performed.

The twenty *CAB* samples fulfilled the specifications at Tyson Fresh meats (Joslin, IL., USA) [23], 370 kg. After chilling for 24 h at 2 °C, loins were removed from the carcasses and shipped to the UIUC facilities under refrigeration (5 °C). Once there, loins were vacuum packaged and aged for 7 days at 4 °C and 95% RH. Afterwards, loins were cut into 2.54-centimeter-thick striploins and trimmed to a maximum of 0.6 cm subcutaneous fat thickness, vacuum packaged, and kept frozen until further sensory analysis was performed. At the same time, half of the loins were shipped in a frozen state to UPNA, where, upon arrival, they were cut into 2.54-centimeter-thick striploins, vacuum packaged, and kept frozen until further analysis was performed (Figure 1).

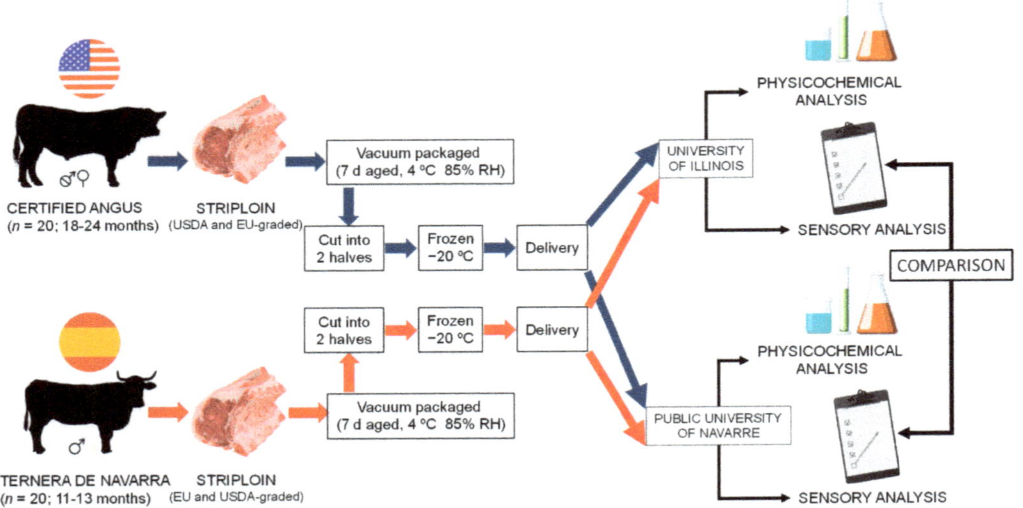

Figure 1. Illustration for experimental design of current study.

Either for physicochemical analysis or for sensory assessment in any location, the samples were thawed for 24 h at 4 °C prior to testing and were immediately analyzed.

2.2. Marbling Score and Longissimus Dorsi Area

In both countries these measurements were performed on the hanging carcass at the respective harvesting plant. The carcass evaluations performed followed the official guidelines of the EU beef classification system [24] and the USDA grading standards [15], in each respective country.

The marbling (visible intramuscular fat) was considered as the intermingling or dispersion of fat within the lean muscle. Graders evaluated the amount and distribution of marbling in the ribeye muscle at the cut surface after the carcass had been ribbed between the 12th and 13th ribs. For their evaluation, marbling scores were divided into 100 subunits. In general, however, marbling scores were divided in tenths within each degree of marbling (e.g., Slight 90, Small 00, Small 10, etc.).

The REA was measured at the last rib. It was measured with an USDA plastic grid of equally spaced dots representing a scale of 0.654 cm^2. The interior dots that were completely within the perimeter of the ribeye muscle were counted. The boundary dots that were on the perimeter of the ribeye muscle were also counted, but the calculated vale was divided by two and the results added to the previous interior dot number.

2.3. Warner Bratzler Shear Force

Steaks were thawed at 4 °C overnight prior to shear force evaluation. Steaks were cooked on a 180 °C preheated grill, turning the steak every 4 min until an internal temperature of 70 °C was reached on an open heart-grill (GR3000, JATA S.A, Tudela, Spain). The final temperature was monitored using copper-constantan thermocouples (Data Acquisition System was a portable digital thermometer Hanna in Spain). After cooking, steaks were cooled for 2 h at 5 °C and then, seven cores with a cross section of 1 cm × 1 cm × 3 cm were cut from each steak parallel to the longitudinal axis of the muscle fiber's orientation. Cores that were not uniform in size, had obvious connective tissue defects, or were not representative of the sample were discarded. Shear tests were performed with a TA-XT2i texture analyzer (Stable Micro System Ltd., Surrey, UK). The equipment was fitted with a 30-kilogram load cell and a Warner–Bratzler with a V slot shear attachment, fulfilling the requirements for this type of test [25]. A 5-kilogram weight was used for system force calibration. Test speed of the shear was set at 10 mm/s. Peak shear force values for the seven cores were averaged for each sample. The measurement of the maximum force (kg) as a function of knife movement (mm) and the compression to shear a sample of meat was considered as the hardness (toughness) of meat. Data were collected via Texture Expert v.2.0. software (Stable Micro Systems Ltd., Surrey, UK).

2.4. Proximate Analysis

The following chemical constituents were determined on thawed samples. In UIUC laboratories, moisture contents were determined using the ISO standard [26]. One steak from each striploin was trimmed of surrounding adipose and connective tissues and homogenized individually using a Cusinart Food Processor (Model DLC 5-TX, Cuisinart, Stamford, CT, USA). Duplicate 10-gram samples of each homogenized steak were weighed, placed in an aluminum film, and covered with filter paper. Each sample was oven dried (110 °C for 48 h) and weighed to determine moisture content. The additional homogenized tissue was frozen at −20 °C to be used for collagen quantification.

For the determination of intramuscular fat content in UPNA, LDL steaks were thawed, trimmed of connective and surrounding adipose tissues, and cut into small pieces for grinding (grinder Moulinex 800W Dpa 251, Groupe Seb Iberica Ltd., Barcelona, Spain). The fat analysis was performed in duplicate with the use of heat and solvents, which is derived from the Soxhlet type extraction [27]. Briefly, 6 g of ground meat was digested with boiling HCl 3 N for one hour. For that, a heating plate (Combiplac, JP Selecta, Barcelona, Spain) was used. Then, samples were filtered through filter paper (Albet 242Ø), and the filter with the fat dried at 70 °C for at least 12 h. After that, fat was extracted by using a Soxhlet system with ethyl ether as the solvent. The fat content was measured by the gravimetric differences of the round bottom flask that collected the extraction solvent during the Soxhlet cycles and correlated with the starting weighted meat quantity. The results were expressed as a percentage of fresh meat.

2.5. Total Collagen Quantification

The total collagen determination in US laboratories was conducted using a method based on AOAC [28] with modifications for a microplate assay. Duplicate 4-gram samples were hydrolyzed in 30 mL of sulfuric acid for 36 h at 105 °C. Hydrolysates were brought to a total volume of 100 mL with water and filtered. Hydroxyproline quantitation was performed in a deep 96-well plate, where 50 µL of each hydrolysate sample was added to 750 µL of water. Then, an oxidant solution (400 µL of cloramine T) was added to each well, and the plate was incubated at room temperature for 20 min. Afterwards, a color reagent (400 µL of Ehrlich's/pDMAB solution) was added to each well, and the plate was heated in a water bath at 60 °C for 15 min. Samples were read on a plate reader against a hydroxyproline standard curve at 557 nm to determine the hydroxyproline concentration in the sample, then multiplied by 8 to calculate the percent collagen in the sample.

2.6. Soluble Collagen Quantification

In UPNA laboratories, the soluble collagen content was obtained using the method described by Hill [29]. Four grams of meat sample were dissolved in 1/4-strength Ringer's solution in a 77 °C water bath for 70 min. After that, the sample was centrifuged, and the upper phase of each sample was hydrolyzed with HCl (50%) by boiling in a reflux for 7 h [30]. Then, the pH was adjusted in a range between 6–7, and the volume was brought to 100 mL with distilled water. The hydrolyzed solution was filtered by gravity through an Albert filter and kept at 5 °C until the hydroxyproline content determination. Hydroxyproline quantitation was performed in a glass tube where 1000 µL of each hydrolysate sample was added. Then, 500 µL of oxidant solution (cloramine T prepared in sodium acetate and isopropanol) was added to each tube, and they were incubated at room temperature for 20 min. Thereafter, 500 µL of the color reagent (Ehrlich's/pDMAB prepared in isopropanol and percloric acid) was added to each tube and heated in a water bath at 65 °C for 15 min. Samples were read on a UV spectrophotometer against a hydroxyproline standard curve (0–14.4 ppm) at 550 nm to determine the hydroxyproline concentration in the sample, then multiplied by 7.52 [31] to calculate the percent of soluble collagen in the samples.

2.7. Sensory Assessment

As described in Section 2.2, a total of 40 striploins steaks were procured either in the USA (*CAB*) or Spain (*CTNA*) and were sensorially tested in both research centers (UPNA in Spain and UIUC in the USA) by six panelists. In the United States, the training, cookery, and sensory evaluation followed the guidelines of the American Meat Science Association [32].

The rigorous methodology of Gorraiz et al. [33] was used in UPNA to train the descriptive panels and to select the attributes to evaluate the meat's sensory quality. In UPNA, panelists were highly trained in beef tasting with an average of 10 years' experience. All were rewarded with a voucher for buying beef.

Steaks were thawed overnight at 4 °C prior to each sensory evaluation session. Steaks were cooked in 180 °C preheated open-hearth electric grills (model 450 Farberware Cookware, Fairfield, CA, USA used at UIUC; and model GR3000, JATA S.A, Tudela, Spain used at UPNA) inside aluminum foil, turning the steak every 4 minutes until an internal temperature of 70 °C was reached. Every steak was then trimmed of any external connective tissue, cut into 3.5 cm × 1 cm × 1 cm rectangle samples, wrapped in coded aluminum foil and kept in a warmer, provided with sand, until the analysis was performed. Samples were coded with a randomized 3-digit number, and they were presented in a random order to the panelist in order to avoid the carry-over effect. Seven tasting sessions were conducted to obtain at least six judgments per steak. Briefly, in each tasting session 3 *CTNA* and 3 *CAB* samples were presented to each judge, except for the first session when two samples per certified beef were tasted.

Panelists were seated in individual booths under red lighting and were provided with water to cleanse the palate. Panelists evaluated samples for tenderness, juiciness, and beef flavor as described by Gorraiz et al. [33] on a 15-centimeter unstructured line scale with an anchor in the middle of the scale where 0 = extremely tough, extremely dry, no beef flavor, and no off-flavor and 15 = extremely tender, extremely juicy, intense beef flavor, and intense off-flavor. In addition to these attributes, the Spanish panelists were asked to evaluate liver- and fat-like flavor, and the US panelists were asked to assess the perception of off-flavor.

2.8. Statistical Analysis

For the physicochemical characterization and sensory assessment, the experimental unit was the animal (n = 40), and the fixed effect was the certified beef program. Mean values were compared using the Student's t-test for paired samples (within each certified beef program/country). Significant statistical differences were declared at $p \leq 0.05$. The sensory quality of *CTNA* and *CAB* samples was also investigated using a discriminant analysis (test of equality of means of Lambda Wilks groups; $p \leq 0.05$). All data were analyzed using IBM SPSS Statistics, version 25.0 for Windows (IBM Corporation, Armonk, NY, USA).

3. Results and Discussion

The descriptive statistics and mean comparisons for the marbling scores and REAs, physicochemical, and sensory attributes are presented in Tables 1–3, respectively. It was noted that the marbling scores showed a large standard deviation (SD) within each certified beef program. As this determination depends on the graders´ criteria and the carcass' fatness, a high variability may be expected [34]. The REA variability was ca. 20% for the

CTNA and ca. 7% for the *CAB* samples. The other analytical parameters, including the sensory traits, showed an SD lower than 10 percent.

Table 1. Descriptive statistics and mean comparison for degree of marbling and *Longissimus dorsi thoracis* eye area (REA) according to EU classification and USDA grading systems for carcasses derived from bullocks PGI-Certified *Ternera de Navarra* (*CTNA*) and *Certified Angus Beef* (*CAB*) steers, respectively.

Attribute	*CTNA*				*CAB*				SEM	*p*-Value
	Mean	Min	Max	SD	Mean	Min	Max	SD		
Marbling [1,3]	307.5 Traces^{07}	230.0 Devoid^{30}	410.0 Slight^{10}	45	837.5 Slightly abundant^{73}	730.0 Moderate^{30}	1090.0 Abundant^{90}	86	15.80	<0.0001
Marbling [2,4]	102.5 Traces^{07}	30.0 Practically devoid^{30}	210.0 Slight^{10}	50	739.0 Slightly abundant^{75}	620.0 Moderate^{30}	920.0 Abundant^{90}	93	16.49	<0.0001
REA (cm^2) [1,3]	104.47	83.87	136.77	14.09	77.89	66.32	98.64	7.62	2.53	<0.0001
REA (cm^2) [2,4]	141.96	106.54	181.41	20.79	78.36	69.51	99.02	7.35	3.48	<0.0001

[1] Tested in the USA; [2] Tested in Spain; [3] Following the USDA beef grading standards [15]; [4] Following the EU beef classification system [24].

Table 2. Descriptive statistics and mean comparison for physicochemical traits of striploins derived from PGI-Certified *Ternera de Navarra* (*CTNA*) and *Certified Angus Beef* (*CAB*), respectively.

Attribute	*CTNA* Striploins				*CAB* Striploins				SEM	*p*-Value
	Mean	Min	Max	SD	Mean	Min	Max	SD		
Fat (%) [2]	0.85	0.39	1.33	0.25	5.84	3.12	8.92	1.70	0.27	<0.0001
Moisture(%) [1]	75.43	72.92	76.34	0.82	65.41	58.02	67.91	2.51	0.41	<0.0001
Total Collagen(mg/g) [1]	2.52	1.45	3.31	0.48	2.69	2.09	3.33	0.40	0.10	0.2485
Soluble Collagen (mg/g) [2]	0.41	0.27	0.78	0.15	0.43	0.24	0.71	0.11	0.29	0.6488
Shear Force (kg) [2]	5.90	2.20	8.25	1.54	3.86	2.84	5.46	0.55	0.25	<0.0001

[1] Tested in the USA; [2] Tested in Spain.

Table 3. Descriptive statistics and mean comparison of sensory descriptive ratings for striploins derived from PGI-Certified *Ternera de Navarra* (*CTNA*) and *Certified Angus Beef* (*CAB*) as tested by trained panels in Spain and the USA, respectively.

Attribute	*CTNA* Striploins				*CAB* Striploins				SEM	*p*-Value
	Mean	Min	Max	SD	Mean	Min	Max	SD		
Tenderness [1]	8.14	5.70	10.10	1.36	9.92	7.63	11.90	1.26	0.29	<0.0001
Juiciness [1]	9.20	6.80	10.77	1.12	10.06	8.12	11.78	0.87	0.22	0.0099
Beef flavor [1]	6.20	4.33	7.93	0.88	7.99	6.55	10.47	1.01	0.21	<0.0001
Off-flavor [1]	0.88	0.17	1.75	0.50	0.32	0.07	0.78	0.22	0.09	<0.0001
Tenderness [2]	8.72	5.46	9.72	0.86	9.86	7.77	11.90	0.96	0.22	<0.0001
Juiciness [2]	6.05	3.66	9.19	1.34	7.39	5.69	9.48	1.19	0.28	0.0018
Beef flavor [2]	6.25	3.77	7.69	1.00	7.28	5.90	8.70	0.77	0.18	0.0004
Liver-like flavor [2]	3.77	2.63	5.04	0.75	3.23	2.12	5.38	0.79	0.17	0.0353
Fat flavor [2]	2.83	1.77	3.66	0.57	4.27	2.80	6.86	1.01	0.18	<0.0001

[1] Tested in the USA; [2] Tested in Spain.

3.1. Marbling Score and REA

The mean marbling score evaluated in both laboratories for the striploins of different experimental groups showed greater values for the *CAB* striploins than for the *CTNA* counterpart (Table 1), with highly significant differences ($p < 0.0001$). These results were expected due to the different genetic background and livestock practices performed in the two countries of origin. Additionally, because of the program's specifications (Table S1. Supplementary Materials). The vast majority of US cattle are grain fed and for their carcasses to be branded as *CAB*, they must have a marbling score of modest or higher [11]. On the other hand, *CTNA* bullocks were subjected to shorter periods of grain feeding and the requirement for this type of certified beef is that the intramuscular fat percentage should be around 2%, which is usually achieved through the feeding protocol of the Protected Designation of Origin program [8].

The *CTNA* carcasses showed larger REAs than those of the *CAB* ($p < 0.0001$), with around 120 and 78 cm^2, respectively. In general, the REA values are directly proportional to the carcass weights reported for the experimental groups, which averaged 330 and 370 kg for the *CTNA* and the *CAB* samples, respectively. The REA value of 106 cm^2 was reported by Alberti et al. [35] for *Pirenaica* carcasses. REA area values ranging from 64.8 to 111.9 cm^2 were reported by Nelson et al. [12] for Angus, Hereford–Angus cross, and Northern animals. The *CAB*'s REA value obtained in this work is within the above-mentioned range.

3.2. Warner–Bratzler Shear Force

WBSF showed lower mean values (Table 2) for the *CAB* striploins steaks as compared to those from *CTNA* (i.e., 3.86 vs. 5.90 kg). ($p < 0.0001$). Accordingly, the *CNTA* striploins were classified as "tough" according to a tenderness classification [30], as the averaged WBSF value was above 5.7 kg, whereas the *CAB* striploins were considered as intermediate in tenderness [36].

The mean WBSF values obtained for *Pirenaica* beef by Panea et al. [37] was 3.8 kg ($n = 55$); whereas, for *CAB*, the mean WBSF value reported by Nelson et al. [12] was 4.15 kg. Both precedent reports [12,37] support the values obtained in this study (5.9 and 3.86 kg for the *CTNA* and *CAB* samples, respectively). It can be claimed that breed had a strong impact on the WBSF values in accordance with the results obtained by López-Pedrouso et al. [38], who studied this textural trait in three different Spanish beef breeds.

3.3. Proximate Analysis

The mean fat content of the *CAB* samples was higher ($p < 0.0001$) than that of the CNTA striploins, which corresponds well with the statistical differences found in the marbling scores (Table 1). The mean values of the moisture content of the *CAB* striploins were about 10% lower ($p < 0.0001$; Table 1) than that of the *CNTA* counterparts. It is most likely that the explanation for these results is the inverse relationship between the fat and moisture contents in proximate composition; this is, as the fat content was higher in the CAB samples, there was a concurrent decrease in the moisture content [16,39,40].

3.4. Total and Soluble Collagen

The total collagen values did not vary with the certified beef program ($p = 0.2485$). The results for the *CTNA* samples are in accordance with those reported by Sañudo et al. [41] (3.5 and 0.27 mg/g for total and soluble collagen, respectively). The latter authors also evaluated these two parameters for other beef Spanish breeds and reported ranges for total collagen and soluble collagen of 4.1–2.8 and 0.27–0.54 mg/g, respectively. The values obtained in the present study were within those ranges [41].

A higher soluble collagen content typically indicates more tender meat [29], which was not supported by the WBSF values for the two certified beef programs. The possible transformation of soluble collagen into an insoluble collagen after meat cooling would explain the low values. Wheeler et al. [42] reported that WBSF was greater (indicating tougher meat) in steaks that exhibited lower marbling scores in *Bos taurus* cattle. The same study also reported a decreased variation in WBSF as the marbling score increased, which may explain the observed differences in the WBSF values obtained for each beef program.

3.5. Sensory Assessment

The sensory panelists in the USA described both types of striploins with medium-high scores for tenderness, juiciness, and beef flavor. Panelists found the *CAB* to be juicier ($p = 0.0099$), more tender ($p < 0.0001$), and more flavorful ($p < 0.0001$) than the *CTNA* samples (Table 3). A similar assessment was provided by the UPNA panel, which defined the *CAB* steaks as juicier ($p = 0.0018$), more tender ($p < 0.0001$), and more flavorful ($p = 0.0004$) than those of CTNA (Table 3). The marbling and the intramuscular fat percentage values obtained for the CAB samples may explain the higher ratings for juiciness because higher fat levels in meat produces a higher initial juiciness sensation in the mouth by increasing mouth lubrication and leading to a perceived improvement in tenderness [43]. It is a well-known relationship between marbling level and sensory attributes, including flavor perception, regardless of breed. In such a way, a higher intramuscular fat percentage is clearly linked to overall liking [43–46]. As fat is the primary driver of beef flavor and acceptability, the results presented herein support that, irrespective of the panelist's country of origin; the *CAB* samples (containing 5.84% fat) were perceived as more tender and more desirable in beef flavor compared to the *CTNA* samples (containing only 0.85% fat).

In a sensory trial run in the USA, consumers rated corn-fed domestic beef higher in flavor, juiciness, and tenderness than Australian grass-fed beef, suggesting that the palatability characteristics of corn-fed beef were perceived as more desirable by US consumers [18]. The study reported fat levels of 8.82 and 6.12% for the US and Australian samples, respectively. A different study reported that US consumers did not detect differences in beef flavor when tasting meat from different breeds produced under similar livestock practices [42]. These findings, along with the present results, would suggest that the animal's diet and the production practices had a greater impact on beef flavor than the marbling score. The beef flavor was detected with higher intensity in the CAB striploins than in the CNTA counterparts by trained panels in both countries. Similar sensory results in both testing places showed that panelists in the USA and in Spain were appropriately trained to detect differences between the samples and provided consistent assessments. The results obtained were in accordance with the assessments obtained from previous consumer studies, in which *CAB* received higher palatability scores as compared to non-US grass-fed beef [18,20,47].

Trained US panelists in the current study perceived less off-flavor in the *CAB* samples as compared to the *CTNA* counterparts (0.32 cm vs. 0.88 cm). The off-flavors detected for the *CAB* samples were described as "buttery" or "metallic", whereas the off-flavors for the *CTNA* samples were described as "grassy", "metallic", "acidic", "bloody", and "serumy." Although there was a significant difference in off-flavor incidence between the two certified beef samples, the mean off-flavor rating was less than 1 on a 15-point scale in any case, indicating a very low incidence of off-flavors in both types of certified beef.

Since the descriptors used in each testing place were slightly different due to differences in terminology and translation difficulties from English to Spanish, it was not possible to make a direct comparison between off-flavor and liver-like flavor, even though several panelists in the USA described the off-flavor in the *CTNA* samples as "livery" or "metallic". These differences in the experiment may be partially attributed to the different tasting experiences and expectations of the panelists in these two different countries. Although both panels were highly trained, the US panelists may have been less sensitive to a liver-like flavor than the Spanish panelists. For US consumers, an intramuscular fat level below 3% is not acceptable [48]. As indicated by Arshad et al. [47], there are several factors influencing beef flavor, particularly fatty acids. The mineral composition and fatty acid profile affect the development of meat off-flavors, such as a liver-like flavor [49]. Another possible reason might be that the descriptors used in both sensory evaluations were not clear enough for panelists to accurately describe the differences between the samples. We must acknowledge this difficulty as a weakness of our work because the vocabulary should had been defined before testing to obtain a reliable and robust sensory assessment from both panels. Furthermore, despite their training, there may have been some inherent bias for panelists to prefer certain flavors more than others. Dransfield et al. [50] conducted trained sensory panels at five European locations using beef from eight different countries. These researchers reported that, even though the perceived differences in tenderness and juiciness were correlated among the panels, some variation was partly attributed to regional differences in reference to juiciness and tenderness [50]. A follow-up study [51], which used similar panelists for evaluating the effects of cattle breed and postmortem aging, reported similar findings. Panels in five European countries were able to distinguish differences in tenderness and juiciness fairly consistently, but there were differences in evaluations of beef flavor [51].

The values of the common descriptors to both panels (juiciness, tenderness, and beef flavor) were used in discriminant analysis. Two canonical functions were obtained and explained 99.8% of the global variability of data. The first function was associated with tenderness (negatively) and juiciness (positively). The second function was associated with beef flavor (positively). Function one could be defined as the "texture factor" and function two as the "flavor factor". The results plotted in Figure 2 show that it is possible to differentiate the two types of meat with only three sensory attributes (juiciness, tenderness, and beef flavor). The *CAB* samples are in the upper quadrants. On the contrary, the *CTNA* samples are in the lower quadrants. The Spanish panel broadly differentiated both meats according to beef flavor. The two panels discriminated both meat types according to their beef texture.

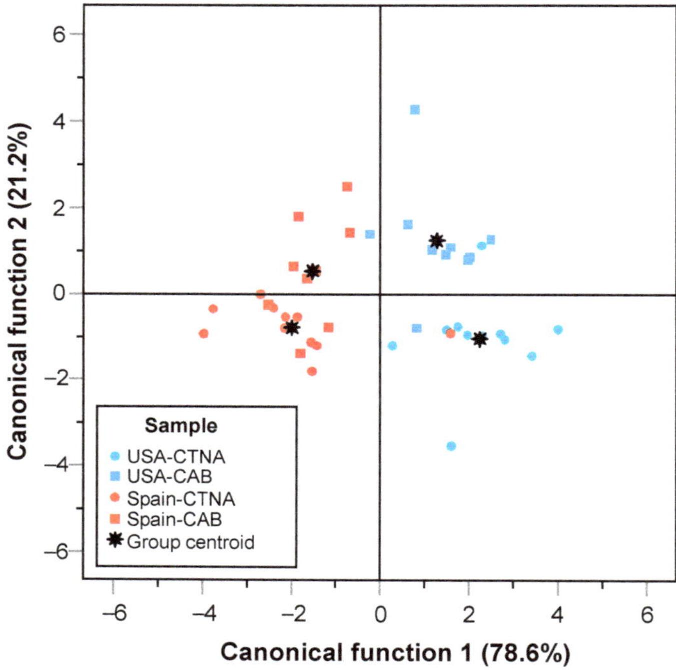

Figure 2. Spatial location of the PGI-Certified *Ternera de Navarra* (*CTNA*) and *Certified Angus Beef* (*CAB*) samples according to common attributes assessed in the USA (UIUC) and Spain (UPNA). Function 1 (texture factor) was associated with tenderness and juiciness, and function 2 (flavor factor) was associated with beef flavor.

4. Conclusions

Based on these results, US *Certified Angus Beef* striploins showed a clear advantage in textural quality as indicated by their lower WBSF and more desirable tenderness ratings, notwithstanding the similarities between CAB and CTNA in total and soluble collagen contents. The noticeable differences between the samples from both certified beef programs in the intramuscular fat and moisture percentages are explained by their inherent genetic and management backgrounds. The same applies to the larger REA and lower marbling scores of the *CTNA* striploins in comparison with those of *CAB*.

In terms of sensory quality, the panels in the two countries concur that the *CAB* striploins outperformed the *CTNA* samples in juiciness, tenderness, and flavor.

The relevant numerical differences found between the rating assigned by the Spanish and the US panelists suggest training differences, or difficulties encountered in using the appropriate terminology for defining each sensory attribute. Furthermore, the lack of product knowledge (i.e., consumption habits) may have been another reason for such differences, despite the blind sensory evaluation.

Supplementary Materials: The following are available online at https://www.mdpi.com/article/10.3390/foods10071474/s1, Table S1: Comparison of minimum specifications between USDA *Certified Angus Beef* and PGI-Certified *Ternera de Navarra*.

Author Contributions: Conceptualization, M.J.B. and T.R.C.; methodology, K.I. and C.L.C.; software, F.C.I.; formal analysis, M.T.M.-A., K.I., F.C.I. and C.L.C.; investigation, M.T.M.-A.; writing—original draft preparation, M.J.B. and C.L.C.; writing—review and editing, F.C.I., K.I. and T.R.C.; supervision, M.J.B. and M.T.M.-A.; project administration, M.J.B. and T.R.C.; funding acquisition, M.J.B. and T.R.C. All authors have read and agreed to the published version of the manuscript.

Funding: The research was supported by Spanish Ministry of Science and Innovation (Project AGL-2009-13303-CO2-01).

Institutional Review Board Statement: Ethical review and approval were waived for this study due to the compliance with national and/or European Regulations on animal husbandry and slaughter.

Data Availability Statement: All data from the research conducted are available on request from the corresponding author.

Acknowledgments: This project was carried out as part of an ongoing collaborative research effort between the University of Illinois at Urbana-Champaign and the Public University of Navarre, Pamplona, Spain entitled "Beef Quality Attributes for Consumers and its Relationship with Beef Carcass Classification Systems in Spain and the USA". The authors would also like to thank the anonymous reviewers of this paper and to the editor for their careful reading of the paper and their constructive comments.

Conflicts of Interest: Christine Leick Cord is employed at the Nestle Purina PetCare Company; however, this was not the situation when this research was performed. The authors declare no conflict of interest.

References

1. Whitnall, T.; Pitts, N. Global trends in meat consumption. *Agric. Commod.* **2019**, *9*, 96–99.
2. OECD/FAO. OECD/FAO Meat. In *OECD-FAO Agricultural Outlook 2020–2029*; OECD Publishing: Rome, Italy, 2020; pp. 162–173.
3. Henchion, M.M.; McCarthy, M.; Resconi, V.C. Beef quality attributes: A systematic review of consumer perspectives. *Meat Sci.* **2017**, *128*, 1–7. [CrossRef]
4. Petrescu, D.C.; Vermeir, I.; Petrescu-Mag, R.M. Consumer understanding of food quality, healthiness, and environmental impact: A cross-national perspective. *Int. J. Environ. Res. Public. Health* **2020**, *17*, 169. [CrossRef]
5. Morales, R.; Aguiar, A.P.S.; Subiabre, I.; Realini, C.E. Beef acceptability and consumer expectations associated with production systems and marbling. *Food Qual. Prefer.* **2013**, *29*, 166–173. [CrossRef]
6. Lee, B.; Yoon, S.; Choi, Y.M. Comparison of marbling fleck characteristics between beef marbling grades and its effect on sensory quality characteristics in high-marbled Hanwoo steer. *Meat Sci.* **2019**, *152*, 109–115. [CrossRef]
7. Barjolle, D.; Lehmann, B.; Chappuis, J.-M.; Dufour, M. *Protected Designation of Origin and Institutions (France, Spain and Italy)*; European Association of Agricultural Economists: Parma, Italy, 1997; pp. 483–502.
8. Ministerio de Agricultura, Pesca y Alimentación. Orden de 26 de diciembre de 2001 por la que se ratifica el Reglamento de la Indicación Geográfica Protegida "Ternera de Navarra o Nafarroako Aratxea" y de su Consejo Regulador. *Bol. Of. Estado* **2002**, *13*, 1755–1763.
9. Berian, M.J.; Sanchez, M.; Carr, T.R. A comparison of consumer sensory acceptance, purchase intention, and willingness to pay for high quality United States and Spanish beef under different information scenarios. *J. Anim. Sci.* **2009**, *87*, 3392–3402. [CrossRef]
10. Lizaso, G.; Beriain, M.J.; Horcada, A.; Chasco, J.; Purroy, A. Effect of intended purpose (dairy/beef production) on beef quality. *Can. J. Anim. Sci.* **2011**, *91*, 97–102. [CrossRef]
11. USDA Certified Angus Beef. G-1 specification. In *Standards and Specifications Division*; US Department of Agriculture: Washington, DC, USA, 2020.
12. Nelson, J.L.; Dolezal, H.G.; Ray, F.K.; Morgan, J.B. Characterization of certified angus beef steaks from the round, loin, and chuck. *J. Anim. Sci.* **2004**, *82*, 1437–1444. [CrossRef]
13. Ministerio de Agricultura, Pesca y Alimentación. I.G.P. Ternera de Navarra/Nafarroako Aratxea. Available online: https://www.mapa.gob.es/es/alimentacion/temas/calidad-diferenciada/dop-igp/carnes/IGP_Ternera_Navarra.aspx (accessed on 18 June 2021).

14. García-González, R.; García Serrano, A.; Revilla, R. Comparación del regimen alimentario de vacas pardo alpinas y pirenaicas en un puerto del Pirineo occidental. In Proceedings of the Actas XXXII Reunion Científica de la S.E.E.P. Sociedad Española para el Estudio de los Pastos: Pamplona, Spain, 1992; pp. 299–305.
15. *USDA Carcass Beef Grades and Standards*; US Department of Agriculture: Washington, DC, USA, 2017; p. 17.
16. Frank, D.; Ball, A.; Hughes, J.; Krishnamurthy, R.; Piyasiri, U.; Stark, J.; Watkins, P.; Warner, R. Sensory and flavor chemistry characteristics of Australian beef: Influence of intramuscular fat, feed, and breed. *J. Agric. Food Chem.* **2016**, *64*, 4299–4311. [CrossRef]
17. Joo, S.-T.; Hwang, Y.-H.; Frank, D. Characteristics of Hanwoo cattle and health implications of consuming highly marbled Hanwoo beef. *Meat Sci.* **2017**, *132*, 45–51. [CrossRef] [PubMed]
18. Sitz, B.M.; Calkins, C.R.; Feuz, D.M.; Umberger, W.J.; Eskridge, K.M. Consumer sensory acceptance and value of domestic, Canadian, and Australian grass-fed beef steaks. *J. Anim. Sci.* **2005**, *83*, 2863–2868. [CrossRef] [PubMed]
19. Banović, M.; Grunert, K.G.; Barreira, M.M.; Fontes, M.A. Beef quality perception at the point of purchase: A study from Portugal. *Food Qual. Prefer.* **2009**, *20*, 335–342. [CrossRef]
20. Killinger, K.M.; Calkins, C.R.; Umberger, W.J.; Feuz, D.M.; Eskridge, K.M. A comparison of consumer sensory acceptance and value of domestic beef steaks and steaks from a branded, Argentine beef program. *J. Anim. Sci.* **2004**, *82*, 3302–3307. [CrossRef] [PubMed]
21. Ellies-Oury, M.-P.; Hocquette, J.-F.; Chriki, S.; Conanec, A.; Farmer, L.; Chavent, M.; Saracco, J. Various statistical approaches to assess and predict carcass and meat quality traits. *Foods* **2020**, *9*, 525. [CrossRef]
22. European Commission. Council Regulation (EC) No 1099/2009 of 24 September 2009 on the protection of animals at the time of killing. *Off. J. Eur. Union* **2009**, *L 303*, 1–30.
23. Tyson Foods Inc. Angus Beef. In *Tyson Fresh Meats*; Tyson Foods Inc.: Dakona Dunes, SD, USA, 2021.
24. European Commission. Commission Regulation (EC) No 1249/2008 of 10 December 2008 laying down detailed rules on the implementation of the Community scales for the classification of beef, pig and sheep carcasses and the reporting of prices thereof. *Off. J. Eur. Union* **2008**, *L 337*, 1–35.
25. Novaković, S.; Tomašević, I. A comparison between Warner-Bratzler shear force measurement and texture profile analysis of meat and meat products: A review. *IOP Conf. Ser. Earth Environ. Sci.* **2017**, *85*, 012063. [CrossRef]
26. ISO. *ISO 1442:1997 Meat and Meat Products—Determination of Moisture Content (Reference Method)*; ISO: Geneva, Switzerland, 1997.
27. AOAC Fat (crude) in meat and meat products (Official method of analysis, no. 991.36). In *Official Methods of Analysis of AOAC International*; AOAC International: Gaithersburg, MD, USA, 2007.
28. AOAC Hydroxyproline in meat and meat products (Official method of analysis, no 990.26). In *Official Methods of Analysis of AOAC International*; AOAC International: Gaithersburg, MD, USA, 1993.
29. Hill, F. The solubility of intramuscular collagen in meat animals of various ages. *J. Food Sci.* **1966**, *31*, 161–166. [CrossRef]
30. Bergman, I.; Loxley, R. Two improved and simplified methods for the spectrophotometric determination of hydroxyproline. *Anal. Chem.* **1963**, *35*, 1961–1965. [CrossRef]
31. Cross, H.R.; Carpenter, Z.L.; Smith, G.C. Effects of intramuscular collagen and elastin on bovine muscle tenderness. *J. Food Sci.* **1973**, *38*, 998–1003. [CrossRef]
32. *AMSA Research Guidelines for Cookery, Sensory Evaluation, and Instrumental Tenderness Measurements of Meat*, 2nd ed.; American Meat Science Association: Champaign, IL, USA, 2015.
33. Gorraiz, C.; Beriain, M.J.; Chasco, J.; Iraizoz, M. Descriptive analysis of meat from young ruminants in mediterranean systems. *J. Sens. Stud.* **2000**, *15*, 137–150. [CrossRef]
34. Liu, J.; Chriki, S.; Ellies-Oury, M.-P.; Legrand, I.; Pogorzelski, G.; Wierzbicki, J.; Farmer, L.; Troy, D.; Polkinghorne, R.; Hocquette, J.-F. European conformation and fat scores of bovine carcasses are not good indicators of marbling. *Meat Sci.* **2020**, *170*, 108233. [CrossRef] [PubMed]
35. Albertí, P.; Sañudo, C.; Lahoz, F.; Olleta, J.; Tena, R.; Jaime, S.; Panea, B.; Campo, M.; Pardos, J. Producción y rendimiento carnicero de siete razas bovinas españolas faenadas a distintos pesos. *Inf. Téc. DGA* **2001**, *101*, 1–16.
36. Stewart, S.M.; Lauridsen, T.; Toft, H.; Pethick, D.W.; Gardner, G.E.; McGilchrist, P.; Christensen, M. Objective grading of eye muscle area, intramuscular fat and marbling in Australian beef and lamb. *Meat Sci.* **2020**, 108358. [CrossRef]
37. Panea, B.; Olleta, J.L.; Sañudo, C.; del Mar Campo, M.; Oliver, M.A.; Gispert, M.; Serra, X.; Renand, G.; del Carmen Oliván, M.; Jabet, S.; et al. Effects of breed-production system on collagen, textural, and sensory traits of 10 European beef cattle breeds. *J. Texture Stud.* **2018**, *49*, 528–535. [CrossRef] [PubMed]
38. López-Pedrouso, M.; Rodríguez-Vázquez, R.; Purriños, L.; Oliván, M.; García-Torres, S.; Sentandreu, M.Á.; Lorenzo, J.M.; Zapata, C.; Franco, D. Sensory and physicochemical analysis of meat from bovine breeds in different livestock production systems, pre-slaughter handling conditions, and ageing time. *Foods* **2020**, *9*, 176. [CrossRef] [PubMed]
39. Thompson, J.M. The effects of marbling on flavour and juiciness scores of cooked beef, after adjusting to a constant tenderness. *Aust. J. Exp. Agric.* **2004**, *44*, 645–652. [CrossRef]
40. Thompson, J.M.; Polkinghorne, R.; Hwang, I.H.; Gee, A.M.; Cho, S.H.; Park, B.Y.; Lee, J.M.; Thompson, J.M.; Polkinghorne, R.; Hwang, I.H.; et al. Beef quality grades as determined by Korean and Australian consumers. *Aust. J. Exp. Agric.* **2008**, *48*, 1380–1386. [CrossRef]

41. Sañudo, C.; Campo, M.; Lasalle, P.; Olleta, J.; Panea, B. Calidad instrumental de la carne de bovino de siete razas españolas. *Arch. Zootec.* **1998**, *47*, 397–402.
42. Wheeler, T.L.; Cundiff, L.V.; Koch, R.M. Effect of marbling degree on beef palatability in *Bos taurus* and *Bos indicus* cattle. *J. Anim. Sci.* **1994**, *72*, 3145–3151. [CrossRef]
43. Matthews, K.R. Saturated fat reduction in butchered meat. In *Reducing Saturated Fats in Foods*; Talbot, G., Ed.; Woodhead Publishing Series in Food Science, Technology and Nutrition; Woodhead Publishing: Cambridge, UK, 2011; pp. 195–209. ISBN 978-1-84569-740-2.
44. Corbin, C.H.; O'Quinn, T.G.; Garmyn, A.J.; Legako, J.F.; Hunt, M.R.; Dinh, T.T.N.; Rathmann, R.J.; Brooks, J.C.; Miller, M.F. Sensory evaluation of tender beef strip loin steaks of varying marbling levels and quality treatments. *Meat Sci.* **2015**, *100*, 24–31. [CrossRef]
45. Frank, D.; Kaczmarska, K.; Paterson, J.; Piyasiri, U.; Warner, R. Effect of marbling on volatile generation, oral breakdown and in mouth flavor release of grilled beef. *Meat Sci.* **2017**, *133*, 61–68. [CrossRef]
46. Legako, J.F.; Dinh, T.T.N.; Miller, M.F.; Adhikari, K.; Brooks, J.C. Consumer palatability scores, sensory descriptive attributes, and volatile compounds of grilled beef steaks from three USDA Quality Grades. *Meat Sci.* **2016**, *112*, 77–85. [CrossRef]
47. Arshad, M.S.; Sohaib, M.; Ahmad, R.S.; Nadeem, M.T.; Imran, A.; Arshad, M.U.; Kwon, J.-H.; Amjad, Z. Ruminant meat flavor influenced by different factors with special reference to fatty acids. *Lipids Health Dis.* **2018**, *17*, 1–13. [CrossRef]
48. Miller, R.; A, T.; Kerry, J.; Ledward, D. *Factors Affecting the Quality of Raw Meat*; Woodhead Publishing: Cambridge, UK, 2002; pp. 27–63. ISBN 978-1-85573-583-5.
49. Jenschke, B.E.; Hodgen, J.M.; Calkins, C.R. Fatty acids and minerals affect the liver-like off-flavor in cooked beef. *Neb. Beef Cattle Rep.* **2007**, *85*, 84–85.
50. Dransfield, E.; Rhodes, D.N.; Nute, G.R.; Roberts, T.A.; Boccard, R.; Touraille, C.; Buchter, L.; Hood, D.E.; Joseph, R.L.; Schon, I.; et al. Eating quality of European beef assessed at five research institutes. *Meat Sci.* **1982**, *6*, 163–184. [CrossRef]
51. Dransfield, E.; Nute, G.R.; Roberts, T.A.; Boccard, R.; Touraille, C.; Buchter, L.; Casteels, M.; Cosentino, E.; Hood, D.E.; Joseph, R.L.; et al. Beef quality assessed at European research centres. *Meat Sci.* **1984**, *10*, 1–20. [CrossRef]

Article

Microbial Growth Study on Pork Loins as Influenced by the Application of Different Antimicrobials

David A. Vargas, Markus F. Miller, Dale R. Woerner and Alejandro Echeverry *

Department of Animal and Food Sciences, Texas Tech University, Lubbock, TX 79409, USA; andres.vargas@ttu.edu (D.A.V.); mfmrraider@aol.com (M.F.M.); dale.woerner@ttu.edu (D.R.W.)
* Correspondence: alejandro.echeverry@ttu.edu; Tel.: +1-806-834-8733

Abstract: The use of antimicrobials in the pork industry is critical in order to ensure food safety and, at the same time, extend shelf life. The objective of the study was to determine the impact of antimicrobials on indicator bacteria on pork loins under long, dark, refrigerated storage conditions. Fresh boneless pork loins ($n = 36$) were split in five sections and treated with antimicrobials: Water (WAT), Bovibrom 225 ppm (BB225), Bovibrom 500 ppm (BB500), Fit Fresh 3 ppm (FF3), or Washing Solution 750 ppm (WS750). Sections were stored for 1, 14, 28, and 42 days at 2–4 °C. Mesophilic and psychrotrophic aerobic bacteria (APC-M, APC-P), lactic acid bacteria (LAB-M), coliforms, and *Escherichia coli* were enumerated before intervention, after intervention, and at each storage time. All bacterial enumeration data were converted into log10 for statistical analysis, and the Kruskal–Wallis test was used to find statistical differences ($p < 0.05$). Initial counts did not differ between treatments, while, after treatment interventions, treatment WS750 did not effectively reduce counts for APC-M, APC-P, and coliforms ($p < 0.01$). BB500, FF3, and WS750 performed better at inhibiting the growth of indicator bacteria when compared with water until 14 days of dark storage.

Keywords: indicator bacteria; chlorine dioxide; rhamnolipids; 1,3-Dibromo-5.5-dimethyl hydantoin; interventions

Citation: Vargas, D.A.; Miller, M.F.; Woerner, D.R.; Echeverry, A. Microbial Growth Study on Pork Loins as Influenced by the Application of Different Antimicrobials. *Foods* **2021**, *10*, 968. https://doi.org/10.3390/foods10050968

Academic Editor: Gary Dykes

Received: 24 March 2021
Accepted: 26 April 2021
Published: 28 April 2021

1. Introduction

In 2017, the United States produced almost 52 billion pounds of red meat, of which 25.6 billion pounds of the total were pork [1]. The United States Department of Agriculture (USDA) data show that, in 2018, the per capita consumption of pork was close to 50.8 pounds per year [2]. Pork has always been one of the major meat sources for people, so it is crucial for the industry to ensure a safe pork supply [3].

The Center for Disease Control and Prevention (CDC), in 2018, estimated that one out of six Americans get sick, and from those who were sick, 128,000 were hospitalized, and 3000 died of foodborne diseases [4]. Moreover, the contribution of meat to foodborne illnesses caused by bacteria is 23.20% (beef: 13.20%, pork: 9.80%, and game: 0.10%) [4]. Although the contribution of pork in foodborne illnesses caused by bacteria is lower when compared with beef, it remains significant.

The Institute of Food Science and Technology defines shelf life as "*the period of time during which the food product will remain safe; be certain to retain its desired sensory, chemical, physical, microbiological, and functional characteristics; where appropriate, comply with any label declaration of nutrition data, when stored under the recommended conditions*" [5]. In order to determine shelf life of products, there are series of different methods and equipment that can be used in relation with sensory characteristics of a product, such as color, odor, structure, and flavor, and how these attributes change with time. These types of equipment have been developed to obtain an objective measurement at the moment of analyzing sensorial characteristics of a product.

Furthermore, consumers expect that foods are free of foodborne pathogens and have a decently long shelf life, where antimicrobials play a substantial role in order to achieve this demand [6]. Food antimicrobials are classified as preservatives, according to the U.S. Food and Drug Administration, which are any chemicals that, when added to food, tend to prevent or retard deterioration [6]. The most common function of an antimicrobial is to prolong shelf life through the process of killing or inhibiting spoilage microorganisms while maintaining and extending all the organoleptic properties [6]. It is important to consider that antimicrobials are never a substitute for good sanitation practices in food processing plants, since low initial counts will always be ideal. Although antimicrobials extend the lag phase, their effects on the surviving population can be overcome through time [6]. The global economy in which we live leads us to store and transport food and assure that the food arrives in the condition that is expected, and this is where antimicrobials undoubtedly play a role.

The purpose of this study was to determine the impact of selected antimicrobial spray products on the microbial growth of indicator bacteria naturally present on pork loins after long term storage under dark and refrigerated conditions.

2. Materials and Methods

2.1. Sample Collection

The study was repeated three times between January to August of 2019. On each repetition, vacuum packaged boneless pork loins (n = 36) were purchased from a commercial pork processing plant located in Oklahoma and transported within five hours in a cooler covered with ice at 0–4 °C to the Gordon W. Davis Texas Tech University Meat Science Laboratory (Lubbock, Texas, TX, USA). Pork loins were stored under dark conditions (no light) at 0–4 °C and processed 24 h later.

2.2. Treatment Preparation

Treatments were prepared two to three hours before application to the boneless pork loins. For each treatment, three liters of solution was prepared and then stored in a handheld sprayer (Chapin 1-Gallon Plastic Tank Sprayer, Chapin, Batavia, NY, USA). Treatments utilized included: cold water, Bovibrom 225 ppm (1,3-Dibromo-5,5-dimenthylhydantoin; prepared in a mixer provided by Passport Food Safety Solutions, West Des Moines, IA, USA), Bovibrom 500 ppm (prepared the same as Bovibrom 225 ppm), Fit Fresh 3 ppm (chlorine dioxide; prepared following label instructions, Selective Micro Technologies, Dublin, OH, USA), and Natural Washing Solution 750 ppm (rhamnolipid, Jeneil Biosurfactant, Saukville, WI, USA).

2.3. Treatment Application

Pork loins were split into five sections of 8.90 cm in length. Each section was randomly assigned to one of the five treatments. For each treatment, 12 pork loin sections were obtained (n = 180). Interventions were sprayed onto the pork loin sections for 30 s using a handheld sprayer (Chapin 1-Gallon Plastic Tank Sprayer, Chapin, Batavia, NY, USA; Flow rate: 5.98 ± 0.75 mL/s). Then, sections were flipped and sprayed for another 30 s, ensuring coverage of the entire loin surface. After 10 min, treated sections were vacuum packaged using Cryovac bags (Sealed Air, Charlotte, NC, USA) and randomly assigned to one of the four dark storages periods (1, 14, 28, and 42 days) and refrigerated at temperatures ranging between 0 and 4 °C.

2.4. Swab Sample Collection

Buffer peptone water (BPW) pre-hydrated 25 mL swabs (3M, St. Paul, MN, USA), were taken at multiple periods of time during pork processing: before application of intervention, after application of intervention (10 min after finishing interventions), and at the end of each storage time (immediately after opening the bag). For swabs in sections, a 100 cm^2 template was used. The swabs were taken from the fat and the lean portions of the pork loin sections.

2.5. Swabs Sample Processing

After arrival to the laboratory, pre-hydrated swabs were homogenized for two minutes at 230 rpm using an automated stomacher (Steward Laboratory Systems, Davie, FL, USA), serial dilutions with BPW were conducted and plated in Petrifilm (3M, St. Paul, MN, USA) or plates (Thermo Fisher Scientific, Waltham, MA, USA) according to each microorganism.

2.6. Total Aerobic Plate Counts

For Aerobic Plate Counts, the Association of Official Agricultural Chemists 990.12 (AOAC) official method was used. After serial dilutions were performed, Petrifilms were placed on a flat surface, inoculated with 1 mL of sample dilution following product instructions. Petrifilms were left undisturbed for one minute to permit gel to solidify. Petrifilms were incubated for 48 ± 3 h at 35 ± 1 °C for mesophilic bacteria conditions and 72 ± 3 h at 20 ± 1 °C for psychrotrophic bacteria conditions. Enumeration was conducted using 3M Petrifilm Plate Reader (3M, St. Paul, MN, USA) and checked in a standard colony counter following the rules of the official method [7–11].

2.7. Coliforms and Escherichia coli Enumeration

For Coliforms and *Escherichia coli*, the AOAC 991.14 official method was used. After serial dilutions were performed, Petrifilms were placed on a flat surface. Then one mL of sample was inoculated onto the center of the film base and covered with the top film in duplicates. Petrifilms were left undisturbed for one minute to permit gel to solidify. Petrifilms were incubated for 48 ± 3 h at 35 ± 1 °C. Enumeration was conducted at 24 h for coliforms and 48 h for *Escherichia coli* in a standard colony counter following rules of the official method [12,13].

2.8. Lactic Acid Bacteria Enumeration

After serial dilutions, one mL of sample was inoculated on a petri dish and pour plated with 20 mL of Mann–Rogosa–Sharpe Agar (MRS) in duplicates. Plates were placed in BD GasPak EZ Container Systems (Becton Dickinson and Company, Franklin Lakes, NJ, USA) and incubated at 48 ± 3 h at 35 ± 1 °C under microaerophilic conditions (6 to 16% O_2 and 2 to 10% CO_2) using BD GasPak EZ Campy Sachets (Becton Dickinson and Company, Franklin Lakes, NJ, USA), [14,15]. Enumeration was conducted using a Q-Counter (Spiral Biotech Inc, Norwood, MA, USA)

2.9. Statistical Analysis

Pork loin section swabs before and after interventions was a 2 × 2 factorial design (Sampling point × Treatment) with two levels under sampling point (before and after) and 5 levels under treatment (Bovibrom 225 ppm, Bovibrom 500 ppm, Fit Fresh 3 ppm, Washing Solution 750 ppm, Water). Pork loin section swabs at different storage times was a complete randomized design with repeated measures over time. All counts were analyzed using Kruskal–Wallis nonparametric test (R. Version 4.04), followed by pairwise multiple comparison Wilcoxon's test adjusted by Benjamin & Hochber method. Wilcoxon's test was used to identify the significant variation in microbial level on swab samples collected at different sampling points, storage times, and treatments. A p-value of 0.05 or less was selected prior to the analysis to determine significant differences in this study.

3. Results

3.1. Microbiological Analysis (before and after Treatment Application)

From pork loin sections, counts for coliforms, *Escherichia coli*, mesophilic aerobic bacteria (APC-M), psychrotrophic aerobic bacteria (APC-P), and mesophilic lactic acid bacteria (LAB-M) were performed before and after treatment application.

For all analysis, coliforms and *Escherichia coli* counts were below detection limit, <0.25 colony-forming unit (CFU)/cm^2. Due to low initial counts, both coliforms and *Escherichia coli*, no statistical difference was found before and after treatment intervention.

Low initial counts suggested that the plant from which samples were collected has implemented good manufacturing practices and dressing procedures and thus a good control of possible cross-contamination of endogenous sources of pathogens with pork carcasses.

A treatment by sampling point interaction was found for APC-M and APC-P ($p < 0.01$), (Figures 1 and 2). For these microorganisms, initial counts (before intervention) did not differ between treatments, while after intervention, treatment with Washing Solution 750 ppm did not effectively reduced counts for APC-M and APC-P. For mesophilic lactic acid bacteria (LAB-M), no effect was found by treatment or sampling point ($p = 0.69$). After treatment application, counts for Washing Solution 750 ppm for APC-M and APC-P were statistically higher ($p < 0.01$), when compared to the other treatments, suggesting a lower antimicrobial efficiency immediately after intervention.

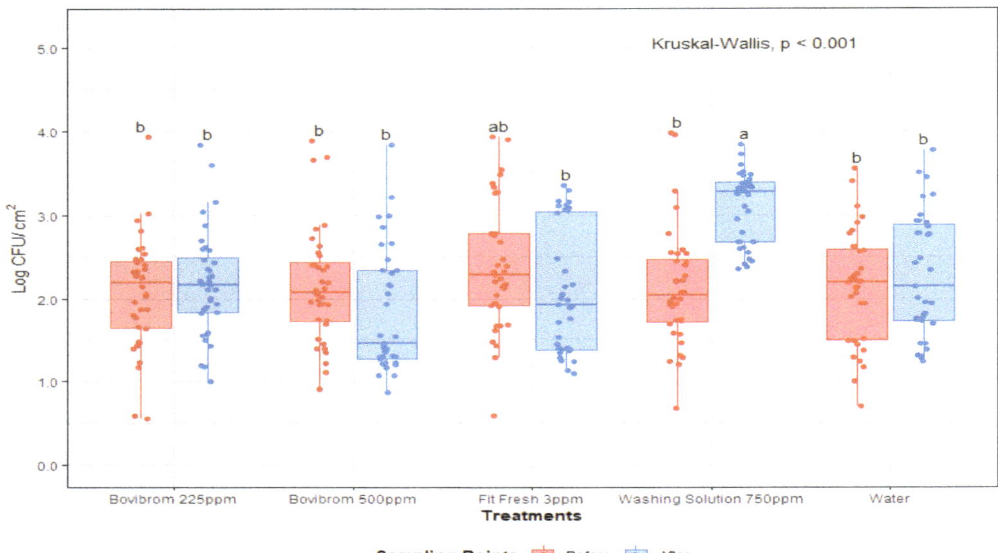

Figure 1. Mesophilic aerobic plate counts (Log CFU/cm^2) before and after treatment application on pork loin sections ($n = 36$ per treatment). In each boxplot, the horizontal line crossing the box represents the median, the bottom and top of the box are the lower and upper quartiles, the vertical top line represents the upper interquartile range, and the vertical bottom line represents the lower interquartile range. Boxes with different letters a,b are significantly different according to Kruskal–Wallis analysis followed by pairwise comparison Wilcoxon's test at $p < 0.05$. The points represent the actual data points. Active ingredients: Bovibrom = 1,3-Dibromo-5,5-dimenthylhydantoin; Fit Fresh = chlorine dioxide; Washing Solution = rhamnolipid.

3.2. Microbiological Analysis (End of Dark Storage Time)

From pork loin sections, enumeration of APC-M, APC-P, coliforms, *Escherichia coli*, and LAB-M were performed at the end of each of four dark storage periods (1 Day, 14 Days, 28 Days, and 42 Days) at refrigerated temperatures 0–4 °C. Similarly, coliform and Escherichia coli initial counts were below detection limit (< 0.25 CFU/cm^2) for the first day of storage. After 14 days, enumeration was above detection limit, and a statistical difference was found between counts at 14, 28, and 42 days (0.09, 0.52, 1.44 Log CFU/cm^2, respectively; largest standard error: 0.09 Log CFU/cm^2).

For APC-M, a dark storage time by treatment interaction was found ($p = 0.05$). As storage time was found to be significant, a statistical analysis was performed per storage time in order to find differences between treatments. Bacterial cells by nature multiply over time; that is why the biological importance lies in the change of treatments over time.

Nonparametric approach tests, also called distribution-free tests, was used in order to analyze the results, because they do not assume that the data follow any specific distribution as parametric tests do. The distribution of the data (Figure 3) suggests neither samples follow a normal distribution, or the sample size was big enough. A Kruskal–Wallis test, the test used when assumptions of ANOVA are not met, was performed to find differences between treatments over the four dark storage times.

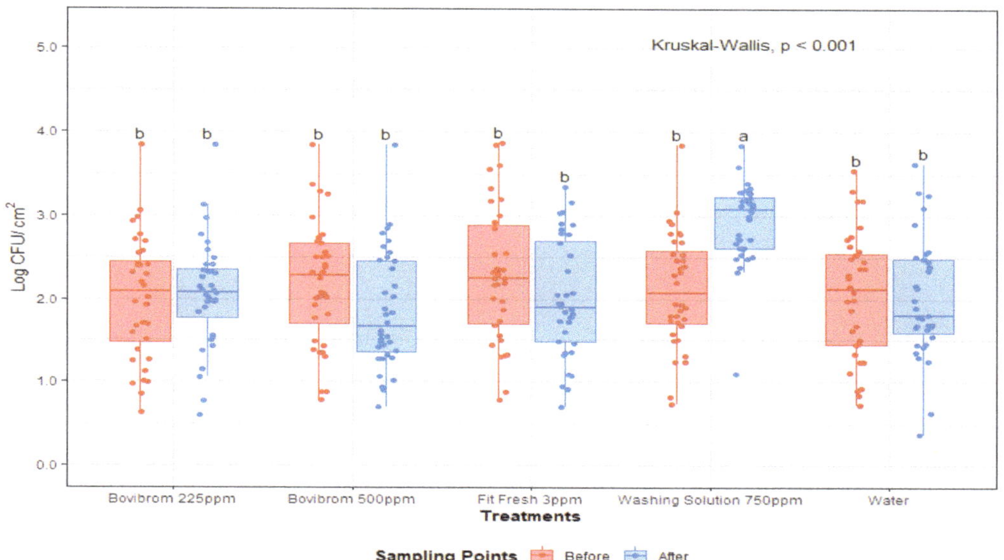

Figure 2. Psychrotrophic aerobic plate counts (Log CFU/cm^2) before and after treatment application on pork loin sections (n = 36 per treatment). In each boxplot, the horizontal line crossing the box represents the median, the bottom and top of the box are the lower and upper quartiles, the vertical top line represents the upper interquartile range, and the vertical bottom line represents the lower interquartile range. Boxes with different letters a,b are significantly different according to Kruskal–Wallis analysis followed by pairwise comparison Wilcoxon's test at $p < 0.05$. The points represent the actual data points. Active ingredients: Bovibrom = 1,3-Dibromo-5,5-dimenthylhydantoin; Fit Fresh = chlorine dioxide; Washing Solution = rhamnolipid.

After one day of dark storage, counts were not different between treatments (p = 0.08). There were differences for Bovibrom 225 ppm, Bovibrom 500 ppm, Fit Fresh 3 ppm, and Washing Solution 750 ppm counts at 14 days of storage time when compared with Water ($p < 0.01$). Then, after 28 days of storage time, all treatments presented similar values when compared with Water. For all treatments, there was no immediate effect on the microbial load of pork loin sections (Day 1), but these results suggest that there was a residual effect of all the treatments by the fact that, at Day 14, all counts were lower when compared with Water. Moreover, these results show that this residual effect is lost by the time the pork samples reached 28 days of refrigerated storage time.

Further, for APC-P (Figure 4), a dark storage time by treatment interaction was found ($p = 0.03$). At Day 1 of dark storage, counts were not different between treatments. There were differences for Bovibrom 500 ppm, Fit Fresh 3 ppm, and Washing Solution 750 ppm counts at 14 days of storage time when compared with Water, but in later storage times, all treatments presented similar values compared with Water. Similar results were obtained for APC-M and APC-P related with the residual effect, found by using this type of interventions on pork loins. For mesophilic lactic acid bacteria (LAB-M), no significant effect was found for treatment or dark storage ($p = 0.45$).

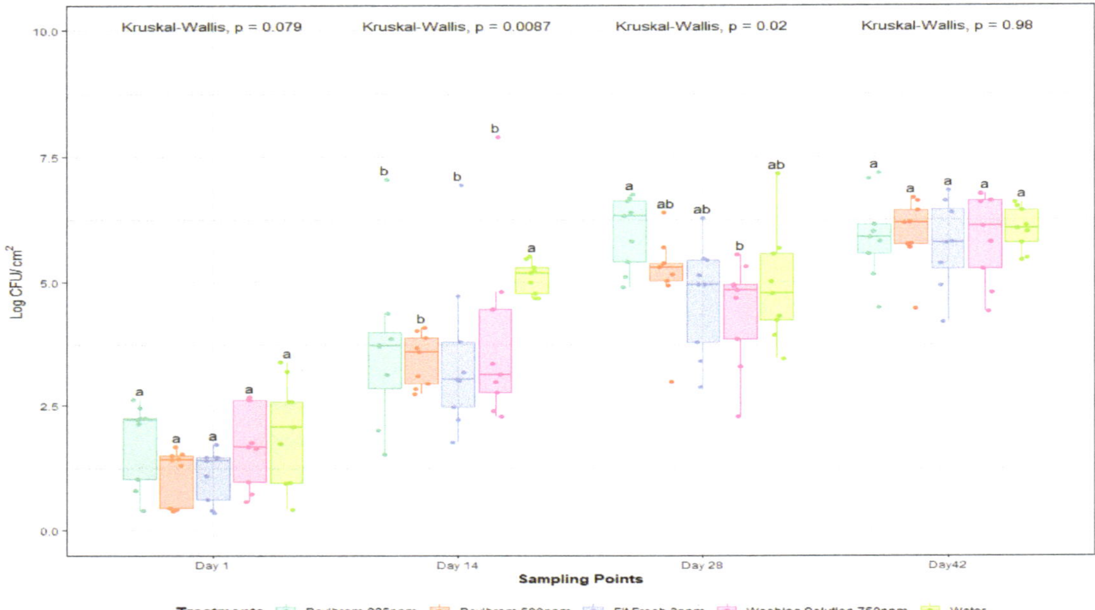

Figure 3. Mesophilic aerobic plate counts (Log CFU/cm^2) after 1, 14, 28, and 42 days of dark storage time on pork loin sections (n = 45 per dark storage time). In each boxplot, the horizontal line crossing the box represents the median, the bottom and top of the box are the lower and upper quartiles, the vertical top line represents the upper interquartile range, and the vertical bottom line represents the lower interquartile range. For each sampling point day, boxes with different letters a,b are significantly different according to Kruskal–Wallis analysis followed by pairwise comparison Wilcoxon's test at $p < 0.05$. The points represent the actual data points. Active ingredients: Bovibrom = 1,3-Dibromo-5,5-dimenthylhydantoin; Fit Fresh = chlorine dioxide; Washing Solution = rhamnolipid.

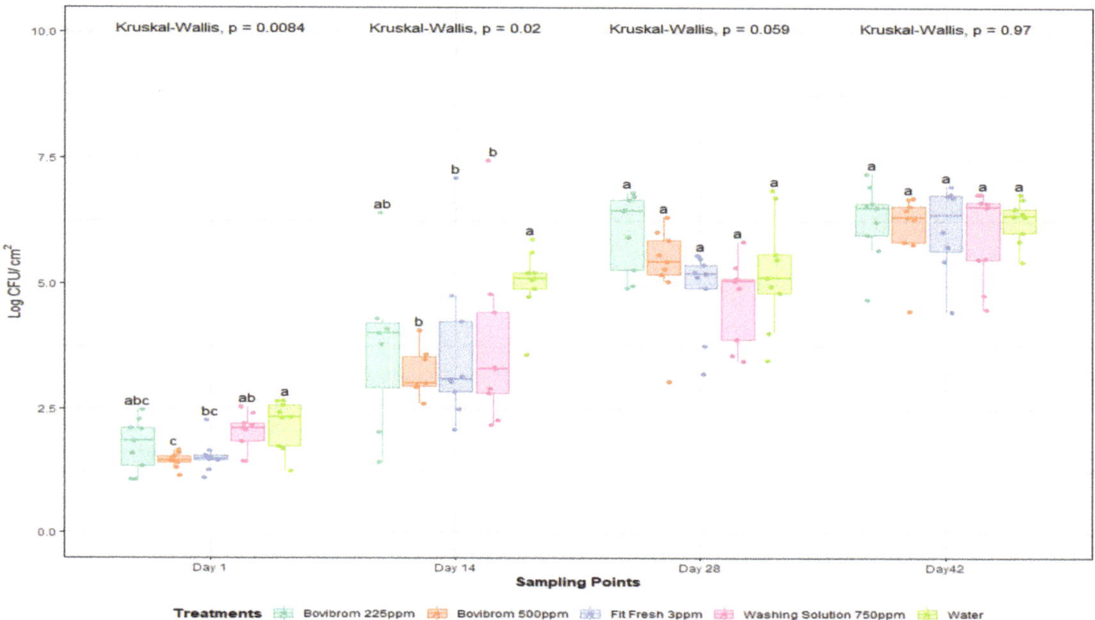

Figure 4. Psychrotrophic aerobic plate counts (Log CFU/cm^2) after 1, 14, 28, and 42 days of dark storage time on pork loin sections (n = 45 per dark storage time). In each boxplot, the horizontal line crossing the box represents the median, the bottom and top of the box are the lower and upper quartiles, the vertical top line represents the upper interquartile range, and the vertical bottom line represents the lower interquartile range. For each sampling point day, boxes with different letters a,b are significantly different according to Kruskal–Wallis analysis followed by pairwise comparison Wilcoxon's test at $p < 0.05$. The points represent the actual data points. Active ingredients: Bovibrom = 1,3-Dibromo-5,5-dimenthylhydantoin; Fit Fresh = chlorine dioxide; Washing Solution = rhamnolipid.

4. Discussion

Each treatment has different active ingredients; therefore, the overall spectrum, the mode of action, and the efficacy against microorganisms are highly dependent on the chemical and physical properties of the antimicrobial [6]. Treatment with Washing Solution 750 ppm is a biosurfactant within the glycolipid category known as rhamnolipids, which are produced mainly by *Pseudomonas aeruginosa* [16,17]. Rhamnolipids is an amphipathic surface-active molecule composed of ß-hydroxy fatty acid connected to a rhamnose sugar molecule used for a broad range of applications, such as antimicrobial agents [18,19]. Its antimicrobial activity is related primary by damaging the cytoplasmic membrane, causing an increase in its permeability due to the release of lipopolysaccharides from the outer membrane [20–22].

In a study where rhamnolipids were tested at different concentrations for Gram-positive and Gram-negative bacteria, the antimicrobial effect was completely indistinguishable for Gram-negative bacteria at all concentrations tested, while Gram-positive bacteria were inhibited at most concentrations, explaining the lower antimicrobial efficiency when compared to other antimicrobials that have a larger overall spectrum of action [22,23].

In a pork chop shelflife study using organic (citric or ascorbic) acid applications and vacuum packaging system, psychrotrophic enumeration was performed in order to see the effect of these interventions in storage time up to 14 days. Results showed that, despite of the intervention and packaging system, psychrotrophic bacteria were still capable of growing over the storage period [24]; however, our study presented a clear decrease in log counts for APC-M and APC-P after 14 days of storage time when compared to the Water application. Evidence in this study suggests that antimicrobial interventions are effective during the 14-day period, but once the antimicrobial effects are depleted, the remaining bacteria have less competition to multiply, increasing the rate of growth up to the point that counts at 42 days of storage did not differ from water application.

Treatment Bovibrom (225 and 500 ppm) is a commercial name for the active ingredient 1,3-dibromo-5,5-dimenthylhydantoin (DBDMH) [25]. DBDMH is a bromanine polymer, which hydrolyzes to hypobromous acid (HOBr) in presence of water [26]. This hypobromous acid has the same biocide property as hypochlorous acid (HOCl), and they both combine with organic compounds to form bromanines or chloramine, respectively; however, bromanines are more potent than chloramines and, therefore, show more effectiveness in the presence of organic matter [6]. Halogen's mechanism of action is not well defined, but theories such as interference in cell metabolism by oxidation of SH groups (sulfhydryl group) essential for bacterial enzymes due to pH or oxidation of purine and pyrimidine bases causing mutation are the most well-known and accepted [27]. Further, treatment with Fit Fresh 3 ppm is a commercial name for the active ingredient that in, presence of water, is known as chlorine dioxide. Chlorine dioxide differs from normal chlorine compounds, as it does not form HOCl, but it presents similarities in the antimicrobial activity due to the oxidation-reduction potential [6]. The mechanisms of action are not well defined, but it is theorized that protein synthesis and disruption of the outer membrane to be highly responsible for their antimicrobial activity [6].

Counts performed at the end of each dark storage time suggested that Fit Fresh, Bovibrom 500 ppm and Washing Solution 750 ppm performed better until 14 days of storage despite the wide spectrum of chlorine dioxide and hypobromous acid when compared with rhamnolipids. In addition, there is a clear increase in counts for all microorganisms as dark storage time increased, which is not surprising, because increased storage time in a vacuum bag will result in increased bacterial proliferation [28]. The use of antimicrobials is mainly to inhibit the growth of microorganisms by extending the lag phase of their lifecycle, and according to literature found, a meat product around 6 log CFU/cm^2 is at a level in which it could be considered spoiled, even though microbial loads is not the only attribute to be considered for shelf life [29]. During the dark storage period, treatments with Bovibrom 500 ppm, Fit Fresh 3 ppm, and Washing Solution 750 ppm presented values below this limit, and it is not until 42 days of storage that the pork section approached this limit, suggesting that the increase of shelf life, considering microbiological characteristics using these antimicrobials, is accomplished. When compared to other studies where, after 28 days of aging this limit was reached, our study delayed reaching the 6-log population limit until 42 days of storage [30].

5. Conclusions

The purpose of the study was to determine the shelf life of pork loins with the application of different antimicrobials evaluating growth on common microbial indicators. The antimicrobials Bovibrom 500 ppm, Fit Fresh 3 ppm, and Washing Solution 750 ppm performed the best for maintaining reduced microbial counts when compared to Water in pork loins after 14 days of dark storage under refrigerated conditions 0–4 °C.

Author Contributions: Conceptualization, D.A.V., A.E., D.R.W., and M.F.M.; Methodology, D.A.V. and A.E.; Validation, D.A.V. and A.E.; Formal analysis, D.A.V.; Investigation, D.A.V. and A.E.; Data Curation, D.A.V.; Writing—original draft preparation, D.A.V.; Writing—review and editing, D.A.V. and A.E.; Supervision, A.E.; Funding acquisition, A.E. All authors have read and agreed to the published version of the manuscript.

Funding: This research was funded by Passport Food Safety Solutions/Church & Dwight Co., but they were not associated with its execution. The authors would like to thank them for their support and technical assistance.

Institutional Review Board Statement: Not applicable.

Informed Consent Statement: Not applicable.

Data Availability Statement: The data presented in this study are available on request from the corresponding author.

Conflicts of Interest: The authors declare no conflict of interest. The funders of this research had no role in the collection, analyses, or interpretation of data nor in the writing of this manuscript.

References

1. NAMI. *Fact Sheet U.S. Meat and Poultry Production & Consumption: An Overview*; NAMI: NW Washington, DC, USA, 2016.
2. USDA. Per Capita Consumption of Pork in the United States from 2015 to 2028 (In Pounds). Available online: https://www.statista.com/statistics/183616/per-capita-consumption-of-pork-in-the-us-since-2000/ (accessed on 29 September 2019).
3. Baer, A.A.; Miller, M.J.; Dilger, A.C. Pathogens of interest to the pork industry: A review of research on interventions to assure food safety. *Compr. Rev. Food Sci. Food Saf.* **2013**, *12*, 183–217. [CrossRef]
4. CDC. *Pathogens Causing the Most Illnesses, Hospitalizations, and Deaths Each Year*; CDC: Atlanta, GA, USA, 2018.
5. Institute of Food Science and Technology. *Shelf-Life of Foods: Guidelines for Its Determination and Prediction*, 1st ed.; Institute of Food Science and Technology, Ed.; Institute of Food Science and Technology: London, UK, 1993; ISBN 9780905367118.
6. Davidson, M.; Sofos, J.N.; Branen, A.L. *Antimicrobials in Food*; Taylor & Francis Group: London, UK, 2005; p. 721, ISBN 0-8247-4037-8.
7. AOAC. *AOAC Official Method 990.12 Aerobic Plate Count in Foods*; AOAC: Rockville, MD, USA, 2002; Volume 242.
8. Dormedy, E.S.; Brashears, M.M.; Cutter, C.N.; Burson, D.E. Validation of acid washes as critical control points in hazard analysis and critical control point systems. *J. Food Prot.* **2000**, *63*, 1676–1680. [CrossRef] [PubMed]
9. Goepfert, J.M. The Aerobic Plate Count, Coliform and Escherichia coli Content of Raw Ground Beef at the Retail Level. *J. Milk Food Technol.* **1976**, *39*, 175–178. [CrossRef]
10. Jay, J. A review of aerobic and psychrotrophic plate count procedures for fresh meat and poultry products. *J. Food Prot.* **2002**, *65*, 1200–1206. [CrossRef] [PubMed]
11. Ercolini, D.; Russo, F.; Nasi, A.; Ferranti, P.; Villani, F. Mesophilic and psychrotrophic bacteria from meat and their spoilage potential in vitro and in beef. *Appl. Environ. Microbiol.* **2009**, *75*, 1990–2001. [CrossRef] [PubMed]
12. AOAC. *AOAC Official Method 991.14 Coliform and Escherichia coli Counts in Foods*; AOAC: Rockville, MD, USA, 2002; Volume 635.
13. Tortorello, M.L. Indicator Organisms for Safety and Quality—Uses and Methods.pdf. *J. AOAC Int.* **2003**, *86*, 1208. [CrossRef] [PubMed]
14. Moraes, P.M.; Perin, L.M.; Ortolani, M.B.T.; Yamazi, A.K.; Viçosa, G.N.; Nero, L.A. Protocols for the isolation and detection of lactic acid bacteria with bacteriocinogenic potential. *LWT Food Sci. Technol.* **2010**, *43*, 1320–1324. [CrossRef]
15. Pothakos, V.; Devlieghere, F.; Villani, F.; Björkroth, J.; Ercolini, D. Lactic acid bacteria and their controversial role in fresh meat spoilage. *Meat Sci.* **2015**, *109*, 66–74. [CrossRef] [PubMed]
16. Kaeppeli, O.; Guerra-Santos, L. Process for the Production of Rhamnolipids. U.S. Patent US4628030A, 9 August 1984.
17. Randhawa, K.K.S.; Rahman, P.K.S.M. Rhamnolipid biosurfactants-past, present, and future scenario of global market. *Front. Microbiol.* **2014**, *5*, 454. [CrossRef]
18. Xu, Q.; Nakajima, M.; Liu, Z.; Shiina, T. Biosurfactants for microbubble preparation and application. *Int. J. Mol. Sci.* **2011**, *12*, 462–475. [CrossRef] [PubMed]
19. Magalhães, L.; Nitschke, M. Antimicrobial activity of rhamnolipids against Listeria monocytogenes and their synergistic interaction with nisin. *Food Control* **2013**, *29*, 138–142. [CrossRef]
20. Rahman, P.K.S.M.; Gakpe, E. Production, characterization, and applications of biosurfactants. *Biotechnology* **2008**, *7*, 360–370. [CrossRef]
21. Parry, A.; Parry, N.; Peilow, C.; Stevenson, P. Combinations of Rhamnolipids and Enzymes for Improved Cleaning. WIPO Patent WO2012010406A1, 4 July 2011.
22. Díaz De Rienzo, M.A.; Stevenson, P.; Marchant, R.; Banat, I.M. Antibacterial properties of biosurfactants against selected Gram-positive and -negative bacteria. *FEMS Microbiol. Lett.* **2016**, *363*, fnv224. [CrossRef] [PubMed]
23. Devendra, H.; Venkata, N.; Smita, S.; Vayalam, P. Rhamnolipid mediated disruption of marine Bacillus cereus biofilms. *Colloids Surf. B Biointerfaces* **2010**, *81*, 242–248. [CrossRef]
24. Huang, N.Y. Retail Shelf-Life of Pork Dipped in Organic Acid before Modified. *J. Food Sci.* **2005**, *70*, 382–387. [CrossRef]
25. Kalchayanand, N.; Arthur, T.; Bosilevac, J.; Schmidt, J.; Wang, R.; Shackelford, S.; Wheeler, T. *Efficacy of Commonly Used Antimicrobial Interventions on Shiga Toxin-Producing Escherichia coli Serotypes O45, O121, and Non-MDR and MDR Salmonella Inoculated Fresh Beef Final Report*; Clay Center: Charleston, WV, USA, 2011.
26. Sun, G.; Allen, L.C.; Luckie, E.P.; Wheatley, W.B.; Worley, S.D. Disinfection of Water by N-Halamine Biocidal Polymers. *Ind. Eng. Chem. Res.* **1995**, *34*, 4106–4109. [CrossRef]
27. Estrela, C.; Estrela, C.R.A.; Barbin, E.L.; Spanó, J.C.E.; Marchesan, M.A.; Pécora, J.D. Mechanism of action of sodium hypochlorite. *Braz. Dent. J.* **2002**, *13*, 113–117. [CrossRef] [PubMed]
28. Hodges, J.H.; Cahill, V.R.; Ockerman, H.W. Effect of Vacuum Packaging on Weight Loss, Microbial Growth and Palatability of Fresh Beef Wholesale Cuts. *J. Food Sci.* **1974**, *39*, 143–146. [CrossRef]
29. Banwart, G.J. *Basic Food Microbiology*; The AVI Publishing Company, Inc.: Westport, CN, USA, 1981.
30. Holmer, S.F.; McKeith, R.O.; Boler, D.D.; Dilger, A.C.; Eggert, J.M.; Petry, D.B.; McKeith, F.K.; Jones, K.L.; Killefer, J. The effect of pH on shelf-life of pork during aging and simulated retail display. *Meat Sci.* **2009**, *82*, 86–93. [CrossRef] [PubMed]

Article

Evaluation of Immersion and Spray Applications of Antimicrobial Treatments for Reduction of *Campylobacter jejuni* on Chicken Wings

Sara V. Gonzalez, Ifigenia Geornaras, Mahesh N. Nair and Keith E. Belk *

Center for Meat Safety & Quality, Department of Animal Sciences, Colorado State University, Fort Collins, CO 80523-1171, USA; sara.gonzalez@colostate.edu (S.V.G.); gina.geornaras@colostate.edu (I.G.); mahesh.narayanan_nair@colostate.edu (M.N.N.)
* Correspondence: keith.belk@colostate.edu; Tel.: +1-970-491-5826

Abstract: The decontamination efficacy of antimicrobial treatments against *Campylobacter jejuni* on chicken wings was evaluated. Chicken wings surface-inoculated with *C. jejuni* (3.9 log colony-forming units [CFU]/mL) were left untreated (control) or were treated by immersion (5 s) or in a spray cabinet (4 s) with water, a sulfuric acid and sodium sulfate blend (SSS; pH 1.2), formic acid (1.5%), peroxyacetic acid (PAA; 550 ppm), or PAA (550 ppm) that was pH-adjusted (acidified) with SSS (pH 1.2) or formic acid (1.5%). All evaluated immersion and spray chemical treatments effectively ($p < 0.05$) lowered *C. jejuni* populations on chicken wings. Spray application of chemical treatments resulted in immediate pathogen reductions ranging from 0.5 to 1.2 log CFU/mL, whereas their application by immersion lowered initial pathogen levels by 1.7 to 2.2 log CFU/mL. The PAA and acidified PAA treatments were equally ($p \geq 0.05$) effective at reducing initial *C. jejuni* populations, however, following a 24 h refrigerated (4 °C) storage period, wings treated with acidified PAA had lower ($p < 0.05$) pathogen levels than samples that had been treated with PAA that was not acidified. Findings of this study should be useful to the poultry industry in its efforts to control *Campylobacter* contamination on chicken parts.

Keywords: *Campylobacter jejuni*; antimicrobials; decontamination; poultry; chicken wings; application method

Citation: Gonzalez, S.V.; Geornaras, I.; Nair, M.N.; Belk, K.E. Evaluation of Immersion and Spray Applications of Antimicrobial Treatments for Reduction of *Campylobacter jejuni* on Chicken Wings. *Foods* **2021**, *10*, 903. https://doi.org/10.3390/foods10040903

Academic Editor: Huerta-Leidenz Nelson and Markus F. Miller

Received: 18 March 2021
Accepted: 16 April 2021
Published: 20 April 2021

1. Introduction

The Foodborne Diseases Active Surveillance Network (FoodNet) of the Centers for Disease Control and Prevention's (CDC) Emerging Infections Program reported that in 2019, *Campylobacter* was the leading bacterial cause of foodborne illness with an incidence rate of 19.5 cases per 100,000 population [1]. Specifically, out of 25,866 total cases of foodborne illness that were laboratory-diagnosed in that year, 9731 were due to infection with *Campylobacter* [1]. Individuals with *Campylobacter* infections, however, do not always seek medical treatment and even if they do, cases may remain undiagnosed [2]. Therefore, when underreporting and underdiagnosis are factored in, estimates indicate that *Campylobacter* spp. are actually responsible for 1.5 million diarrheal illnesses each year in the United States [2,3]. The most common species of *Campylobacter* associated with human campylobacteriosis cases is *C. jejuni*, and is responsible for at least 80% of *Campylobacter* enteric infections [4,5].

Campylobacter infections are primarily associated with consumption of unintentionally undercooked contaminated poultry products [6,7]. Moreover, *Campylobacter* in poultry is the number one pathogen-food combination in terms of annual illness burden, with a total of 608,231 infections and an estimated cost of more than $1.2 billion [8]. In an effort to reduce the incidence of foodborne illness cases from poultry products, slaughter facilities are required by the U.S. Department of Agriculture's (USDA) Food Safety and

Inspection Service (FSIS) to identify points during slaughter and processing where physical and/or chemical interventions can be applied to reduce pathogen contamination levels [9]. In the United States, FSIS Directive 7120.1 [10] provides poultry processors with a list of antimicrobials that are approved for use as decontamination treatments of poultry products. Peroxyacetic acid (PAA), which is currently the most widely used antimicrobial intervention in U.S. poultry processing facilities, is approved for use up to a maximum concentration of 2000 ppm [10,11]. Also approved are various organic and inorganic acids, cetylpyridinium chloride, chlorine, acidified sodium chlorite, trisodium phosphate, and a blend of sulfuric acid and sodium sulfate (SSS; also referred to as AFTEC 3000 or Amplon in the literature) [10,11]. SSS can be used as a spray, immersion, or wash treatment of poultry products at concentrations that would achieve a targeted pH range of 1.0 to 2.2 [10].

Performance standards for *Salmonella* and *Campylobacter*, established by FSIS, are used to assess the effectiveness of decontamination interventions used by a facility, in limiting or reducing pathogen contamination [12]. Since more than 85% of poultry meat in the United States is sold as parts, FSIS includes in its testing program sampling sites for both pathogens in the cut-up room to test poultry parts [13,14]. The current performance standards for the maximum acceptable *Campylobacter*-positives for chicken are 15.7% of broiler carcasses, 9.6% of comminuted products, and 7.7% of parts [13,15]. Thus, the poultry industry is reevaluating current antimicrobial interventions used for pathogen control and is looking for novel decontamination treatments to apply to meet the strict regulations for poultry [14,16].

There are numerous published studies on the antimicrobial effects of various chemical treatments against *Salmonella* populations on whole chicken carcasses and parts [11,14,17–19]. In comparison, however, fewer research studies have reported on the effect of such treatments against *Campylobacter*, and in particular, on chicken parts. Additionally, regardless of poultry product type and pathogen, studies investigating the decontamination efficacy of chemical treatments that combine two or more modes of action are also limited. Therefore, the objectives of this study were to (i) evaluate the antimicrobial effects of SSS, formic acid, PAA, and PAA that was pH-adjusted with SSS or formic acid (hereafter referred to as "acidified PAA"), when applied to chicken wings inoculated with *C. jejuni*, and (ii) determine the antimicrobial efficacy of the treatments as a result of applying the test solutions by immersion or spraying. Additionally, the antimicrobial effects against inoculated populations were evaluated immediately after treatment application (0 h) and after 24 h of storage at 4 °C.

2. Materials and Methods

2.1. Bacterial Strains and Inoculum Preparation

The inoculum consisted of a mixture of six *C. jejuni* strains of poultry origin (Table 1). Working cultures of the strains were maintained at 4 °C on plates of Campy Cefex Agar, Modified (mCCA; Hardy Diagnostics, Santa Maria, CA, USA) that were held within anaerobic containers (AnaeroPack Rectangular Jar; Mitsubishi Gas Chemical America, New York, NY, USA) with a microaerophilic environment generating gas pack (mixture of approximately 6 to 12% O_2 and 5 to 8% CO_2; AnaeroPack-MicroAero sachet, Mitsubishi Gas Chemical America).

Table 1. *Campylobacter jejuni* strains used in the study.

Strain ID	Origin	Source
FSIS21822450	Chicken drumsticks	USDA-FSIS-OPHS [a]
FSIS21822588	Chicken drumsticks	USDA-FSIS-OPHS
FSIS11815850	Ground chicken	USDA-FSIS-OPHS
CVM N55886	Chicken wings	FDA-CVM [b]
CVM N56299	Chicken wings	FDA-CVM
CVM N16C024	Chicken breast	FDA-CVM

[a] U.S. Department of Agriculture, Food Safety and Inspection Service, Office of Public Health Science. [b] U.S. Food and Drug Administration, Center for Veterinary Medicine.

The *C. jejuni* strains were individually cultured and subcultured in 10 mL of Bolton broth (Hardy Diagnostics) incubated at 42 °C for 48 h under microaerophilic conditions (Oxoid CampyGen sachet, Thermo Scientific, Basingstoke, UK). Cultures of the six strains were then combined and centrifuged (6000× *g*, 15 min, 25 °C; Sorvall Legend X1R centrifuge, Thermo Scientific, Waltham, MA, USA). Resulting cell pellets were washed twice with 10 mL of phosphate-buffered saline (PBS; pH 7.4; Sigma-Aldrich, St. Louis, MO, USA), and the final washed cell pellet comprising all six strains was resuspended in 60 mL of PBS. This cell suspension (ca. 7 log colony-forming units [CFU]/mL concentration) was then diluted 10-fold in PBS, and the diluted inoculum (ca. 6 log CFU/mL concentration) was used to inoculate the chicken wings. The concentration of the *C. jejuni* inoculum (undiluted and diluted) was determined by plating serial dilutions onto mCCA.

2.2. Inoculation of Chicken Wings

Fresh (i.e., not frozen) skin-on whole chicken wings were purchased from a wholesale food distributor. Wings were stored at 2 °C and were used for the study within six days of receipt. Two trials (repetitions) of the study were conducted on two separate days. On the first day of each trial, wings were randomly assigned to a control treatment or one of six treatments to be applied by immersion or spraying. For each antimicrobial treatment and application method, six samples were placed on trays lined with ethanol-sterilized aluminum foil and were inoculated under a biological safety cabinet. A 0.1 mL (100 µL) aliquot of the diluted *C. jejuni* inoculum was deposited, with a micropipette, on one side of each wing and then spread over the entire surface with a sterile disposable spreader. After a 10 min bacterial cell attachment period, samples were turned over, with sterile forceps, and were inoculated on the second side using the same procedure. The second inoculated side was also left undisturbed for 10 min to allow for inoculum attachment. The target inoculation level was 3 to 4 log CFU/mL of wing rinsate.

2.3. Antimicrobial Treatment of Chicken Wings

Inoculated wings were left untreated, to serve as controls, or they were treated by immersion or a spray application with water, SSS (pH 1.2; Amplon, Zoetis, Florham Park, NJ, USA), formic acid (1.5%; BASF Corporation, Florham Park, NJ, USA), PAA (550 ppm; Actrol Max, Kroff, Pittsburgh, PA, USA), PAA (550 ppm) acidified with SSS (pH 1.2; SSS-aPAA), or PAA (550 ppm) acidified with formic acid (1.5%; FA-aPAA). The water treatment was included to determine the rinsing effect of the immersion and spray treatments. Antimicrobial treatment solutions were prepared according to the manufacturers' instructions, and the pH of solutions was measured (Orion Star A200 Series pH meter and Orion RossUltra pH electrode, Thermo Scientific, Schaumburg, IL, USA). Average pH values of the SSS, formic acid, and PAA solutions were 1.2, 2.9, and 3.2, respectively. For the SSS-aPAA and FA-aPAA solutions, average pH values were 1.2 and 2.8, respectively. The PAA concentration was verified using a hydrogen peroxide and peracetic acid test kit (LaMotte Company, Chestertown, MD, USA).

For immersion application of the test solutions, inoculated wings were individually immersed for 5 s in 500 mL of the solution in a Whirl-Pak bag (1627 mL; Nasco, Fort

Atkinson, WI, USA). A different Whirl-Pak bag and fresh, unused solution was used to immersion-treat each sample. Spray application of the water and chemical treatments was performed using a custom-built spray cabinet (Birko/Chad Equipment, Olathe, KS, USA) fitted with two 0.38 L/min FloodJet spray nozzles (Spraying Systems Co., Glendale Heights, IL, USA) positioned above the product belt. The inoculated wings were placed on a cutting board on top of the ladder-style conveyor belt of the cabinet and were sprayed with the test solution at a pressure of 69 to 83 kPa and a product contact time of 4 s.

Immersion- and spray-treated wings were placed on sterile wire racks for 5 min to allow excess solution to drip off samples before microbiological analysis or refrigerated storage. For each trial, three of the six samples per treatment were analyzed for *C. jejuni* populations following treatment application (0 h analysis), and the three remaining samples were placed in individual 710 mL Whirl-Pak bags (Nasco) and analyzed after a 24 ± 1 h storage period at 4 °C.

2.4. Microbiological Analysis

At each sampling time (0 h and 24 h), untreated (control) and treated samples were analyzed for *C. jejuni* populations. For microbial analysis of 0 h samples, wings were placed in a Whirl-Pak bag (710 mL) containing 150 mL of neutralizing buffered peptone water (nBPW; Acumedia-Neogen, Lansing, MI, USA) [20]. For the 24 h samples, which were already in Whirl-Pak bags, 150 mL of nBPW was aseptically poured into each bag. Sample bags containing individual wings were vertically shaken by hand with a strong downward force, 60 times, to recover cells from the wing surface. Rinsates were serially diluted (1:10) in buffered peptone water (Difco, Becton Dickinson and Company, Sparks, MD, USA) and appropriate dilutions were surface-plated, in duplicate, onto pre-warmed (42 °C) mCCA plates. Plates were placed into anaerobic containers (AnaeroPack Rectangular Jar) with an appropriate number of microaerophilic environment generating gas packs (AnaeroPack-MicroAero), per manufacturer instructions, and were incubated at 42 °C for 48 ± 1 h. Three uninoculated and untreated chicken wings were also analyzed on each of the inoculation and treatment application days, for natural microflora counts (on Tryptic Soy Agar [Acumedia-Neogen]; 25 °C for 72 h) and for any naturally-present *Campylobacter* populations (on mCCA) on the chicken wings used in the study. The detection limit of the microbiological analysis was 1 CFU/mL.

2.5. Statistical Analysis

The study was designed as a 7 (treatments) × 2 (sampling times) factorial for each solution application method (immersion, spraying), blocked by trial day. It was repeated on two separate days, and three samples were analyzed per treatment and sampling time (0 h and 24 h) in each trial (i.e., a total of six samples per treatment and sampling time). For each solution application method, recovered *C. jejuni* populations were statistically analyzed across all treatments within each sampling time (0 h, 24 h), and across the two sampling times for each antimicrobial treatment. Bacterial populations were expressed as least squares means for log CFU/mL of wing rinsate under the assumption of a log-normal distribution of plate counts. Data were analyzed using the emmeans package [21] in R (version 3.5.1). Means were separated with Tukey adjustment using a significance level of $\alpha = 0.05$.

3. Results

3.1. Untreated Chicken Wings

Aerobic microbial populations of the uninoculated and untreated chicken wings used for the study ranged from 2.6 to 4.3 log CFU/mL, with a mean of 3.6 ± 0.7 log CFU/mL. Naturally occurring *Campylobacter* populations were not detected (<1 CFU/mL) in five of the six uninoculated and untreated wings analyzed, while the remaining sample had a *Campylobacter* count of 1 CFU/mL. As such, bacterial populations recovered with the mCCA culture medium from inoculated control (untreated) and treated samples (Tables 2 and 3) were those of the inoculum strains.

Table 2. Mean (n = 6) *Campylobacter jejuni* populations (log colony-forming units [CFU]/mL ± standard deviation [SD]) for inoculated (six-strain mixture; 3 to 4 log CFU/mL) chicken wings that were left untreated (control) or were immersion-treated (5 s, 500 mL of solution per sample) with various treatment solutions.

Treatment	Mean *C. jejuni* Populations (log CFU/mL ± SD)	
	0 h	24 h
Control	3.9 ± 0.1 ^{a,z}	3.7 ± 0.3 ^{a,z}
Water	3.4 ± 0.1 ^{b,z}	3.2 ± 0.2 ^{b,z}
SSS (pH 1.2)	2.2 ± 0.1 ^{c,z}	1.6 ± 0.2 ^{c,y}
Formic acid (1.5%)	2.1 ± 0.2 ^{cd,z}	1.2 ± 0.1 ^{cd,y}
PAA (550 ppm)	1.7 ± 0.3 ^{d,z}	1.4 ± 0.4 ^{c,z}
SSS-aPAA	1.7 ± 0.3 ^{d,z}	0.9 ± 0.2 ^{de,y}
FA-aPAA	1.8 ± 0.2 ^{cd,z}	<0.6 ± 0.5 ^{e,y} *

SSS: sulfuric acid and sodium sulfate blend; PAA: peroxyacetic acid; SSS-aPAA: PAA (550 ppm) acidified with SSS (pH 1.2); FA-aPAA: PAA (550 ppm) acidified with formic acid (1.5%). $^{a–e}$ Least squares means in the same column without a common superscript letter are different ($p < 0.05$). $^{y–z}$ Least squares means in the same row without a common superscript letter are different ($p < 0.05$). * One of the six samples analyzed had a *C. jejuni* count that was below the microbial analysis detection limit of 1 CFU/mL; therefore, the mean is reported as < (less than) the mean.

Table 3. Mean (n = 6) *Campylobacter jejuni* populations (log colony-forming units [CFU]/mL ± standard deviation [SD]) for inoculated (six-strain mixture; 3 to 4 log CFU/mL) chicken wings that were left untreated (control) or were spray-treated (4 s, 69 to 83 kPa) with various treatment solutions.

Treatment	Mean *C. jejuni* Populations (log CFU/mL ± SD)	
	0 h	24 h
Control	3.9 ± 0.1 ^{a,z}	3.7 ± 0.3 ^{a,y}
Water	3.6 ± 0.1 ^{b,z}	3.5 ± 0.2 ^{ab,z}
SSS (pH 1.2)	3.4 ± 0.2 ^{bc,z}	3.3 ± 0.2 ^{bc,z}
Formic acid (1.5%)	3.2 ± 0.2 ^{cd,z}	3.0 ± 0.2 ^{cd,y}
PAA (550 ppm)	3.0 ± 0.2 ^{de,z}	2.8 ± 0.2 ^{d,z}
SSS-aPAA	2.8 ± 0.1 ^{e,z}	2.4 ± 0.5 ^{e,y}
FA-aPAA	2.7 ± 0.1 ^{e,z}	2.1 ± 0.4 ^{e,y}

SSS: sulfuric acid and sodium sulfate blend; PAA: peroxyacetic acid; SSS-aPAA: PAA (550 ppm) acidified with SSS (pH 1.2); FA-aPAA: PAA (550 ppm) acidified with formic acid (1.5%). $^{a–e}$ Least squares means in the same column without a common superscript letter are different ($p < 0.05$). $^{y–z}$ Least squares means in the same row without a common superscript letter are different ($p < 0.05$).

Immersion and spray application methods of the test solutions were evaluated on the same experiment day; therefore, the same set of untreated inoculated samples were used as controls for both application methods (Tables 2 and 3). The inoculation level of *C. jejuni* on the wings following the inoculation procedure, as determined by microbial analysis of untreated inoculated samples, was 3.9 log CFU/mL, and similar pathogen levels were recovered from untreated wings stored aerobically at 4 °C for 24 h (Tables 2 and 3).

3.2. Chicken Wings Treated by Immersion Application of Antimicrobial Treatments

C. jejuni populations recovered from immersion-treated wings immediately after treatment (0 h) and after 24 h of refrigerated (4 °C) storage are shown in Table 2. Compared to the untreated control, all six immersion treatments effectively ($p < 0.05$) reduced initial (0 h) inoculated *C. jejuni* populations (3.9 log CFU/mL), with reductions ranging from 0.5 (water) to 2.2 (PAA, and SSS-aPAA) log CFU/mL. Moreover, pathogen counts recovered from wings that had been treated with any of the five tested chemical solutions were 1.2 (SSS) to 1.7 (PAA, SSS-aPAA) log CFU/mL lower ($p < 0.05$) than the pathogen counts of samples that had been treated with water. No ($p \geq 0.05$) differences in efficacy against *C. jejuni* were observed at the 0-h sampling time between the SSS, formic acid, and FA-aPAA. Additionally, formic acid, PAA, and the two acidified PAA treatments were equally ($p \geq 0.05$) effective against *C. jejuni* immediately following their application, reducing initial populations by 1.8 (formic acid) to 2.2 (PAA, and SSS-aPAA) log CFU/mL.

Within each immersion treatment, pathogen counts of samples analyzed after the refrigerated storage period were similar (water, PAA; $p \geq 0.05$) or lower (SSS, formic acid, SSS-aPAA, FA-aPAA; $p < 0.05$) than the counts of corresponding 0-h samples (Table 2). More specifically, at the 24-h sampling time, pathogen counts of wings that had been treated with SSS, formic acid, SSS-aPAA, or FA-aPAA were 0.6, 0.9, 0.8, and >1.2 log CFU/mL lower ($p < 0.05$), respectively, than counts of the corresponding treatments at the 0-h sampling time. Furthermore, it was observed that within the 24-h sampling point, *C. jejuni* counts of wings that had been treated with SSS-aPAA or FA-aPAA were lower (by 0.5 and >0.8 log CFU/mL, respectively; $p < 0.05$) than the counts of samples that had been treated with non-acidified PAA.

3.3. Chicken Wings Treated by Spray Application of Antimicrobial Treatments

Results for the spray-treated wings are presented in Table 3. Spray application of the treatments lowered ($p < 0.05$) initial *C. jejuni* populations (3.9 log CFU/mL) by 0.3 (water) to 1.2 (FA-aPAA) log CFU/mL. No ($p \geq 0.05$) differences in efficacy against the pathogen were noted between the water treatment and SSS treatment. Additionally, formic acid and PAA had similar ($p \geq 0.05$) immediate (0 h) antimicrobial effects, reducing ($p < 0.05$) initial pathogen populations by 0.7 and 0.9 log CFU/mL, respectively. At the 0-h sampling time, surviving *C. jejuni* populations of wings treated with SSS-aPAA or FA-aPAA were lower ($p < 0.05$) than those of samples treated with SSS or formic acid (by 0.6 and 0.5 log CFU/mL, respectively). No ($p \geq 0.05$) differences in antimicrobial efficacy were obtained at 0 h between PAA and the two acidified PAA treatments.

C. jejuni counts recovered from wings treated with formic acid, SSS-aPAA or FA-aPAA, and stored at 4 °C (24 h), were 0.2, 0.4, and 0.6 log CFU/mL lower ($p < 0.05$), respectively, than the counts obtained for these treatments at the 0-h sampling time (Table 3). However, for samples that had received the water, SSS or PAA treatment, pathogen levels recovered after 24 h of storage were similar ($p \geq 0.05$) to those obtained immediately after treatment application. Lastly, as seen for the immersion application method (Table 2), although no statistical differences ($p \geq 0.05$) were observed between the PAA and either of the acidified PAA treatments at the 0-h sampling point, after refrigerated storage, pathogen counts of SSS-aPAA and FA-aPAA spray-treated samples were lower (by 0.4 and 0.7 log CFU/mL, respectively; $p < 0.05$) than those of wings that were spray-treated with PAA (Table 3).

4. Discussion

Multiple intervention strategies, including the use of chemical antimicrobial treatments, are used by the U.S. poultry processing industry to reduce the prevalence of *Campylobacter* and *Salmonella* on whole carcasses and parts [11]. These chemical decontamination treatments are applied as sprays and/or immersion (dip) treatments at pre- and post-chill stages of processing [11,22]. In the current study, SSS, formic acid, PAA, and two acidified PAA treatments (SSS-aPAA and FA-aPAA) were evaluated for their antimicrobial effects against *C. jejuni* populations on chicken wings. Overall, all of the chemical treatments were effective ($p < 0.05$) in reducing initial pathogen levels, and under the experimental conditions of the study, greater reductions were obtained when the wings received the treatment by immersion (5 s) than as a spray (4 s) (Tables 2 and 3).

The antimicrobial effects of SSS against various foodborne pathogens have been previously evaluated, mostly on beef products [23–30] but also on poultry carcasses and parts by a few investigators [18,31,32]. Scott et al. [18] reported a 1.2 log CFU/mL reduction of inoculated (5.5 log CFU/mL) *Salmonella* populations on chicken wings that were immersed for 20 s in a pH 1.1 solution of SSS. In another study [31], immersion of turkey drumsticks in SSS (pH 1.3) for 30 s lowered inoculated (7–8 log CFU/g) *Salmonella* Reading and *Salmonella* Typhimurium populations by 2.2 and 2.4 log CFU/g, respectively. In the current study, *C. jejuni* levels on wings were reduced ($p < 0.05$) by 1.7 and 0.5 log CFU/mL immediately following immersion or spray treatment with SSS (pH 1.2), respectively (Tables 2 and 3). To our knowledge, there has only been one other published study that has investigated the antimicrobial efficacy of SSS against *Campylobacter* on poultry. In this particular study [32], a 1.5 log CFU/chicken reduction of naturally occurring *Campylobacter* spp. populations was reported when post-chilled whole carcasses were immersed in SSS (pH 1.4) for 15 s.

Published reports on the use of formic acid as a decontamination treatment of poultry are limited. Riedel et al. [33] observed a 1.6 log CFU/mL reduction of *C. jejuni* inoculated on chicken skin that was immersed for 1 min in 2% formic acid. In the present study, the antimicrobial efficacy of 1.5% formic acid against initial populations of *C. jejuni* was similar ($p \geq 0.05$) to that of SSS, regardless of the application method (Tables 2 and 3). Specifically, reductions of 1.8 and 0.7 log CFU/mL were obtained for wings immersion- or spray-treated with formic acid, respectively.

As previously mentioned, PAA is currently one of the most commonly used antimicrobials in U.S. poultry slaughter and processing facilities, and its effectiveness in reducing pathogen contamination on poultry-associated products has been extensively reported [11,14,16,18,31,32,34–39]. Naturally occurring *Campylobacter* spp. levels were reduced by 2.2 log CFU/chicken when post-chilled whole carcasses were subjected to a 15 s dip in 750 ppm PAA [32]. Nagel et al. [35] also evaluated PAA as a post-chill immersion (20 s) treatment of whole carcasses and reported 1.9 and 2.0 log CFU/mL reductions of inoculated (ca. 5 log CFU/mL) *C. jejuni* populations with 400 ppm and 1000 ppm PAA, respectively. In another study [39], 200 ppm PAA applied as an immersion (60 s) or spray (62 s) treatment lowered *C. jejuni* levels of chicken carcasses by 1.4 and 0.6 log CFU/mL, respectively. PAA was also recently evaluated as a decontamination treatment of skinless, boneless chicken breast fillets [38]. Specifically, breast fillets inoculated with *Campylobacter coli* populations (4.9 log CFU/mL) were reduced by 0.9 and 0.8 log CFU/mL when they were immersed (3.5 L, 4 s) or sprayed (15 mL/s, 5 s) with 500 ppm PAA [38].

While the antimicrobial effects of PAA have been extensively investigated, there are only a few recently published studies on the use of pH-adjusted (acidified) PAA as a decontamination treatment of meat and poultry products [30,31]. In our study, no differences ($p \geq 0.05$) were obtained between PAA and the acidified PAA treatments (SSS-aPAA and FA-aPAA) with regard to reducing initial (0 h) levels of *C. jejuni* contamination, irrespective of whether the treatments were applied by immersion or in the spray cabinet. Specifically, the three PAA-containing treatments reduced 0 h pathogen populations by 2.1 to 2.2 log CFU/mL in immersion-treated samples, and 0.9 to 1.2 log CFU/mL in spray-treated samples (Tables 2 and 3). After refrigerated storage (4 °C, 24 h), however, differences ($p < 0.05$) were noted between recovered pathogen populations from wings that had been treated (immersion or spray) with PAA and those that received one of the acidified PAA treatments. While 0 h *C. jejuni* populations of PAA-treated wings remained relatively unchanged ($p \geq 0.05$) following the 24-h storage period, pathogen levels of 24 h samples that had received either of the acidified PAA treatments were lower ($p < 0.05$; by 0.8 to >1.2 log CFU/mL for immersion-treated samples, and 0.4 to 0.6 log CFU/mL for spray-treated samples) than the populations recovered from the corresponding treatments at 0 h. Acidification of PAA, regardless of the acidifier (i.e., SSS or formic acid), combines two mechanisms of action. PAA is an oxidizing agent that disrupts bacterial cell walls and essential enzyme functions [40,41], and formic acid and SSS cause cytoplasmic acidification which results in the accumulation of protons that leads to the cell using its energy to try to re-establish the intracellular pH [42–44]. Therefore, the combination of hurdles of the acidified PAA coupled with the subsequent low-temperature storage conditions probably impeded recovery of sub-lethally injured cells and likely explains the further reduction of *C. jejuni* levels in the 24-h acidified PAA-treated samples. Evidence of sub-lethal cell injury was also observed for wings that were immersed in SSS or formic acid (i.e., without PAA) (Table 2). Scott et al. [18] and Riedel et al. [33] also reported further reductions of pathogen populations following refrigerated storage of SSS- and formic acid-treated samples, respectively.

Two previous studies have evaluated the antimicrobial effects of acidified PAA treatments [30,31]. Similar to the 0 h results of our study, Olson et al. [31] reported no differences between *Salmonella* reductions obtained immediately following treatment (30 s immersion) of turkey drumsticks with 500 ppm PAA or PAA (500 ppm) acidified with SSS (pH 1.3). In contrast to the findings of our study, subsequent storage (4 °C, 24 h) did not result in further reductions of *Salmonella* populations on samples treated with SSS-acidified PAA [31]. Acidified PAA solutions have also been evaluated as spray treatments (10 s, 103 kPa) of prerigor beef carcass surface tissue for reduction of nonpathogenic *Escherichia coli* surrogates for Shiga toxin-producing *E. coli* and *Salmonella* [30]. The authors of this study reported that acidification of PAA (350 ppm or 400 ppm) with 2% acetic acid or pH 1.2 SSS did not ($p \geq 0.05$) enhance the immediate antimicrobial effects of non-acidified PAA (350 ppm or 400 ppm) [30].

5. Conclusions

Results of this investigation demonstrated that all tested chemical interventions (SSS, formic acid, PAA, SSS-aPAA and FA-aPAA) were effective ($p < 0.05$) in reducing *C. jejuni* populations on chicken wings, with greater immediate reductions obtained when the treatments were applied by immersion than by spraying. Acidification of PAA (550 ppm) with pH 1.2 SSS or 1.5% formic acid did not enhance the immediate (0 h) bactericidal effects of non-acidified PAA; however, the combination of hurdles of the acidified PAA treatments and the subsequent chilled storage conditions (4 °C, 24 h) likely prevented recovery of sub-lethally injured bacterial cells. As a result, chicken wings treated with SSS-aPAA or FA-aPAA and stored at 4 °C for 24 h had the lowest pathogen levels. Further research should be conducted to evaluate the efficacy of SSS-aPAA and FA-aPAA in reducing *C. jejuni* contamination on chicken parts when applied under conditions found in commercial processing facilities.

Author Contributions: All of the authors contributed significantly to the study. Conceptualization, K.E.B.; study design, I.G., M.N.N., K.E.B.; supervision, I.G., M.N.N., K.E.B.; performed the experiments, S.V.G.; data collection and analysis, S.V.G.; data interpretation, S.V.G., I.G.; writing—original draft preparation, S.V.G.; writing—review and editing, S.V.G., I.G., M.N.N., K.E.B. All authors have read and agreed to the published version of the manuscript.

Funding: This research was funded by Zoetis, Inc. (Parsippany, NJ, USA).

Institutional Review Board Statement: Not applicable.

Informed Consent Statement: Not applicable.

Data Availability Statement: The data presented in this study are available on request from the corresponding author.

Acknowledgments: The authors would like to thank Zoetis Inc. (Parsippany, NJ, USA) for funding the study, for providing the antimicrobial products, and for technical support from James O. Reagan and Stephen Mixon on the preparation of the chemical solutions. The authors would also like to thank Glenn Tillman (U.S. Department of Agriculture, Food Safety and Inspection Service, Office of Public Health Science, Eastern Laboratory, Microbiology Characterization Branch, Athens, GA, USA) and Shaohua Zhao (U.S. Food and Drug Administration, Center for Veterinary Medicine, Laurel, MD, USA) for providing the *C. jejuni* strains used in this study.

Conflicts of Interest: The authors declare no conflict of interest. The funder of the study had no role in the design of the study; in the collection, analyses, or interpretation of data; in the writing of the manuscript; or in the decision to publish the results.

References

1. Tack, D.M.; Ray, L.; Griffin, P.M.; Cieslak, P.R.; Dunn, J.; Rissman, T.; Jervis, R.; Lathrop, S.; Muse, A.; Duwell, M.; et al. Preliminary Incidence and Trends of Infections with Pathogens Transmitted Commonly Through Food-Foodborne Diseases Active Surveillance Network, 10 U.S. Sites, 2016–2019. *MMWR. Morb. Mortal. Wkly. Rep.* **2020**, *69*, 509–514. [CrossRef]
2. Scallan, E.; Hoekstra, R.M.; Angulo, F.J.; Tauxe, R.V.; Widdowson, M.-A.; Roy, S.L.; Jones, J.L.; Griffin, P.M. Foodborne Illness Acquired in the United States—Major Pathogens. *Emerg. Infect. Dis.* **2011**, *17*, 7–15. [CrossRef]
3. Centers for Disease Control and Prevention. *Campylobacter* (Campylobacteriosis). 2020. Available online: https://www.cdc.gov/campylobacter/index.html (accessed on 28 January 2020).
4. Moore, J.E.; Corcoran, D.; Dooley, J.S.G.; Fanning, S.; Lucey, B.; Matsuda, M.; McDowell, D.A.; Mégraud, F.; Millar, B.C.; O'Mahony, R.; et al. *Campylobacter*. *Veter. Res.* **2005**, *36*, 351–382. [CrossRef]
5. Connerton, I.F.; Connerton, P.L. *Campylobacter* Foodborne Disease. In *Foodborne Diseases*, 3rd ed.; Aldsworth, T., Stein, R., Eds.; Academic Press: Cambridge, MA, USA, 2017; pp. 209–221. ISBN 978-0-1238-5008-9.
6. Acheson, D.; Allos, B.M. *Campylobacter jejuni* Infections: Update on Emerging Issues and Trends. *Clin. Infect. Dis.* **2001**, *32*, 1201–1206. [CrossRef] [PubMed]
7. Arritt, F.M.; Eifert, J.D.; Pierson, M.D.; Sumner, S.S. Efficacy of Antimicrobials Against *Campylobacter jejuni* on Chicken Breast Skin. *J. Appl. Poult. Res.* **2002**, *11*, 358–366. [CrossRef]
8. Batz, M.B.; Hoffmann, S.; Morris, J.G. Ranking the Disease Burden of 14 Pathogens in Food Sources in the United States Using Attribution Data from Outbreak Investigations and Expert Elicitation. *J. Food Prot.* **2012**, *75*, 1278–1291. [CrossRef]
9. U.S. Department of Agriculture, Food Safety and Inspection Service. Pathogen Reduction; Hazard Analysis and Critical Control Point (HACCP) Systems; Final Rule. *Fed. Regist.* **1996**, *61*, 38805–38989.
10. U.S. Department of Agriculture, Food Safety and Inspection Service. *Safe and Suitable Ingredients Used in the Production of Meat, Poultry and Egg Products-Revision 55*; 2021. Available online: https://www.fsis.usda.gov/policy/fsis-directives/7120.1 (accessed on 11 April 2021).
11. Thames, H.T.; Sukumaran, A.T. A Review of *Salmonella* and *Campylobacter* in Broiler Meat: Emerging Challenges and Food Safety Measures. *Foods* **2020**, *9*, 776. [CrossRef]
12. U.S. Department of Agriculture, Food Safety and Inspection Service. New Performance Standards for *Salmonella* and *Campylobacter* in Young Chicken and Turkey Slaughter Establishments: Response to Comments and Announcement of Implementation Schedule. *Fed. Regist.* **2011**, *76*, 15282–15290.
13. U.S. Department of Agriculture, Food Safety and Inspection Service. New Performance Standards for *Salmonella* and *Campylobacter* in Not-Ready-to-Eat Comminuted Chicken and Turkey Products and Raw Chicken Parts and Changes to Related Agency Verification Procedures: Response to Comments and Announcement of Implementation Schedule. *Fed. Regist.* **2016**, *81*, 7285–7300.
14. Ramírez-Hernández, A.; Brashears, M.M.; Sanchez-Plata, M.X. Efficacy of Lactic Acid, Lactic Acid–Acetic Acid Blends, and Peracetic Acid To Reduce *Salmonella* on Chicken Parts under Simulated Commercial Processing Conditions. *J. Food Prot.* **2018**, *81*, 17–24. [CrossRef]
15. U.S. Department of Agriculture, Food Safety and Inspection Service. Changes to the *Campylobacter* Verification Testing Program: Revised Performance Standards for *Campylobacter* in Not-Ready-to-Eat Comminuted Chicken and Turkey and Related Agency Procedures. *Fed. Regist.* **2019**, *84*, 38203–38210.
16. Chen, X.; Bauermeister, L.J.; Hill, G.N.; Singh, M.; Bilgili, S.F.; McKee, S.R. Efficacy of Various Antimicrobials on Reduction of *Salmonella* and *Campylobacter* and Quality Attributes of Ground Chicken Obtained from Poultry Parts Treated in a Postchill Decontamination Tank. *J. Food Prot.* **2014**, *77*, 1882–1888. [CrossRef]
17. Laury, A.M.; Alvarado, M.V.; Nace, G.; Alvarado, C.Z.; Brooks, J.C.; Echeverry, A.; Brashears, M.M. Validation of a Lactic Acid– and Citric Acid–Based Antimicrobial Product for the Reduction of *Escherichia coli* O157:H7 and *Salmonella* on Beef Tips and Whole Chicken Carcasses. *J. Food Prot.* **2009**, *72*, 2208–2211. [CrossRef] [PubMed]
18. Scott, B.R.; Yang, X.; Geornaras, I.; Delmore, R.J.; Woerner, D.R.; Reagan, J.O.; Morgan, J.B.; Belk, K.E. Antimicrobial Efficacy of a Sulfuric Acid and Sodium Sulfate Blend, Peroxyacetic Acid, and Cetylpyridinium Chloride against *Salmonella* on Inoculated Chicken Wings. *J. Food Prot.* **2015**, *78*, 1967–1972. [CrossRef] [PubMed]
19. Sukumaran, A.T.; Nannapaneni, R.; Kiess, A.; Sharma, C.S. Reduction of *Salmonella* on Chicken Meat and Chicken Skin by Combined or Sequential Application of Lytic Bacteriophage with Chemical Antimicrobials. *Int. J. Food Microbiol.* **2015**, *207*, 8–15. [CrossRef]
20. Berrang, M.E.; Gamble, G.R.; Hinton, A., Jr.; Johnston, J.J. Neutralization of Residual Antimicrobial Processing Chemicals in Broiler Carcass Rinse for Improved Detection of *Campylobacter*. *J. Appl. Poult. Res.* **2018**, *27*, 299–303. [CrossRef]
21. Lenth, R. Emmeans: Estimated Marginal Means, Aka Least-Squares Means. 2019. R Package Version 1.4.3.01. Available online: https://cran.r-project.org/web/packages/emmeans/index.html (accessed on 2 December 2019).
22. Oyarzabal, O.A. Reduction of *Campylobacter* spp. by Commercial Antimicrobials Applied during the Processing of Broiler Chickens: A Review from the United States Perspective. *J. Food Prot.* **2005**, *68*, 1752–1760. [CrossRef]
23. Geornaras, I.; Yang, H.; Manios, S.; Andritsos, N.; Belk, K.E.; Nightingale, K.K.; Woerner, D.R.; Smith, G.C.; Sofos, J.N. Comparison of Decontamination Efficacy of Antimicrobial Treatments for Beef Trimmings against *Escherichia coli* O157:H7 and 6 Non-O157 Shiga Toxin-Producing *E. coli* Serogroups. *J. Food Sci.* **2012**, *77*, M539–M544. [CrossRef]

24. Geornaras, I.; Yang, H.; Moschonas, G.; Nunnelly, M.C.; Belk, K.E.; Nightingale, K.K.; Woerner, D.R.; Smith, G.C.; Sofos, J.N. Efficacy of Chemical Interventions against *Escherichia coli* O157:H7 and Multidrug-Resistant and Antibiotic-Susceptible *Salmonella* on Inoculated Beef Trimmings. *J. Food Prot.* **2012**, *75*, 1960–1967. [CrossRef]
25. Schmidt, J.W.; Bosilevac, J.M.; Kalchayanand, N.; Wang, R.; Wheeler, T.L.; Koohmaraie, M. Immersion in Antimicrobial Solutions Reduces *Salmonella enterica* and Shiga Toxin–Producing *Escherichia coli* on Beef Cheek Meat. *J. Food Prot.* **2014**, *77*, 538–548. [CrossRef] [PubMed]
26. Acuff, J.C. Evaluation of Individual and Combined Antimicrobial Spray Treatments on Chilled Beef Subprimal Cuts to Reduce Shiga Toxin-Producing *Escherichia coli* Populations. Master's Thesis, Kansas State University, Manhattan, KS, USA, 2017. Available online: https://krex.k-state.edu/dspace/handle/2097/35504 (accessed on 8 February 2020).
27. Scott-Bullard, B.R.; Geornaras, I.; Delmore, R.J.; Woerner, D.R.; Reagan, J.O.; Morgan, J.B.; Belk, K.E. Efficacy of a Blend of Sulfuric Acid and Sodium Sulfate against Shiga Toxin–Producing *Escherichia coli*, *Salmonella*, and Nonpathogenic *Escherichia coli* Biotype I on Inoculated Prerigor Beef Surface Tissue. *J. Food Prot.* **2017**, *80*, 1987–1992. [CrossRef]
28. Yang, X.; Bullard, B.R.; Geornaras, I.; Hu, S.; Woerner, D.R.; Delmore, R.J.; Morgan, J.B.; Belk, K.E. Comparison of the Efficacy of a Sulfuric Acid–Sodium Sulfate Blend and Lactic Acid for the Reduction of *Salmonella* on Prerigor Beef Carcass Surface Tissue. *J. Food Prot.* **2017**, *80*, 809–813. [CrossRef]
29. Muriana, P.M.; Eager, J.; Wellings, B.; Morgan, B.; Nelson, J.; Kushwaha, K. Evaluation of Antimicrobial Interventions against *E. coli* O157:H7 on the Surface of Raw Beef to Reduce Bacterial Translocation during Blade Tenderization. *Foods* **2019**, *8*, 80. [CrossRef]
30. Britton, B.C.; Geornaras, I.; Reagan, J.O.; Mixon, S.; Woerner, D.R.; Belk, K.E. Antimicrobial Efficacy of Acidified Peroxyacetic Acid Treatments Against Surrogates for Enteric Pathogens on Prerigor Beef. *Meat Muscle Biol.* **2020**, *4*. [CrossRef]
31. Olson, E.G.; Wythe, L.A.; Dittoe, D.K.; Feye, K.M.; Ricke, S. Application of Amplon in Combination with Peroxyacetic Acid for the Reduction of Nalidixic Acid–Resistant *Salmonella* Typhimurium and *Salmonella* Reading on Skin-on, Bone-in Tom Turkey Drumsticks. *Poult. Sci.* **2020**, *99*, 6997–7003. [CrossRef]
32. Kim, S.A.; Park, S.H.; Lee, S.I.; Owens, C.M.; Ricke, S.C. Assessment of Chicken Carcass Microbiome Responses During Processing in the Presence of Commercial Antimicrobials Using a Next Generation Sequencing Approach. *Sci. Rep.* **2017**, *7*, 43354. [CrossRef] [PubMed]
33. Riedel, C.T.; Brøndsted, L.; Rosenquist, H.; Haxgart, S.N.; Christensen, B.B. Chemical Decontamination of *Campylobacter jejuni* on Chicken Skin and Meat. *J. Food Prot.* **2009**, *72*, 1173–1180. [CrossRef] [PubMed]
34. Bauermeister, L.J.; Bowers, J.W.J.; Townsend, J.C.; McKee, S.R. The Microbial and Quality Properties of Poultry Carcasses Treated with Peracetic Acid as an Antimicrobial Treatment. *Poult. Sci.* **2008**, *87*, 2390–2398. [CrossRef] [PubMed]
35. Nagel, G.M.; Bauermeister, L.J.; Bratcher, C.L.; Singh, M.; McKee, S.R. *Salmonella* and *Campylobacter* Reduction and Quality Characteristics of Poultry Carcasses Treated with Various Antimicrobials in a Post-Chill Immersion Tank. *Int. J. Food Microbiol.* **2013**, *165*, 281–286. [CrossRef]
36. Purnell, G.; James, C.; James, S.J.; Howell, M.; Corry, J.E.L. Comparison of Acidified Sodium Chlorite, Chlorine Dioxide, Peroxyacetic Acid and Tri-Sodium Phosphate Spray Washes for Decontamination of Chicken Carcasses. *Food Bioprocess. Technol.* **2013**, *7*, 2093–2101. [CrossRef]
37. Kataria, J.; Vaddu, S.; Rama, E.N.; Sidhu, G.; Thippareddi, H.; Singh, M. Evaluating the Efficacy of Peracetic Acid on *Salmonella* and *Campylobacter* on Chicken Wings at Various pH Levels. *Poult. Sci.* **2020**, *99*, 5137–5142. [CrossRef] [PubMed]
38. Kumar, S.; Singh, M.; Cosby, D.E.; Cox, N.A.; Thippareddi, H. Efficacy of Peroxyacetic Acid in Reducing *Salmonella* and *Campylobacter* spp. Populations on Chicken Breast Fillets. *Poult. Sci.* **2020**, *99*, 2655–2661. [CrossRef]
39. Smith, J.; Corkran, S.; McKee, S.R.; Bilgili, S.F.; Singh, M. Evaluation of Post-Chill Applications of Antimicrobials against *Campylobacter jejuni* on Poultry Carcasses. *J. Appl. Poult. Res.* **2015**, *24*, 451–456. [CrossRef]
40. Imlay, J.A. Pathways of Oxidative Damage. *Annu. Rev. Microbiol.* **2003**, *57*, 395–418. [CrossRef] [PubMed]
41. Kitis, M. Disinfection of Wastewater with Peracetic Acid: A Review. *Environ. Int.* **2004**, *30*, 47–55. [CrossRef]
42. Mani-López, E.; García, H.S.; López-Malo, A. Organic Acids as Antimicrobials to Control *Salmonella* in Meat and Poultry Products. *Food Res. Int.* **2012**, *45*, 713–721. [CrossRef]
43. Stratford, M.; Plumridge, A.; Nebe-von-Caron, G.; Archer, D.B. Inhibition of Spoilage Mould Conidia by Acetic Acid and Sorbic Acid Involves Different Modes of Action, Requiring Modification of the Classical Weak-Acid Theory. *Int. J. Food Microbiol.* **2009**, *136*, 37–43. [CrossRef] [PubMed]
44. U.S. Food and Drug Administration. *GRAS Notices, GRN No. 408: Sulfuric Acid and Sodium Sulfate*; 2012. Available online: https://www.accessdata.fda.gov/scripts/fdcc/index.cfm?set=GRASNotices&id=408 (accessed on 7 February 2020).

Article

UV-C LED Irradiation Reduces *Salmonella* on Chicken and Food Contact Surfaces

Alexandra Calle [1,*], Mariana Fernandez [1], Brayan Montoya [2], Marcelo Schmidt [1] and Jonathan Thompson [3]

[1] School of Veterinary Medicine, Texas Tech University, 7671 Evans Dr, Amarillo, TX 79106, USA; Mariana.fernandez@ttu.edu (M.F.); Marcelo.schmidt@ttu.edu (M.S.)
[2] Escuela de Medicina Veterinaria, Universidad Nacional de Costa Rica, Lagunilla, Heredia 40101, Costa Rica; brayan.montoya.torres@est.una.ac.cr
[3] Department of Chemistry, Texas Tech University, MS 1061, Lubbock, TX 79409, USA; jon.thompson@ttu.edu
* Correspondence: alexandra.calle@ttu.edu; Tel.: +1-806-834-4074

Abstract: Ultraviolet (UV-C) light-emitting diode (LED) light at a wavelength of 250–280 nm was used to disinfect skinless chicken breast (CB), stainless steel (SS) and high-density polyethylene (HD) inoculated with *Salmonella enterica*. Irradiances of 2 mW/cm^2 (50%) or 4 mW/cm^2 (100%) were used to treat samples at different exposure times. Chicken samples had the lowest *Salmonella* reduction with 1.02 and 1.78 Log CFU/cm^2 ($p \leq 0.05$) after 60 and 900 s, respectively at 50% irradiance. Higher reductions on CB were obtained with 100% illumination after 900 s (>3.0 Log CFU/cm^2). *Salmonella* on SS was reduced by 1.97 and 3.48 Log CFU/cm^2 after 60 s of treatment with 50% and 100% irradiance, respectively. HD showed a lower decrease of *Salmonella*, but still statistically significant ($p \leq 0.05$), with 1.25 and 1.77 Log CFU/cm^2 destruction for 50 and 100% irradiance after 60 s, respectively. Longer exposure times of HD to UV-C yielded up to 99.999% (5.0 Log CFU/cm^2) reduction of *Salmonella* with both irradiance levels. While UV-C LED treatment was found effective to control *Salmonella* on chicken and food contact surfaces, we propose three mechanisms contributing to reduced efficacy of disinfection: bacterial aggregation, harboring in food and work surface pores and light absorption by fluids associated with CB.

Keywords: UV-C; *Salmonella*; chicken; microbial intervention; food-contact surfaces

Citation: Calle, A.; Fernandez, M.; Montoya, B.; Schmidt, M.; Thompson, J. UV-C LED Irradiation Reduces *Salmonella* on Chicken and Food Contact Surfaces. *Foods* **2021**, *10*, 1459. https://doi.org/10.3390/foods10071459

Academic Editor: Juana Fernández-López

Received: 6 May 2021
Accepted: 22 June 2021
Published: 24 June 2021

1. Introduction

Salmonella sp. is a major public health concern and a common food safety hazard associated with poultry processing [1–4]. Foodborne illness caused by this microorganism is one of the most frequent diseases affecting millions of people worldwide every year. Outbreaks related to *Salmonella* in poultry are very frequent [1,2,5]. A recent report by the Centers for Disease Control and Prevention (CDC), between 2015 and 2017, stated that poultry was associated with 262 outbreaks, 4807 illnesses, 849 hospitalizations and 12 deaths in the United States [6]. *Salmonella* is usually carried by live animals in their gastrointestinal track and transferred to processing environments where end-products can become contaminated [7,8]. Consistently, the presence of *Salmonella* in poultry houses is also very common with up to 100% prevalence among surveyed operations [9]. Efforts to control this pathogen are constantly made by the industry and government [10]. The most typical interventions to reduce *Salmonella* in poultry products involve the application of chemical treatments at different steps of processing, which include the use of organic acids, inorganic compounds, chlorine-based treatments, phosphate-based products, among other chemical compounds [11]. Consumers of poultry seem to have adverse opinions about the use of such chemicals in food [12], creating a challenge for the food industry to control bacterial contaminants. Therefore, poultry facilities would benefit from having alternative technologies to chemical interventions for pathogen control in food and processing environment.

The use of ultraviolet light has been proven to be effective for microbial inactivation by damaging bacterial DNA [13–15]. Pathogens absorb the ultraviolet (UV) light and thymine dimers are formed, blocking transcription and replication, which ultimately lead to cell death [15,16]. There is a growing interest in the use of UV treatments for the inactivation of pathogens in food [17]. The use of UV light in the food industry gained interest after the approval by the Food & Drug Administration (FDA) in 1997 to use UV irradiation as an alternative for microbial control in meat products [18,19]. Today, applications of UV light are commonly used to control pathogens in water, for decontamination of food contact surfaces (bakeries, dairy, and meat plants), and for decontamination of food packaging materials (boxes, bottles, leads, food wrapping films, thermoformable plastics, cartons for liquid foods and others) [19]. Commercially available equipment can be found advertised to disinfect surfaces, but most of the applications pertain to treating drinking water or being used for washing food products.

Several studies have demonstrated the effectiveness of using this technology in a wide variety of food products, such as fresh berries, apple juice, milk, fresh fish, processed meats, and in water [16,20–24]. Similarly, several studies have investigated the use UV light produced by mercury lamps, demonstrating the effectiveness of this technology in a wide variety of food products. However, mercury lamps require high voltage power supplies for operation, and certain lamps produce deep UV radiation of $\lambda < 240$ nm that generates significant quantities of ozone, a very reactive oxidative gas harmful to human health and food quality. UV light-emitting diodes (LED) are increasingly being used as substitutes for mercury lamps for several reasons. UV LEDs are much smaller than mercury lamps and generate less heat. As a result, they may be placed close to food contact surfaces to achieve high irradiance, and presumably more effective inactivation of pathogens. In addition, the emission spectrum of UV LEDs can be tuned to emit UV light specifically of wavelengths between 250–280 nm, which are most effective at driving the photochemical reactions leading to formation of thymine dimers. Considering the need to control *Salmonella* in poultry operations, this research aimed to evaluate the effectiveness of UV-C LED light for the reduction of *Salmonella* sp. applied to the surface of chicken breasts (CB), stainless steel (SS), and high-density polyethylene (HD) using different times and irradiance intensities.

2. Materials and Methods

2.1. Bacterial Cultures

A five-strain *Salmonella* cocktail was prepared with *Salmonella* Thyphimurium ATCC BAA-712, *Salmonella* Newport ATCC 6962 (food poisoning fatality), *Salmonella* Enteritidis ATCC 31194, *Salmonella* Senftenberg ATCC 43845, and *Salmonella* Heidelberg ATCC 8326. Each strain was grown individually by transferring 10 µL from the stock culture into 9-mL of Tryptic Soy Broth (TSB) (EMD Millipore Chemicals; Darmstadt, Germany) and incubating for 24 h at 37 °C. Equal amounts (2 mL) from each grown *Salmonella* suspension were combined into a sterile test tube and homogenized. The bacterial cocktail was freshly prepared prior each repetition. *Salmonella* concentration in the cocktail was confirmed at each repetition of the experiment by conducting serial dilutions and plating onto Trypticase Soy Agar (TSA) (Becton, Dickinson and Company, Franklin Lakes, NJ, USA), followed by incubation for 24 h at 37 °C and subsequent enumeration.

2.2. UV-C LED Light and Surfaces Subjected to Irradiation

The ultraviolet type C (UV-C) light used as the irradiation source for this project was a Klaran class LED acquired from Crystal IS Inc. (Green Island, NY, USA). The UV-C LED had a wavelength range of 250–280 nm, 20 mW power and a viewing angle of 105 degrees. The lamp was operated under forward bias at a maximum 400 mA current, corresponding to 100% irradiance, which is the maximum current recommended by the manufacturer. The average irradiance used in this study was either 2 mW/cm^2 (referred in this experiment as 50% or half irradiance) or 4 mW/cm^2 (referred in this experiment as 100% or full irradiance). Three different surfaces were treated with UV-C LED irradiation: (1) boneless skinless chicken breast (CB), (2) stainless steel (SS) and (3) high density polyethylene (HD). To treat each surface, experiments were carried out on 2 × 2 cm coupons used as the experimental units. SS and HD were selected to be treated with the UV-C light since they are commonly used as food-contact surfaces in the poultry processing industry.

2.3. Chicken Inoculation and Treatment

Chicken breast was obtained boneless and skinless from a local supermarket. Portions of 2 × 2 cm and approximately 4 mm thick were aseptically cut. The upper surface was inoculated with the five-strain *Salmonella* cocktail at a target concentration of ca. 6.0 Log CFU/cm^2. The inoculated CB squares were placed on a tray and set under refrigeration for 30 min to allow for bacterial attachment. Two irradiance conditions, 50 and 100%, were explored. In all cases, the CB squares were irradiated individually under the UV-C LED source. In the first case (50% irradiance), the CB squares were treated for varying times with integrated doses of UV-C radiation corresponding to 0–1.8 J/cm^2. For the second treatment (100% irradiance), the UV-C dose ranged from 0–3.6 J/cm^2. As light intensity scales linearly with drive current, the UV-C irradiance was controlled by metering the drive current of the LED. The exposure times were: 60, 180, 300, 600 and 900 s. An additional control set of samples (inoculated, not irradiated) were considered. Control samples are referred as 0 s.

2.4. Stainless Steel Inoculation and Treatments

Stainless Steel 304 (SS, C 0.08% max., Mn 2.00% max., P 0.045% max., S 0.03% max., Si 0.75% max., Cr 18.00–20.00%, Ni 8.00–12.00%, N 0.10 max., Fe balance), 2 mm thickness was obtained from Agrosuper (Rancagua, Chile). Sterile SS squares were surface inoculated before each experiment. Squares were cleaned, degreased with acetone, flamed with 95% ethanol, stored in a glass container and autoclaved at 121 °C (15 lb/in^2) for 15 min. Sterile SS squares were surface-inoculated by applying a 20 μL aliquot of the five-strain *Salmonella* cocktail on one side of each 4 cm^2 square. A target surface inoculation of 6.5 Log_{10} CFU/cm^2 was attempted. The inoculum was completely spread on the entire surface using a sterile 1-μL loop and then let sit for 30 min under refrigeration to dry and for bacterial attachment. Treatments were performed with both the low and high irradiance cases, applying a spatially averaged irradiance of approximately 2 mW/cm^2 and 4 mW/cm^2, respectively. Irradiation occurred for a period of 15, 30, 45 and 60 s, and additionally for a control set of samples (inoculated, not irradiated). Controls are referred to as 0 s.

2.5. High Density Polyethylene Inoculation and Treatments

Kitchen cutting boards (approx. 1 cm thick) were obtained from the microbiology research lab, which had been previously used to chop meat samples. The cutting boards were intentionally chosen as used to mimic scratched surfaces from processing facilities. Prior to the study, the HD board was cut into 2 × 2 cm squares (4 cm^2). HD squares were treated and inoculated following the same procedures as with SS. Both the full irradiance (100%) and half irradiance (50%) cases were considered. Irradiation times included trials for: 30, 60, 90, 120, 150, 180, 300, 600 and 900 s. Additionally, control samples (inoculated, not irradiated) were tested and referred as 0 s.

2.6. Analysis of Chicken Rinse Fluid

Fluids associated with CB were analyzed to evaluate whether they could offer a protective coat effect for bacteria by absorbing ultraviolet light. The extent of light absorption by CB juices was studied by ultraviolet-visible (UV-VIS) absorption spectroscopy. A portion of chicken breast was placed in a plastic bag and 10 mL of deionized water was added to wash the surface of the chicken. A 3 mL portion of the deionized water was collected and placed into a 1 cm path length quartz cuvette, and the full UV—VIS absorbance spectrum was recorded against a deionized water blank on an Agilent photodiode array spectrometer with 1 nm spectral resolution.

2.7. Microbial Analysis

CB portions were placed immediately after the treatment into 9-mL Buffered Peptone Water (BPW) (BD BBL™, Franklin Lakes, NJ, USA) tubes and thoroughly homogenized. Serial dilutions were conducted to facilitate enumeration followed by spread plating on Xylose-Lysine-Tergitol 4 (XLT4) (BD Difco™ Franklin Lakes, NJ, USA). Inoculated XLT4 plates were incubated for 24 h at 37 °C. SS and HD squares exposed to the LED UV-C treatment were transferred immediately after the exposure time to sterile conical tubes (50 mL capacity, Corning™ Falcon™) containing 10 mL of phosphate-buffered saline solution (PBS, Sigma-Aldrich®, Saint Louis, MO, USA), then mixed by vortex motion for 60 s to transfer the bacterial cells from the surface to the saline solution. The number of viable bacteria in the saline solution was determined by serially diluting with BPW, spread plating on XLT4 plates and incubating for 24 h at 37 °C. For each surface, colonies were enumerated upon incubation, and final counts were reported as CFU/cm^2 considering the size of CB, SS, and HD coupons of 4 cm^2. Control samples were also enumerated following the corresponding protocol.

2.8. Electron Micrographs

Scanning electron microscope (SEM) images were taken by the Texas Tech College of Arts and Sciences Microscopy (CASM). Samples were provided to CASM frozen at −80 °C with the bacterial cells suspended in sterile water. CB, SS, and HD squares with bacterial cells were dried frozen and coated with Iridium (Ir). SEM imaging were obtained with an electron microscope Zeiss Crossbeam 540 FIB-SEM.

2.9. Statistical Analysis

Each surface (CB, SS and HD) treated with the UV-C was subjected to two different treatment combinations that included irradiance and exposure time. Analyses of variance were used to test the effect of time periods (illumination time; measured in seconds) on *Salmonella* reduction (Log CFU/cm^2) under two levels of irradiance exposure (irradiance), and on three specific surfaces conditions (i.e., chicken breast, stainless steel, and high-density polyethylene. Three experimental repetitions were conducted and a total of six separate ANOVAs were conducted. Each model revealed a significant ($\alpha = 0.05$) UV-C illumination time effect on Log CFU/cm^2. Multiple comparisons were calculated using Bonferronni correction to determine differences at each level of the illumination time variable. All statistical analyses were conducted with STATA (StataCorp. 2019. Stata Statistical Software: Release 16. College Station, TX, USA: StataCorp LLC.).

3. Results and Discussion

LED UV-C treatment was applied to inactivate *Salmonella* sp. deposited on three different surfaces: chicken breast (CB), type 304 stainless steel (SS) and high-density polyethylene (HD). For all samples tested, two irradiance intensities were tested, 2 mW/cm^2 (50%) and 4 mW/cm^2 (100%). Illumination times between 0 and 900 s (0 and 15 min) were explored. An overview of the findings per treatment is summarized in Tables 1–3 and discussed below. The UV-C wavelengths used during the experiments were in the range 250–280 nm, which are considered safe for food products according to the FDA permitted levels of 253.7 nm [25]; however, this regulation only refers to the use of mercury lamps and not LED lamps.

Table 1. *Salmonella* reduction on chicken breast.

Illumination Time (s)	Irradiance [1] (mW/cm^2)	UV [4] Dose (J/cm^2)	Bacterial Count (Log CFU/cm^2)	St. Dev. [5]	Reduction [2] (Log CFU/cm^2)	Bacterial Reduction (%) [3]
0	2	0	6.21	0.16	-	-
60	2	0.12	5.20	0.48	1.01	90.2
180	2	0.36	4.89	0.81	1.32	95.2
300	2	0.60	4.64	0.67	1.57	97.3
600	2	1.20	4.36	0.70	1.85	98.6
900	2	1.80	4.43	0.70	1.78	98.3
0	4	0	6.26	0.11	-	-
60	4	0.24	4.21	0.77	2.05	99.1
180	4	0.72	3.99	0.86	2.27	99.5
300	4	1.2	3.67	0.63	2.59	99.7
600	4	2.4	3.89	0.44	2.37	99.6
900	4	3.6	3.25	0.53	3.01	99.9

[1] Irradiance of 2 and 4 mW/cm^2 are equivalent to 50 and 100%, respectively. [2] Reduction based on the initial attachment at time 0. [3] Percentage calculated using actual values of colony forming units (CFU) before log transformation. [4] Ultraviolet. [5] Standard Deviation.

Table 2. *Salmonella* reduction on stainless steel.

Illumination Time (s)	Irradiance [1] (mW/cm^2)	UV Dose (J/cm^2)	Log CFU/cm^2	St. Dev.	Reduction [2] (Log CFU/cm^2)	Bacterial Reduction (%) [3]
0	2	0	3.4	0.61	-	-
15	2	0.03	2.1	0.72	1.3	93.7
30	2	0.06	1.94	0.83	1.46	95.6
45	2	0.09	1.87	0.72	1.53	96.3
60	2	0.12	1.43	0.41	1.97	98.7
0	4	0	6.27	0.49	-	-
15	4	0.06	4.91	0.56	1.36	95.6
30	4	0.12	3.78	1.5	2.49	99.7
45	4	0.18	3.47	0.65	2.8	99.8
60	4	0.24	2.79	1.76	3.48	99.9

[1] Irradiance of 2 and 4 mW/cm^2 are equivalent to 50 and 100%, respectively. [2] Reduction based on the initial attachment at time 0. [3] Percentage calculated using actual values of colony forming units (CFU) before log transformation. St. Dev. refers to standard deviation and UV refers to ultraviolet.

3.1. Boneless Skinless Chicken Breast (CB)

Results for reduction of *Salmonella* on CB are reported in Table 1. For the CB treated with 50% irradiance, initial bacterial attachment was estimated to be 6.21 ± 0.16 Log CFU/cm^2. Significant ($p \leq 0.05$) reductions of *Salmonella* were obtained after each of the treatment times (60, 180, 300, 600 and 900 s) compared to the starting inoculation level. After 60 s of exposure, *Salmonella* decreased by 1.02 Log CFU/cm^2, which was significant at $p \leq 0.05$. Upon completion of a 900 s irradiance, a total *Salmonella* reduction of 1.78 Log CFU/cm^2 ($p \leq 0.05$) was achieved.

On the other hand, when CB was treated with 100% irradiance the reduction of *Salmonella* was enhanced. Considering the initial attachment level of *Salmonella* observed in the samples (6.26 ± 0.11 Log CFU/cm^2), significant ($p \leq 0.05$) reductions were also obtained after each treatment time relative to the *Salmonella* level before treatments. Data show a total reduction of >3.0 Log CFU/cm^2 during the total exposure time (900 s). Based on the data obtained at the different time points, the rate of reduction of *Salmonella* occurred most efficiently within the first 60 s of UV illumination. During this time, *Salmonella* was reduced by 2.05 Log CFU/cm^2, which was significant at $p \leq 0.05$. After that first minute of UV-C exposure, *Salmonella* was reduced only by an additional 0.96 Log CFU/cm^2 total, which was still significant at $p \leq 0.05$. Comparable results were found by McLeod et al. [26]. In their investigation using 254 nm wavelength, skinless chicken fillets were exposed for 5,

10, 30, 60 and 300 s. After the first 60 s of treatment, they were able to observe a *Salmonella* reduction of 1.5 Log CFU/cm^2. However, when the exposure was 300 s, a 2.4 Log CFU/cm^2 reduction was achieved.

Table 3. *Salmonella* reduction on high density polyethylene.

Illumination Time (s)	Irradiance [1] (mW/cm^2)	UV Dose (J/cm^2)	Log CFU/cm^2	St. Dev.	Reduction [2] (Log CFU/cm^2)	Bacterial Reduction (%) [3]
0	2	0	6.58	0.16	-	-
30	2	0.06	5.67	0.28	0.91	87.7
60	2	0.12	5.33	0.12	1.25	94.4
90	2	0.18	5.28	0.17	1.3	95.0
120	2	0.24	5.13	0.16	1.45	96.5
150	2	0.30	5.05	0.25	1.53	97.0
180	2	0.36	4.57	0.47	2.01	99.0
300	2	0.6	2.75	1.68	3.83	99.99
600	2	1.2	2.04	1.68	4.54	99.997
900	2	1.8	1.84	1.46	4.74	99.998
0	4	0	5.20	0.15	-	-
30	4	0.12	3.97	0.24	1.23	94.1
60	4	0.24	3.43	0.24	1.77	98.3
90	4	0.36	2.82	0.29	2.38	99.6
120	4	0.48	2.42	0.07	2.78	99.8
150	4	0.60	2.40	0.00	2.8	99.8
180	4	0.72	0 *	0 *	5.2	99.999
300	4	1.20	0 *	0 *	5.2	99.999
600	4	2.40	0 *	0 *	5.2	99.999
900	4	3.60	0 *	0 *	5.2	99.999

[1] Irradiance of 2 and 4 mW/cm^2 are equivalent to 50 and 100%, respectively. [2] Reduction based on the initial attachment at time 0. [3] Percentage calculated using actual values of colony forming units (CFU) before log transformation. St. Dev. refers to standard deviation and UV refers to ultraviolet. * No colonies recovered.

There are three hypothesis we propose for this outcome. First, the porosity of the chicken surface could play an important role in the reduced effect of the UV-C against *Salmonella*. An image of the boneless skinless chicken breast sample used during this research was obtained via electron microscopy (Figure 1). The chicken surface may visually appear smooth; however, cracks, crevices, and/or pores could protect bacterial cells from light exposure [26]. A single *Salmonella* cell is 2–5 µm long by 0.5–1.5 µm wide [27], thus the micro holes in the chicken breast, as observed in the electron micrograph (Figure 1A), could harbor bacterial cells. Some cells may become trapped or sequestered within the irregular and porous surface of the chicken, which may affect the effectiveness of the UV-C to evenly cover and reach the entire surface area of the sample as seen in Figure 1B,C. As described by Lagunas-Solar et al. [28], complex surface properties of foods bring a challenge; microorganisms located in pores and crevices of a food surface can be shaded from light, and thus remain unaffected. The use of UV light can be more effective to reduce microorganisms on foods with smooth surfaces such as fresh whole fruits, vegetables, hard cheese, and smooth-surface meat slices [29].

The second hypothesis for reduced efficacy of UV illumination after the initial minute is that the fluid on the surface associated with CB absorbs ultraviolet radiation, reducing the light intensity and thereby reducing the rate of bacterial deactivation. To support this premise, the UV-VIS absorption spectrum of the fluid associated with the chicken breast samples was obtained. It was found that the fluid strongly absorbs ultraviolet light below 300 nm, and this light absorption will lower the intensity of light interacting with *Salmonella*, offering a protective or shielding effect to prevent deactivation (see Section 3.5). The presence of fluids between bacterial cells and the light, most likely affects the efficacy of the treatment as the liquid may absorb the light [26,30].

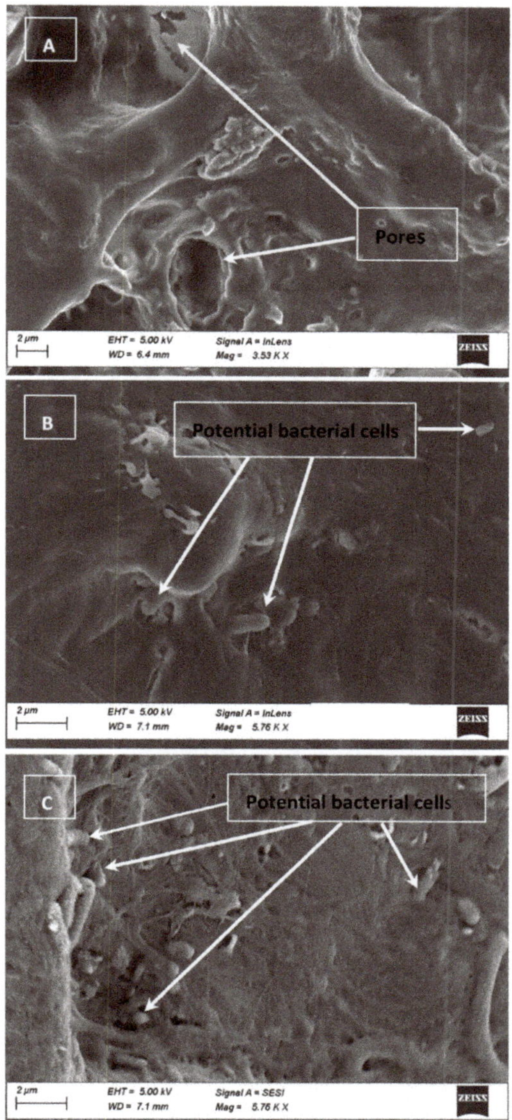

Figure 1. Electron micrograph of chicken breast samples illustrating the porous nature of the chicken (**A**). *Salmonella*, being only 2–5 μm long by 0.5–1.5 μm wide may enter pores on the surface of the chicken and be sheltered from full illumination (see (**B**,**C**)).

The third hypothesis involves the tendency for cells to aggregate into clusters. When illuminated, cells near the surface of the cluster (nearest to LED) may absorb the UV radiation and be inactivated. However, cells located beneath the top layer may be shaded from full illumination and protected. The premise for this mechanism of bacterial cell protection during UV irradiation has been presented recently in the literature [31].

3.2. Stainless Steel (SS)

The reduction of *Salmonella* was evaluated at intervals of 15 s during a period of 60 s, and the summary of the findings is presented in Table 2. For experiments with 50% irradiance, the initial attachment level of *Salmonella* on the SS squares was only 3.4 ± 0.61 Log CFU/cm^2. Results indicate that a rapid reduction (1.3 Log CFU/cm^2) of *Salmonella* occurred after the first 15 s of exposure to UV light, which was statistically different ($p \leq 0.05$) from the starting level. When the total exposure time of 60 s was applied, a reduction of 1.97 Log CFU/cm^2 ($p \leq 0.05$) was observed.

In the case of experiments with 100% irradiance, SS was inoculated with an average load of 6.26 ± 0.49 Log CFU/cm^2 of *Salmonella*. A loss of approx. 1.36 Log CFU/cm^2 *Salmonella* occurred within the first 15 s of illumination, and nearly 2.49 Log CFU/cm^2 was reduced after 30 s of exposure to the UV LED. These reductions were statistically significant at $p \leq 0.05$. The reduction of *Salmonella* over time on SS did not plateau and level-off, but rather a decrease in numbers continued through the entire duration of the experiment. The highest reduction was observed after 60 s of UV-C exposure (3.48 Log CFU/cm^2). Lim and Harrison [14] evaluated the effect of UV-C (254 nm) in reducing *Salmonella* contamination on 3 × 5 cm stainless steel coupons. They obtained reductions of 2.75 and 3.51 Log CFU/coupon of 15 cm^2 after treatment times of 5 and 30 s, respectively. Bae and Lee [32] exposed stainless steel for longer periods and found reductions of 1.25 and 2.02 Log CFU/coupon of 5 × 2 cm after 30 min and 1 h, respectively. While it appears that their investigation suggests a low effectiveness of the UV treatment, it is important to mention that their group used a UV 253.7 nm wavelength with intensity of 0.236 ± 0.013 mW/cm^2, which was much lower than the irradiance used in the current research (2 or 4 mW/cm^2). Consistent reductions were also observed by Sommers et al. [30]. Their findings indicate a *Salmonella* reduction of 5 Log CFU/coupon on stainless steel when inoculated coupons were exposed to UV-C at a dose of 400 mJ/cm^2. When inoculated coupons were treated with 50 mJ/cm^2, the pathogens were reduced by only 1.86–3.05 Log CFU/coupon. Kim et al. [31] found that UV-C intensities of 250 or 500 μW/cm^2 decreased three target microorganisms (*L. monocytogenes*, *S. typhimurium*, and *E. coli* O157:H7) on stainless steel surfaces. A UV-C dose of 90 mJ/cm^2 reduced the three pathogens by >4 Log CFU/coupon; however, a dose of 15 mJ/cm^2 decreased the pathogens by 2.43–4.38 Log CFU/sample. These doses and times were considerably higher (1, 2, and 3 min) compared to those use in the current research.

Based on the above cited investigations, it may be possible to increase the rate of *Salmonella* destruction by increasing exposure times. To investigate whether the porosity of the SS surface causes harboring of cells, electron micrographs of the SS coupons used during the present experiment were obtained (Figure 2). The images show minor surface imperfections. Although the depth can't be determined, the apparent size of the crevice may not be large enough to harbor *Salmonella* cells (Figure 2A). An agglomeration of cells was also observed (Figure 2B), forming horizontal and vertical layers of cells (Figure 2C).

3.3. High Density Polyethylene (HD)

Salmonella reduction was observed when HD was treated with UV-C, as presented in Table 3. For the treatment of HD with 50% irradiance, the initial inoculation level was 6.58 ± 0.16 Log CFU/cm^2. After 30 s of exposure, *Salmonella* was reduced by nearly 1 Log CFU/cm^2 ($p > 0.05$); however, only after 150 s of irradiation was a significant ($p \leq 0.05$) reduction in *Salmonella* obtained. Disinfection on the HD surface followed a different temporal pattern compared to CB, as statistically significant reduction of *Salmonella* continued to be achieved after even several minutes of illumination. This result suggests the *Salmonella* on HD surfaces may not experience the shielding effect proposed for the CB samples (*vide supra*).

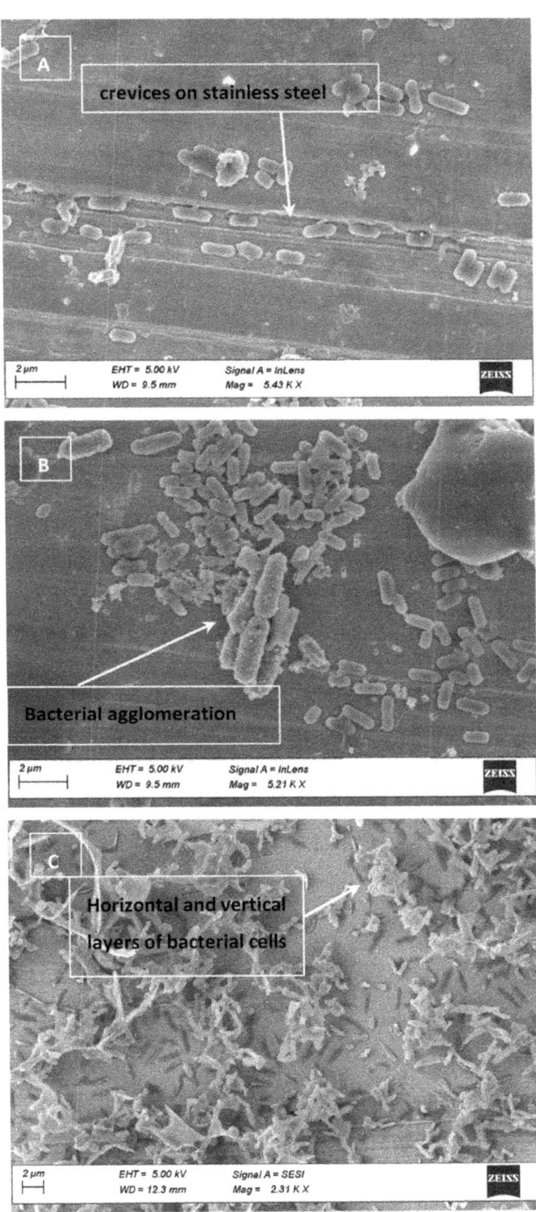

Figure 2. Electron micrograph of stainless steel (SS) inoculated with *Salmonella* showing imperfections on SS (**A**) and agglomeration of the cells on the surface (**B**). Vertical and horizontal agglomerations were observed when high volumes of cells are present (**C**).

When HD squares were treated with 100% irradiance, *Salmonella* was also effectively reduced. The initial attachment level of the microorganism was 5.20 ± 0.15 Log CFU/cm^2. Experimental data showed a statistically significant ($p \leq 0.05$) reduction of 1.77 Log CFU/cm^2 during the first 60 s of exposure, with approximately 1.23 Log CFU/cm^2 reduction ($p > 0.05$)

occurring within the initial 30 s. Lim and Harrison [14] obtained similar results when they exposed 35 cm high density polyethylene coupons inoculated with *Salmonella*. After 5 and 30 s of treatment with UV-C light (254 nm), the reduction of *Salmonella* was 2.93 and 4.32 Log CFU/coupon of 15 cm^2, respectively. In 2011, Haughton et al. [33] treated nine different food contact surfaces (black & white polypropylene, polystyrene, aluminum, polyethylene-polypropylene blend, polyolefin, polyvinyl chloride, stainless steel, polyethylene) with UV-C does ranging from 0–192 mJ/cm^2. The authors found that *C. jejuni*, *E. coli*, and *Salmonella* could be reduced by >2 Log CFU/cm^2 on all surfaces during treatment. However, substantial differences in disinfection efficacy were noted for different materials. For the polyethylene cutting board tested, a UV-C dose of <20 mJ/cm^2 was effective at inactivating the bacteria to levels below the limit of detection. Bae and Lee [32] obtained reductions of 1.62 and 1.18 Log CFU/coupon of 5 × 2 cm after 30 min and 1 h of UV treatment. Although the authors reported statistically significant reductions relative to the level of *Salmonella* before treatments, they were considerably lower than the reductions found in the present research.

Longer exposure times yielded much higher inactivation levels of *Salmonella*. After 180 s of treatment, no *Salmonella* was recovered from the samples in any of the repetitions. To confirm the inactivation of *Salmonella*, samples exposed to the UV-C treatment during 180, 300, 600 and 900 s were enriched in 10-mL BPW, incubated overnight at 37 °C, and streaked onto XLT4. After 24 h of incubation at 37 °C no *Salmonella* colonies were recovered. Sommers, et al. [34] inoculated both stainless steel and HDPE surfaces with *F. tularensis* in food exudate prior to treating with UV-C. These authors found that exposure to 500 mJ/cm^2 reduced the pathogen level by >4 Log CFU/coupon for both surfaces. However, their treatment was at a higher UV dose, and it is possible that *F. tularensis* is less sensitive to UV treatment compared to *Salmonella*.

Electron micrograph of the HD coupons used during these experiments were obtained. As depicted in Figure 3, deep crevices were observed (Figure 3A,B), which could potentially hide bacterial cells. These crevices may be associated with use of the board for cutting. Clumping of cells was observed (Figure 3C). Thus, reductions obtained when the UV dose was >0.72 J/cm^2, indicate that the quality of the surface may not have affected bacterial survival.

Only in the case of HD did the UV dose seem to show consistent reductions regardless of what combinations of irradiance and time were used to achieve the given dose. For example, when the UV dose was 0.12 J/cm^2, a reduction of 1.25 and 1.23 Log CFU/cm^2 was observed at 50 and 100% illumination, respectively. Similar cases were observed with 0.24, 0.36, and 1.2 J/cm^2 as observed in Table 3. This appears to follow the Bunsen–Roscoe reciprocity law, which suggests that the effectiveness of the irradiation is achieved regardless of what combination of time and irradiation rate is used to reach a certain UV dose exposure (short exposure with high irradiance or long exposure time with high irradiance) [30].

3.4. Cell Clumping and SEM Images

SEM images obtained from the inoculated surfaces and chicken are presented in Figures 1–3. On CB and HD, pores were large enough to shelter *Salmonella* cells. Electron micrographs of SS show clear scratches that are long but apparently not wide or deep enough to harbor *Salmonella*. As an important finding, SEM images with bacterial cells inoculated on the CB, SS and HD show vertical and horizontal accumulations of cells (clumping or aggregation). The conglomeration of cells may also cause shading and shielding effects, protecting those cells that are below the top layer, as previously mentioned [31]. It can be hypothesized that with a larger concentration of cells on the surfaces, UV-C light penetration could represent a challenge. This possibility should consider the fact that chicken and food contact surfaces could also carry other microorganisms that could potentially cause shielding.

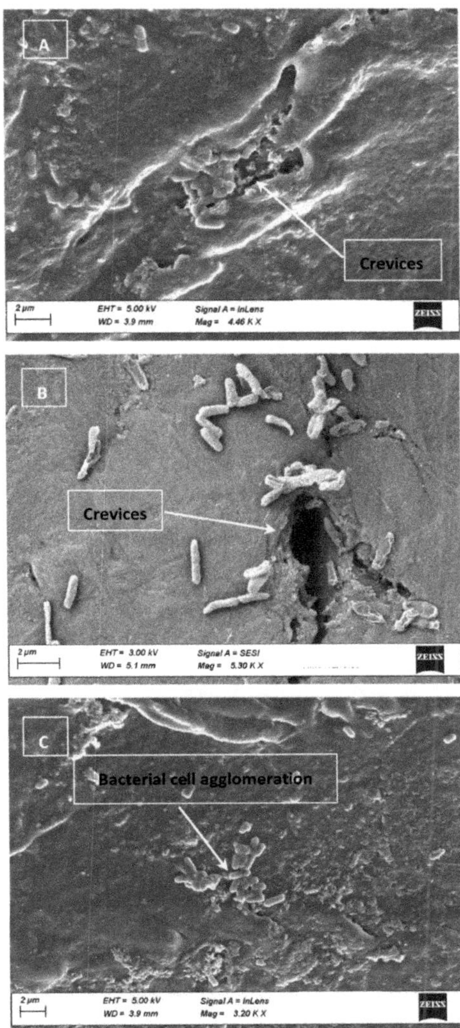

Figure 3. (**A–C**) Electron micrograph of three different high-density polyethylene (HD) samples used during the experiments. Based on the scale indicated in the micrograph, the crevice highlighted in (**B**) appears to be large enough to harbor *Salmonella* cells (rod shaped). (**C**) depicts bacterial cell agglomerates present on the surface of the HD.

3.5. Absorption of UV Light by Chicken Rinse Fluid

An extremely high absorption of light by the fluid present on the chicken at wavelengths lower than 300 nm was observed. The resulting absorbance spectrum is depicted in Figure 4. The data suggests that <0.01% of light below 290 nm was transmitted through the 1 cm path sample used. The fluid associated with the chicken breast absorbs UV light very strongly, and bacteria immersed within this fluid are likely protected or sheltered from photochemical damage caused by irradiation by the LED. This effect may cause the observed rapid initial reduction in bacterial load followed by leveling off between 4–5 Log CFU/cm^2. Bacterial cells not well immersed within the fluid may experience full illumination from the LED and resultant deactivation, while other bacterial cells more immersed within the fluid/broth are sheltered by the fluid's absorption of light and are protected.

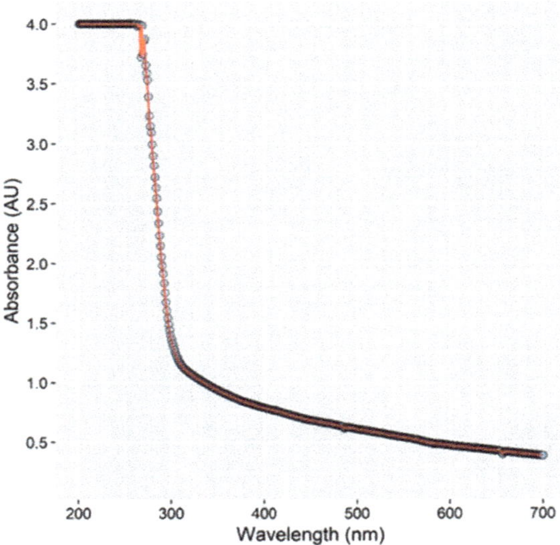

Figure 4. Absorption spectrum of fluid removed from surface of chicken breast (CB). The graph represents the light absorbance of the chicken fluid vs. the wavelength of light. As observed, below 300 nm, the absorption of light increases, which could potentially shelter *Salmonella* and prevent deactivation.

The effectiveness of UV-C light on disinfecting liquids is known to be dependent on the type of fluid [35]. A low transmittance of UV light is common in fluids other than water due to their tendency to scatter and/or absorb UV light [12]. When liquids have low transmissivity due to the presence of organic compounds, soluble solutes or particulate matter, UV-C disinfection can be challenging [36]. As a point of reference, the penetration depth of some fluid foods (the distance at which 90% of the light is absorbed) is 0.67, 0.25, 0.22, 0.10 and 0.01 mm for clear apple cider, apple cider, liquid sucrose, orange juice, and egg whites, respectively [12].

The commercial availability of deep UV-C LEDs has led to an emergence of potential applications in the food processing industry [37,38]. Due to their advantages, LED lamps are now being implemented in systems for water disinfection; however, other uses are currently rare. One exciting application is in the disinfection of food products and food contact surfaces while on the production line. LED devices could be more robust, durable, and portable compared to mercury lamps because there are no glass tubes that may break and contaminate workstations with mercury.

4. Conclusions

This study demonstrated the effectiveness of UV-C LED at reducing *Salmonella* on chicken breast samples and common food contact surfaces such as stainless steel and high-density polyethylene. At a minimum, a 1 Log CFU/cm^2 reduction for CB was noted in trials, with up to 3 Log CFU/cm^2 being reached. Further reductions seemed to be limited by the remaining *Salmonella* in the sample being shaded from the UV-C light. This is believed to occur by *Salmonella* sheltering within pores on the CB surface or behind neighboring bacterial cells, absorption of UV light by fluid present on the CB, or both effects simultaneously. *Salmonella* was also reduced on both food contact surfaces, yielding reductions up to 3.5 and 5.2 Log CFU/cm^2 on stainless steel and high-density polyethylene, respectively. An increase in irradiance yielded higher reductions of *Salmonella* on food and food contact surfaces with up to 99.999% in the case of HD.

A clumping cell factor, when large number of bacteria are accumulated on the surfaces, should be considered. Electron micrographs showed formation of layers of *Salmonella* that extended horizontally and accumulated vertically, which could protect cells beneath the top layer.

UV-C LED illumination could be an effective means to deactivate *Salmonella*, especially for nonporous surfaces which are not UV light absorbing.

By doubling the irradiance (mW/cm^2) from 50 to 100%, the UV dose (J/cm^2) deposited on each surface was also increased or duplicated. Larger UV doses were directly correlated with the *Salmonella* reduction (Log CFU/cm^2) attained on each surface tested; however, such reduction did not necessarily double. In other words, *Salmonella* reductions were consistent with the intensity of exposure but not exactly proportional to the increase in the UV dose.

The majority of research studies investigating the effect of UV treatments to control bacterial pathogens from food or food contact surfaces focus on the use of conventional mercury UV lamps. Since the present investigation found the effectiveness of using UV-C LED light for food and environmental surface treatment, findings could be relevant particularly to the poultry industry. The advantages of UV-C LEDs over chemical treatments and conventional mercury UV should be highlighted when considering UV-C LEDs as an alternative for pathogen control. UV-C LEDs do not contain mercury, are environmentally friendly, robust, durable, energy efficient, and their full illumination power can be reached more rapidly, without time delay for warm-up [39].

Author Contributions: Conceptualization, A.C. and J.T.; methodology, A.C.; formal analysis, M.S.; investigation, M.F., B.M.; resources, A.C.; writing—original draft preparation, A.C. and J.T.; supervision, A.C.; funding acquisition, A.C. All authors have read and agreed to the published version of the manuscript.

Funding: This research received no external funding.

Institutional Review Board Statement: Not applicable.

Informed Consent Statement: Not applicable.

Data Availability Statement: Project data has been reported within this manuscript.

Acknowledgments: The authors want to acknowledge ICFIE laboratory personnel at Texas Tech University for their support during this research project, and Ilan Arvelo for overseeing some experiments conducted by research students. Special acknowledgment to Bo Zhao from CASM for processing samples and capturing SEM images. Acknowledgement of contribution of effort in no way formally implies endorsement of statements made within this manuscript.

Conflicts of Interest: The authors declare no conflict of interest.

References

1. USDA; FSIS. Serotypes Profile of Salmonella Isolates from Meat and Poultry Products January 1998 through December 2014. 2016. Available online: https://www.fsis.usda.gov/wps/wcm/connect/3866026a-582d-4f0e-a8ce-851b39c7390f/Salmonella-Serotype-Annual-2014.pdf?MOD=AJPERES (accessed on 17 April 2021).
2. Tack, D.M.; Ray, L.; Griffin, P.M.; Cieslak, P.; Dunn, J.; Rissman, T.; Jervis, R.; Lathrop, S.; Muse, A.; Duwell, M.; et al. Preliminary Incidence and Trends of Infections with Pathogens Transmitted Commonly Through Food—Foodborne Diseases Active Surveillance Network, 10 U.S. Sites, 2016–2019. *MMWR Morb. Mortal. Wkly. Rep.* **2020**, *69*, 509–514. [CrossRef]
3. Grant, A.; Parveen, S.; Schwarz, J.; Hashem, F.; Vimini, B. Reduction of Salmonella in ground chicken using a bacteriophage. *Poult. Sci. J.* **2017**, *96*, 2845–2852. [CrossRef]
4. Cox, N.A.; Richardson, L.J.; Cason, J.A.; Buhr, R.J.; Vizzier-Thaxton, Y.; Smith, D.P.; Fedorka-Cray, P.J.; Romanenghi, C.P.; Pereira, L.V.; Doyle, M.P. Comparison of neck skin excision and whole carcass rinse sampling methods for microbiological evaluation of broiler carcasses before and after immersion chilling. *J. Food Prot.* **2010**, *73*, 976–980. [CrossRef] [PubMed]
5. Antunes, P.; Mourão, J.; Campos, J.; Peixe, L. Salmonellosis: The role of poultry meat. *Clin. Microbiol. Infect.* **2016**, *22*, 110–121. [CrossRef]

6. Centers for Disease Control and Prevention (CDC). National Center for Emerging and Zoonotic Infectious Diseases (NCEZID). December 2018. Available online: https://wwwn.cdc.gov/norsdashboard/ (accessed on 17 April 2019).
7. Whyte, P.; Collins, J.D.; McGill, K.; Monahan, C.; O'Mahony, H. Quantitative investigation of the effects of chemical decontamination procedures on the microbiological status of broiler carcasses during processing. *J. Food Prot.* **2001**, *64*, 179–183. [CrossRef]
8. Capita, R.; Prieto, M.; Alonsi-Calleja, C. Sampling methods for microbiological analysis of red meat and poultry carcasses. *J. Food Prot.* **2004**, *67*, 1303–1308. [CrossRef] [PubMed]
9. Velasquez, C.; Macklin, K.; Kumar, S.; Bailey, M.; Ebner, P.; Oliver, H.; Martin-Gonzalez, F.; Singh, M. Prevalence and antimicrobial resistance patterns of Salmonella isolated from poultry farms in southeastern United States. *Poult. Sci. J.* **2018**, *97*, 2144–2152. [CrossRef] [PubMed]
10. U.S. Department of Agriculture (USDA); Food Safety Inspection Service (FSIS). Roadmap to Reducing Salmonella. Driving Change through Science-Based Policy. 2020. Available online: https://www.fsis.usda.gov/sites/default/files/media_file/2020-12/FSISRoadmaptoReducingSalmonella.pdf (accessed on 20 April 2021).
11. Loretz, M.; Stephan, R.; Zweifel, C. Antimicrobial activity of decontamination treatments for poultry carcasses: A literature survey. *Food Control* **2010**, *21*, 791–804. [CrossRef]
12. Koutchma, T. Advances in Ultraviolet Light Technology for Non-thermal Processing of Liquid Foods. *Food Bioproc. Technol.* **2009**, *2*, 138–155. [CrossRef]
13. Hessling, M.; Gross, A.; Hoenes, K.; Rath, M.; Stangl, F.; Tritschler, H.; Sift, M. Efficient Disinfection of Tap and Surface Water with Single High Power 285 nm LED and Square Quartz Tube. *IEEE Photonics J.* **2016**, *3*, 7. [CrossRef]
14. Hijnen, W.; Beerendonk, E.; Medema, G. Inactivation credit of UV radiation for viruses, bacteria and protozoan (oo)cysts in water. *Water Res.* **2006**, *40*, 3–22. [CrossRef] [PubMed]
15. Lim, W.; Harrison, M.A. Effectiveness of UV light as a means to reduce Salmonella contamination on tomatoes and food contact surfaces. *Food Control* **2016**, *66*, 166–173. [CrossRef]
16. Wang, T.; MacGregor, S.; Anderson, J.; Woolsey, G. Pulsed ultra-violet inactivation spectrum of Escherichia coli. *Water Res.* **2005**, *39*, 2921–2925. [CrossRef] [PubMed]
17. U.S. Food and Drug Administration (FDA). Kinetics of Microbial Inactivation for Alternative Food Processing Technologies–Ultraviolet Light. 2013. Available online: http://www.fda.gov/Food/FoodScienceResearch/SafePracticesforFoodProcesses/ucm103137.htm (accessed on 2 May 2021).
18. Morehouse, K.; Komolprasert, V. Overview of Irradiation of Food and Packaging. ACS Symposium Series 875, Irradiation of Food and Packaging. 2004; Chapter 1; pp. 1–11. Available online: https://www.fda.gov/food/irradiation-food-packaging/overview-irradiation-food-and-packaging (accessed on 31 May 2021).
19. Koutchma, T. UV Light for Processing Foods. *Ozone Sci. Eng.* **2008**, *30*, 93–98. [CrossRef]
20. Koivunen, J.; Heinonen-Tanski, H. Inactivation of enteric microorganisms with chemical disinfectants, UV irradiation and combined chemical/UV treatments. *Water Res.* **2005**, *39*, 1519–1526. [CrossRef]
21. Matak, K.; Churey, J.; Worobo, W.; Sumner, S.; Hovingh, E.; Hackney, C.; Pierson, M. Efficacy of UV Light for the Reduction of Listeria monocytogenes in Goat's Milk. *J. Food Prot.* **2005**, *68*, 2212–2216. [CrossRef]
22. Ozer, N.; Demirci, A. Inactivation of Escherichia coli O157:H7 and Listeria monocytogenes inoculated on raw salmon fillets by pulsed UV-light treatment. *Int. J. Food Sci. Technol.* **2006**, *41*, 354–360. [CrossRef]
23. Keklik, N.; Krishnamurthy, K.; Demirci, A. Microbial Decontamination of Food by Ultraviolet (UV) and Pulsed UV Light. In *Microbial Decontamination in the Food Industry: Novel Methods and Applications*; Elsevier: Amsterdam, The Netherlands, 2012; pp. 344–369.
24. Bialka, K.; Demirci, A.; Puri, V. Modeling the inactivation of Escherichia coli O157:H7 and Salmonella enterica on raspberries and strawberries resulting from exposure to ozone or pulsed UV-light. *J. Food Eng.* **2008**, *85*, 444–449. [CrossRef]
25. U.S. Food and Drug Administration (FDA). Code of Federal Regulations Title 21, Part 179: Irradiation in the Production, Processing and Handling of Food. 2010. Available online: https://www.accessdata.fda.gov/scripts/cdrh/cfdocs/cfcfr/CFRSearch.cfm?CFRPart=179&showFR=1 (accessed on 1 May 2021).
26. McLeod, A.; Hovde Liland, K.; Haugen, J.E.; Sørheim, O.; Myhrer, K.S.; Holck, A.L. Chicken fillets subjected to UV-C and pulsed UV light: Reduction of pathogenic and spoilage bacteria, and changes in sensory quality. *J. Food Saf.* **2018**, *38*, 1–15. [CrossRef]
27. Andino, A.; Hanning, I. Salmonella enterica: Survival, colonization, and virulence differences among serovars. *Sci. World J.* **2015**, *2015*, 520179. [CrossRef]
28. Lagunas-Solar, M.C.; Piña, C.; MacDonald, J.D.; Bolkan, L. Development of pulsed UV light processes for surface fungal disinfection of fresh fruits. *J. Food Prot.* **2006**, *69*, 376–384. [CrossRef] [PubMed]
29. Oms-Oliu, G.; Martín-Belloso, O.; Soliva-Fortuny, R. Pulsed Light Treatments for Food Preservation. A Review. *Food Bioproc. Tech.* **2010**, *3*, 13–23. [CrossRef]
30. Gómez-López, V.; Ragaert, P.; Debevere, J.; Devlieghere, F. Pulsed light for food decontamination: A review. *Trends Food Sci. Technol.* **2007**, *9*, 464–473. [CrossRef]
31. Kim, D.; Kang, D. Effect of surface characteristics on the bactericidal efficacy of UVC LEDs. *Food Control* **2020**, *108*, 106869. [CrossRef]

32. Bae, Y.-M.; Lee, S.-Y. Inhibitory Effects of UV Treatment and a Combination of UV and Dry Heat against Pathogens on Stainless Steel and Polypropylene Surfaces. *J. Food Sci.* **2012**, *77*, M61–M64. [CrossRef] [PubMed]
33. Haughton, P.N.; Lyng, J.G.; Cronin, D.A.; Morgan, D.J.; Fanning, S.; Whyte, P. Efficacy of UV Light Treatment for the Microbiological Decontamination of Chicken, Associated Packaging, and Contact Surfaces. *J. Food Prot.* **2011**, *74*, 565–572. [CrossRef] [PubMed]
34. Sommers, C.H.; Sites, J.E.; Musgrove, M. Ultraviolet light (254 nm) inactivation of pathogens on foods and stainless steel surfaces. *J. Food Saf.* **2010**, *30*, 470–479. [CrossRef]
35. Shama, G. Ultraviolet Light. In *Encyclopedia of Food Microbiology-2*; Batt, C.A., Tortorello, M.L., Eds.; Academic Press: London, UK, 2014; pp. 665–671.
36. Guerrero-Beltran, J.A.; Barbosa-Canovas, G.V. Advantages and Limitations on Processing Foods by UV Light. *Int. J. Food Sci. Technol.* **2004**, *10*, 137–147. [CrossRef]
37. Shin, J.Y.; Kim, S.J.; Kim, D.K.; Kang, D.H. Fundamental characteristics of deep-UV light-emitting diodes and their application to control foodborne pathogens. *Appl. Environ. Microbiol.* **2016**, *82*, 2–10. [CrossRef] [PubMed]
38. D'Souza, C.; Yuk, H.; Hoon Khoo, G.; Zhou, W. Application of Light-Emitting Diodes in Food Production, Postharvest Preservation, and Microbiological Food Safety. *Compr. Rev. Food Sci. Food Saf.* **2015**, *14*, 719–740. [CrossRef]
39. Jarvis, P.; Autin, O.; Goslan, E.H.; Hassard, F. Application of ultraviolet light-emitting diodes (UV-LED) to full-scale drinking-water disinfection. *Water* **2019**, *11*, 1894. [CrossRef]

MDPI

Article

Effect of UV-C Irradiation and Lactic Acid Application on the Inactivation of *Listeria monocytogenes* and Lactic Acid Bacteria in Vacuum-Packaged Beef

Giannina Brugnini [1,2], Soledad Rodríguez [1], Jesica Rodríguez [1] and Caterina Rufo [1,*]

[1] Instituto Polo Tecnológico de Pando, Facultad de Química, Universidad de la República, By Pass de Pando y Ruta 8, Pando 91000, Uruguay; gbrugnini@fq.edu.uy (G.B.); srodriguez@fq.edu.uy (S.R.); jsrodriguez@fq.edu.uy (J.R.)

[2] Graduate Program in Chemistry, Facultad de Química, Universidad de la República, General Flores 2124, Montevideo 11800, Uruguay

* Correspondence: crufo@fq.edu.uy; Tel.: +598-2-292-2021

Abstract: The objective of this study was to test the effect of the combined application of lactic acid (0–5%) (LA) and UV-C light (0–330 mJ/cm^2) to reduce *Listeria monocytogenes* and lactic acid bacteria (LAB) on beef without major meat color (L *, a *, b *) change and its impact over time. A two-factor central composite design with five central points and response surface methodology (RSM) were used to optimize LA concentration and UV-C dose using 21 meat pieces (10 g) inoculated with *L. monocytogenes* (LM100A1). The optimal conditions were analyzed over 8 weeks. A quadratic model was obtained that predicted the *L. monocytogenes* log reduction in vacuum-packed beef treated with LA and UV-C. The maximum log reduction for *L. monocytogenes* (1.55 ± 0.41 log CFU/g) and LAB (1.55 ± 1.15 log CFU/g) with minimal impact on meat color was achieved with 2.6% LA and 330 mJ/cm^2 UV-C. These conditions impaired *L. monocytogenes* growth and delayed LAB growth by 2 weeks in vacuum-packed meat samples throughout 8 weeks at 4 °C. This strategy might contribute to improving the safety and shelf life of vacuum-packed beef with a low impact on meat color.

Keywords: beef; lactic acid; UV-C; *Listeria monocytogenes*; LAB; response surface methodology

Citation: Brugnini, G.; Rodríguez, S.; Rodríguez, J.; Rufo, C. Effect of UV-C Irradiation and Lactic Acid Application on the Inactivation of *Listeria monocytogenes* and Lactic Acid Bacteria in Vacuum-Packaged Beef. *Foods* **2021**, *10*, 1217. https://doi.org/10.3390/foods10061217

Academic Editors: Huerta-Leidenz Nelson and Markus F. Miller

Received: 1 May 2021
Accepted: 19 May 2021
Published: 28 May 2021

1. Introduction

Listeria monocytogenes is a human pathogen that may cause listeriosis, a foodborne infection with a low morbidity and a high mortality rate (20–30%) [1]. The presence of *L. monocytogenes* in raw meat does not cause major public health problems since meat is generally consumed after cooking at temperatures above 70 °C. However, contaminated raw meat when used as raw material for food products that in their production process fail to eliminate the pathogen may present a safety risk [1]. In addition, the presence of *L. monocytogenes* in raw meat constitutes restrictions on international trade.

Contamination of meat with *Listeria monocytogenes* is a consequence of the production process [2]. In addition, *L. monocytogenes* can survive and grow in vacuum-packed meat cuts stored at temperatures between 0 and 4 °C; therefore, different strategies are applied in abattoirs to minimize bacterial contamination [3,4]. Among the different strategies, lactic acid (LA) application is accepted because it does not present risks to consumer health. The maximum concentration of LA allowed is 5% (m/v) [5]. UV-C light irradiation (UV-C) stands out for its low cost, non-generation of potentially hazardous chemical residues, and low carbon footprint [6]. In addition, UV-C irradiation is an FDA-approved intervention for surface decontamination of foods [7] (FDA, 2019a).

The application of between 2 and 4% LA on meat was reported to reduce *L. monocytogenes* counts on beef surface [8,9]. Reductions between 0.79 to 1.31 log CFU/cm^2 were obtained in fresh beef when LA was applied from 1% to 4% [9]. Different levels of

L. monocytogenes reductions have been observed and were associated with factors such as variabilities among strain sensitivity towards stress and forms of LA application [10,11].

UV-C radiation (200 to 280 nm) has been used for decontamination of food surfaces [12]. The ability of UV-C to inactivate *L. monocytogenes* has also been reported, being strain dependent and showing a direct correlation between UV-C dose and *Listeria monocytogenes* reduction [13–15]. In addition, UV-C radiation can penetrate the packaging material usually used on meat and meat products such as transparent polypropylene and polyethylene bags [16] and cause significant *L. monocytogenes* reduction on food [13].

Antimicrobial interventions may affect fresh meat color, which is considered to be the single most important characteristic influencing consumer's purchase decisions [17,18]. Negative effects on meat color are the major problem associated with the use of lactic acid, especially at high concentration [19]. In contrast, UV-C (118–590 mJ/cm^2) on fresh meat does not appear to cause detrimental color changes [15].

Other bacteria present on the beef surface may be affected by LA and UV-C [15,20]. In refrigerated vacuum-packed meat, lactic acid bacteria (LAB) are the ones that develop the most, being responsible for the production of strong lactic acid off-odors when counts reach 10.000.000 UFC/g at the end of shelf life [21–23]. Thus, knowing the effect of LA and UV-C application on LAB may be relevant to improve vacuum packed beef shelf life.

In the past few years, special attention has been given to experimental design and response surface methodology (RSM) to optimize conditions in different systems [24,25]. These modeling tools enable the study of the simultaneous effects of different factors and their interactions on experimental characteristics. This strategy has not been widely used to study the effects of the LA and UV-C combination on vacuum-packed beef. To date, only one report was found describing a similar strategy to study the effects of LA and UV-C on *Salmonella typhimurium* reduction on sliced Brazilian dry-cured loin [26].

In the present work, it was hypothesized that the combined application of LA at low concentrations and UV-C after vacuum packaging might achieve a significant level of reduction in *L. monocytogenes* contamination on beef with a minimal impact on meat color and would contribute to its shelf life by reducing meat LAB counts. To test this hypothesis, a two-factor central composite design and response surface methodology (RSM) were used to optimize the concentration of LA and the UV-C dose applied to vacuum-packed meat that will reduce the amount of *L. monocytogenes* and LAB without significant effects on meat color.

2. Materials and Methods

2.1. Meat Samples

Eye of round *(Semitendinosus Muscle)* cuts were obtained from a local abattoir. Meat samples were not decontaminated prior to the study. Meat was cut by hand into square pieces of 10 g measuring 5 × 5 cm^2. Each piece was individually inoculated with *L. monocytogenes* and treated according to the experimental design.

2.2. *L. monocytogenes* Culture Preparation

A strain of *L. monocytogenes* (LM 100A1) previously isolated and characterized in our laboratory was used for this study [27]. The culture was prepared by growing LM 100A1 overnight at 35 °C, to the stationary phase, in tryptic soy broth (Oxoid Ltd., Hampshire, UK) supplemented with 0.6% yeast extract (Oxoid Ltd., Hampshire, UK). The overnight culture was diluted with butterflied buffer to 6.1 log CFU/mL.

2.3. Preparation of Lactic Acid Solution

The lactic acid solutions were prepared by diluting a concentrated lactic acid solution (85% *m*/*v*) (PURAC®, Corbion, Montevideo, Uruguay) with sterilized distilled water to make 2.5%, 5.0% and 6.0% (*m*/*v*) lactic acid solution. Fresh solutions were prepared prior to each test.

2.4. UV-C Irradiation

The specifications of the UV-C lamp used were: 30 W T6 tubular 254 nm with UV germicidal lamp (Code ZW30S19W-Z894, Cnlight Co., Ltd., Guangdong, China), diameter 19 mm, length 894 mm and UV intensity at one meter of 107 $\mu W/cm^2$. Intensity at the application distance was 3.137 mW/cm^2 measured with a ZED Smart Meter s/N 800,009 (EN61326-1-2013) and the reference sensor D-SICONORM-LP-REF-500 W/m^2.

Before each trial, the UV-C lamp was preheated for 20 min to stabilize the UV-C emission. UV-C treatments did not increase the surface temperature of the meat to greater than 20 °C.

2.5. Experimental Design

A two-factor central composite design with five central points, 2 replicates of factorial points and 2 replicates of axial points were used. The experimental design matrix and all data analysis were performed using Design-Expert® (Version 10, Stat-Ease Inc., Minneapolis, MN, USA). The independent variables were lactic acid concentration (X_1) and UV-C dose (X_2), and the dependent variables or response variables were *L. monocytogenes* (LM) log reduction (Y_1), Lactic Acid Bacteria (LAB) log reduction (Y_2), and Chroma value (Y_3). The design matrix consisted of 21 experimental runs including 8 factorial points and 8 axial points with five replicates at the center point (Table 1). Observed responses were fitted to first order, second order and quadratic models. Models were selected by the Sequential Sum of Square Method and assessed based on statistically significant coefficients and R^2 values using ANOVA technique, with a significance level of $\alpha = 0.05$. For each response variable (Y), a second-order polynomial model equation was defined:

$$Y = \beta_0 + \beta_1 X_1 + \beta_2 X_2 + \beta_{11} X_1^2 + \beta_{22} X_2^2 + \beta_{12} X_1 X_2 \quad (1)$$

where Y is the measured response of the dependent variables, X_1 and X_2 are the independent variables, β_0 is the intercept, β_1 and β_2 are the linear coefficients, β_{11} and β_{22} are the squared coefficients, and β_{12} is the interaction coefficient.

Table 1. Central composite experimental design matrix and observed responses.

Runs	% Lactic Acid (*m*/*v*) (X_1)	UV-C Dose (mJ/cm^2) (X_2)	Reduction LM (Log CFU/g) (Y_1)	Reduction LAB (Log CFU/g) (Y_2)	Chroma Value (Y_3) [a]
1	2.5	398	1.59	2.44	21.48
2	0.0	165	0.62	0.66	26.90
3	2.5	0	0.38	0.45	19.14
4	2.5	0	0.63	0.59	20.25
5	2.5	165	1.20	1.13	18.69
6	5.0	330	1.74	1.14	15.02
7	5.0	330	1.34	1.70	14.51
8	5.0	0	0.58	1.48	17.34
9	2.5	165	1.53	1.11	18.73
10	0.0	0	0.04	0.08	22.57
11	2.5	165	1.43	2.25	21.61
12	2.5	398	1.45	1.67	20.61
13	0.0	165	0.82	0.15	25.24
14	6.0	165	1.49	2.24	16.60
15	0.0	0	−0.04	−0.08	22.77
16	5.0	0	0.85	2.06	16.82
17	2.5	165	1.08	1.39	18.69
18	6.0	165	0.96	1.63	17.43
19	0.0	398	0.92	0.49	23.82
20	0.0	398	0.94	1.12	23.20
21	2.5	165	1.42	1.94	18.73

[a] Mean of three values per sample.

Response surface methodology (RSM) included the generation of 3D response surface and contour plots to study the overall relationships and interactions between independent variables and response factors.

2.6. Sample Treatments

According to the experimental design, 21 pieces of 10 g of meat were inoculated with 5.8 log of CFU of the strain LM100A1. Then, 500 µL of the inoculum were disposed on the meat surface and spread with a bent glass rod.

After 10 min, inoculated meat pieces were treated with 1.5 mL of lactic acid solutions from 0 to 6 (m/v) %. The LA was disposed on the inoculated side, drop by drop covering the entire surface of the meat samples. Then the samples were vacuum packaged in a multi-laminar (EVA, PVDC, PE) thermo-shrinkable bag with a 76% UV-C transmission rate (Cryovac® BB 2620; 50 µm thick, oxygen permeability of 20 cm^3 m^{-2}, 24 h, at 23 °C, and 75% RH; and maximum carbon dioxide permeability of 100 cm^3 m^{-2}, 24 h, at 23 °C, and 75% RH) by use of a vacuum-packaging machine SAMMIC model V-410SGI (Spain).

After packaging, each side of the samples were exposed to 3.137 mW/cm^2, at 7 cm from the emitting lamp, for 53, 105 and 127 s, achieving doses of UV-C of 165, 330 and 398 mJ/cm^2 respectively. The UVC-dose range was selected considering the reported UV-C doses applied on beef that did not affect meat color [15], the application conditions, distance from the lamp and duration of the exposures, to be easily implementable in industrial production lines. A second set of samples was prepared for color measurements.

2.7. Microbiological Analyses

After treatments, samples were homogenized in sterile bags with 90 mL of Butterfield buffer, appropriate dilutions were plated by duplicate on PALCAM Listeria Selective Agar (Oxoid Ltd., Hampshire, UK) incubated at 37 °C for 48 h for *L. monocytogenes* and on MRS Agar (Oxoid Ltd., Hampshire, UK) incubated anaerobically at 30 °C for 72 h for LAB. Colonies were counted and log transformed. Log reductions of *L. monocytogenes* and LAB per gram of meat compared to samples with no treatments were calculated.

2.8. Instrumental Color

At twenty-four hours post treatments, color measurements were performed 30 min after opening the packages. Instrumental lean color (CIE L*: brightness, a*: redness and b*: yellowness) was measured with a Minolta chromameter CR-400 (Konica Minolta Sensing Inc., Tokyo, Japan) using a C illuminant, a 2° standard observer angle and 8 mm aperture size, and calibrated with a white tile before use. Three measurements from each sample were taken and the mean value was calculated. Chroma value was calculated as $C^* = \sqrt{(a^{*2} + b^{*2})}$.

2.9. Optimization and Model Validation

The optimized conditions, lactic acid concentration and UV-C dose, were obtained by applying the following constraints on the response factors: (i) to maximize *L. monocytogenes* log reduction; (ii) to maximize LAB log reduction; and (iii) Chroma value > 20, according to MacDougall, et al. 1982 [28] (values above 20 indicate bright red beef).

To validate the proposed model, three experiments were carried out using the optimized conditions as the checkpoint. Experimental responses (log reduction of LM100A1 per gram and log reduction of LAB per gram) of the checkpoint were compared to the predicted results from the fitted models to evaluate the precision of the polynomial equations.

2.10. Evolution of *L. monocytogenes*, LAB, pH and Instrumental Color of LA/UV-C Treated Meat Vacuum Packed and Stored at 4 °C for 8 Weeks

Three experiments were carried out using the optimized conditions (LA 2.6% (w/v) and UV-C 330 mJ/cm^2). Meat samples were treated as explained in 2.6., untreated samples were used as a control group. Samples were stored at 4 °C and analyzed for LM, LAB at initial time (week 0) and every two weeks until week 8. Instrumental color was measured at week 0 and week 8. An additional set of samples was prepared for pH determination using a Hanna® model 9025c pH meter with a surface electrode (HI1413B). The T-test was used to compare the data for control and LA/UV-C samples. Data were expressed as mean ± standard deviation. Significance was determined at the $p < 0.05$ level.

3. Results

3.1. Central Composite Design and Response Surface

L. monocytogenes log reduction (Y_1) varied from no reduction to 1.74 log CFU/g, and LAB log reduction (Y_2) varied from none to 2.44 log CFU/g. The Chroma value (Y_3) ranged from 14.51 to 26.90 (Table 1). The analysis of variance (ANOVA) indicates that the best-fitted model ($p < 0.0001$) for *L. monocytogenes* log reduction was the quadratic model, and for LAB log reduction and meat color the best fitted model was the linear (Table 2).

Table 2. Summary of the regression analysis of the three responses.

Response	Model	Significance	R^2	Adjusted R^2	Predicted R^2	Adequate Precision
Y_1	Quadratic	<0.0001	0.9038	0.8718	0.8111	16.646
Y_2	Linear	<0.0001	0.5774	0.5304	0.4329	9.563
Y_3	Linear	<0.0001	0.8002	0.7780	0.7314	14.571

For *L. monocytogenes* log reduction (Y_1), the quadratic model had a significance of $p < 0.0001$ and an R^2 value of 0.9038, explaining 90.38% of the variability in the response. The similarity between the R^2 and adjusted R^2 values showed the adequacy of the model to predict the corresponding response. The resulting signal-to-noise ratio, measured by the term "adequate precision" (above 4), indicated that the model could be used to navigate the design space (Table 2). The lack of fit of the quadratic model was not significant (F = 0.26, $p = 0.8522$).

For LAB log reduction (Y_2), the linear model had a significance of $p < 0.0001$, an R^2 value of 0.5774 similar to the adjusted R^2, and a non-significant lack of fit (F = 2.43, $p = 0.0903$). However, the low R^2 value (Table 2) indicates that the model has low precision in the predictions.

The adjusted linear model for meat color (Y_3) had a significance of $p < 0.0001$ with an R^2 value of 0.8002, similar to the adjusted R^2 (Table 2). The lack of fit of this model was significant (F = 6.74, $p = 0.0026$), suggesting that besides LA and UV-C, there are other factors affecting meat color that were not considered in the experimental design.

For *L. monocytogenes* and LAB log reduction, both LA and UV-C were significant factors. For color, UV-C was not a significant factor in our system. The generated equations for each response, including only the terms with statistical significance ($p < 0.05$), were as follows (Equations (2)–(4)):

$$Y_1 = -0.033819 + 0.35890\ X_1 + 6.054 \times 10^{-3}\ X_2 - 0.043525\ {X_1}^2 - 9.17 \times 10^{-6}\ {X_2}^2 \quad (2)$$

$$Y_2 = 0.24423 + 0.24127\ X_1 + 2.04052 \times 10^{-3}\ X_2 \quad (3)$$

$$Y_3 = 23.63579 - 1.41044\ X_1 \quad (4)$$

The 3D response surface plots allow one to visualize the response in the design space (Figure 1). Both LA and UV-C in the ranges studied have a positive effect on LM 100A1 and LAB reduction (Figure 1a,b). For *L. monocytogenes* reduction the 3D response surface plot reflects a curvature according to the quadratic terms in the equation model (Figure 1a). The LAB 3D response surface plot does not present curvature (Figure 1b).

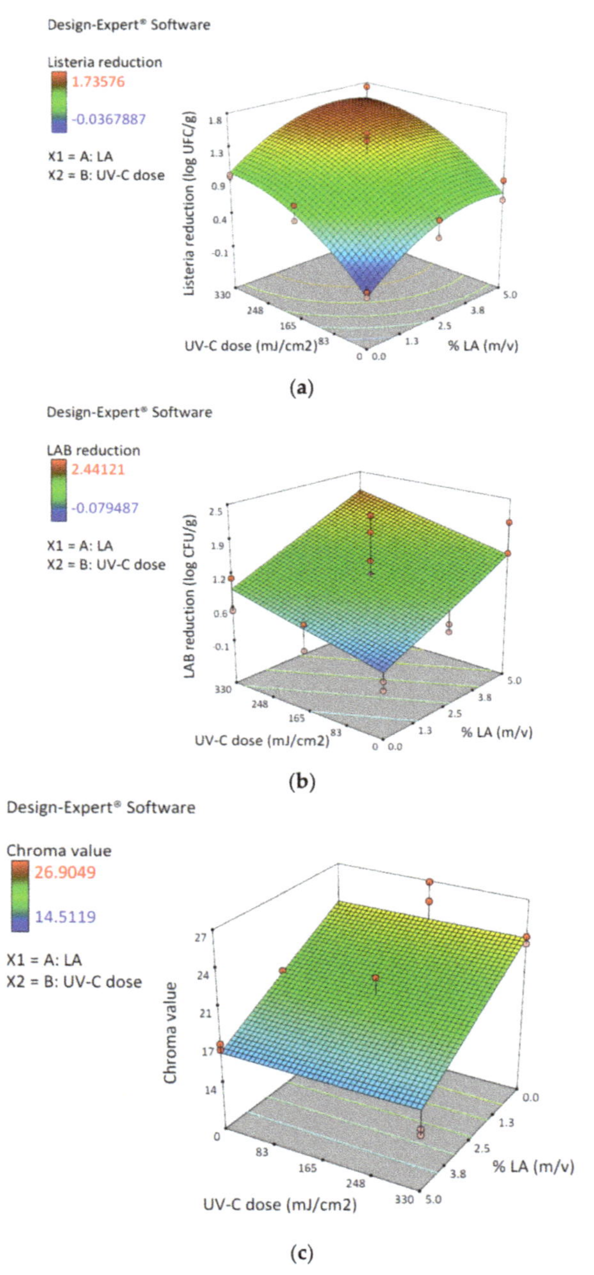

Figure 1. 3D response surface plots generated from the Central Composite design showing: (**a**) effect of lactic acid concentration and UV-C dose on *L. monocytogenes* reduction (log CFU/g); (**b**) effect of lactic acid concentration and UV-C dose on LAB reduction (log CFU/g); (**c**) effect of lactic acid concentration and UV-C dose on meat color (expressed as Chroma value).

For Chroma value according to Equation (4), the 3D response surface plot showed no changes in Chroma value due to UV-C and a negative effect by LA (Figure 1c).

3.2. Optimization and Model Validation

Based on the model generated using the Design Expert software with a desirability factor close to 1, the optimal conditions that satisfy the constraints applied (maximize *L. monocytogenes* reduction; maximize LAB reduction; Chroma value > 20) were: 2.6% lactic acid solution and UV-C dose of 330 mJ/cm^2. Using these conditions, the model predicted a *L. monocytogenes* reduction of 1.55 $\pm$ 0.41 log CFU/g and a LAB reduction of 1.55 $\pm$ 1.15 log CFU/g.

Experimental responses using the optimal conditions to treat meat samples were compared to the predicted results from the fitted models to evaluate the precision of the polynomial equations. The experimental values for *L. monocytogenes* and LAB reduction were 1.24 $\pm$ 0.18 log CFU/g and 1.20 $\pm$ 0.20 log CFU/g respectively. Both *L. monocytogenes* and LAB reduction experimental values were within the 95% CI of the predicted outcome by the models.

3.3. Evolution of *L. monocytogenes*, LAB, pH and Meat Color Treated with 2.6% of LA and 330 mJ/cm^2 of UV-C Dose Vacuum Packed and Stored at 4 °C for 8 Weeks

3.3.1. *L. monocytogenes* and LAB Counts

Application of 2.6% of LA and 330 mJ/cm^2 of UV-C reduced *L. monocytogenes* and LAB initial log counts by 1.2 and 1.3 log compared to control. Treated meat samples had *L. monocytogenes* and LAB log counts significantly lower ($p < 0.05$) than the control samples throughout the 8 weeks (Figure 2a,b). *L. monocytogenes* counts in LA/UV-C treated meat decreased from 3.6 log to 3.0, while in control samples an increase from 4.86 to 7.38 log CFU/g was observed (Figure 2a). LAB counts in treated samples remained constant until week 4 ($p > 0.05$), then an increase was observed at week 6, reaching 6.89 log CFU/g in week 8. In control samples, LAB counts remained unchanged during the first two weeks, and then increased over time up to 7.85 log CFU/g (Figure 2b).

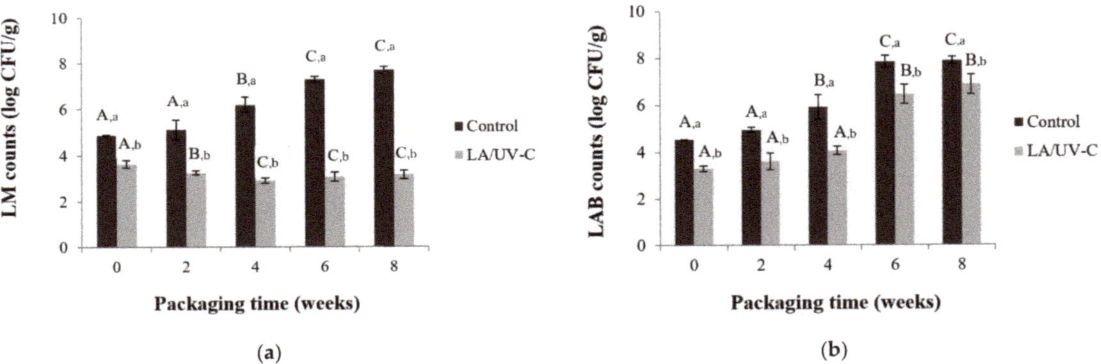

Figure 2. Bacterial count evolution in vacuum packed meats stored at 4 °C (**a**) *L. monocytogenes* (LM) (**b**) Lactic Acid Bacteria (LAB). Light grey and dark grey represent samples treated with 2.6% of LA/330 mJ/cm^2 of UV-C and no treatment control, respectively. Mean $\pm$ SD (n = 3) of the values are presented. Different capital letters indicate significant differences at $p \leq 0.05$ among the means over time for each treatment, and different small letters indicate significant differences at $p \leq 0.05$ between control and treated samples for each time point.

3.3.2. pH Values

LA/UV-C treatment decreased ($p < 0.05$) superficial meat pH from 5.78 to 3.70. Then, the treated sample's pH increased, reaching 5.29 at week 2. After week 2, the superficial pH of treated and untreated samples decreased over time. The pH value of LA/UV-C treated meat was always lower ($p < 0.05$) than the control. The final pH for control samples was 5.55 and for treated samples was 5.07 (Figure 3).

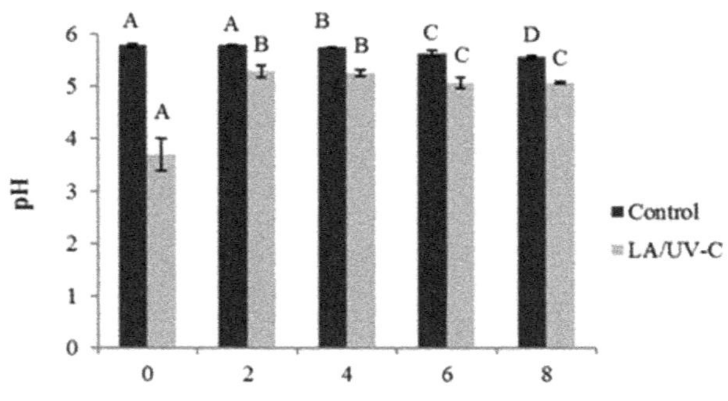

Figure 3. Superficial pH evolution in vacuum packed meats stored at 4 °C. Light grey and dark grey represent samples treated (2.6% of LA and 330 mJ/cm^2 of UV-C) and control, respectively. Mean ± SD (n = 3) of the values are presented. Different letters indicate a significant difference at $p \leq 0.05$ across time for each treatment.

3.3.3. Color Measurements

Changes in CIE L*, a*, b* and Chroma values (C*) at weeks 0 and 8 are shown in Table 3. At the initial time, L*, b* and C* values of LA/UV-C treated and untreated meat did not show variations from each other ($p > 0.05$); the a* value of control meat was higher ($p < 0.05$) than the value of LA/UV-C treated meat. At week 8, treated and untreated meat had non-significant differences in b* and C* values. However, the L* and a* values of LA/UV-C treated meat were lower ($p < 0.05$) than the control.

Table 3. Instrumental color parameters (L*, a*, b*) measured and Chroma value (C*) treated LA/UV-C and control meat samples at initial time and at 8 weeks.

Time (Weeks)		Control	LA/UV-C
0	Lightness (L*)	47.69 ± 2.94 ^{a}	46.71 ± 3.51 ^{a}
	Redness (a*)	20.21 ± 2.85 ^{a}	15.43 ± 1.92 ^{b}
	Yellowness (b*)	13.07 ± 0.77 ^{a}	13.36 ± 0.63 ^{a}
	Chroma (C*)	24.10 ± 2.48 ^{a}	20.44 ± 1.56 ^{a}
8	Lightness (L*)	46.94 ± 2.71 ^{a}	40.95 ± 1.81 ^{b}
	Redness (a*)	13.53 ± 1.63 ^{a}	9.92 ± 0.69 ^{b}
	Yellowness (b*)	11.08 ± 0.27 ^{a}	11.62 ± 0.56 ^{a}
	Chroma (C*)	17.52 ± 1.14 ^{a}	15.28 ± 0.66 ^{a}

Different superscripts within a row show significant results ($p \leq 0.05$). Data recorded as Mean ± Standard Deviation.

For both control and LA/UV-C treated meat, CIE L*, a*, b* and C* values were lower ($p < 0.05$) at 8 weeks compared to values at week 0, except the L* value of control samples that showed no significant change ($p > 0.05$).

4. Discussion

The present study shows for the first time the effect of the combined application of LA and UV-C on *L. monocytogenes* and LAB in vacuum-packed beef. A central composite design and Response Surface Methodology were used to optimize the concentration of LA and the dose of UV-C to reduce the population of *L. monocytogenes* and LAB without major changes in meat color.

The major findings of the present study were: (i) the quadratic model obtained allowed us to predict *L. monocytogenes* log reduction in vacuum-packed beef treated with LA and UVC, (ii) the maximum log reduction for both *L. monocytogenes* (1.55 ± 0.41 log CFU/g) and LAB (1.55 ± 1.15 log CFU/g) with minimal impact on meat color was achieved with the application of 2.6% LA and 330 mJ/cm^2 UV-C, and (iii) under these conditions, there was no increase in *L. monocytogenes* counts over 8 weeks of storage at 4 °C, and LAB growth was delayed by 2 weeks compared to control samples.

In the present study, the quadratic model obtained for predicting inoculated *L. monocytogenes* reduction had a good predictor value (R^2 = 0.9038). Both LA (0–5%) and UVC (0–330 mJ/cm^2) were significant ($p < 0.05$) factors and had independent effects (no significative interaction ($p > 0.05$)). The non-significant interaction between the factors indicated that the effects were additive, observed also for *Salmonella* inactivation in a different food matrix [26]. The reduction in *L. monocytogenes* increased as LA concentration and UV-C dose increased. According to the model, the combination of the maximum levels of LA (5.0%) and UV-C (330 mJ/cm^2) reduced the 5.8 log inoculum by 1.73 log, a fraction of viable microorganisms remained in the sample, indicating the presence of a tailing effect. This is depicted in the 3D Response Surface plot (Figure 1a), where *L. monocytogenes* log reduction had an initial sharp increasing rate and then decreased at higher UV-C doses and LA concentrations. As mentioned before, there are no other studies reporting the combined action of LA and UV-C on *Listeria monocytogenes* in fresh meat, although similar inactivation patterns were observed in fresh beef treated separately with LA or with UV-C [9,15,27]. In this respect, a previous study from our group obtained a reduction of 1.13 log using 2.5% LA. DeGeer et al. 2016, using a 4% LA solution, reduced by 1.3 log a *L. monocytogenes* inoculum of 8 log and, Kalchayanand et al. 2020, using a 590 mJ/cm^2 UV-C dose, reduced a 6 log *L. monocytogenes* inoculum by 0.89 log. The observed tailing effect for *L. monocytogenes* inactivation in meat may be explained by the ability of the meat matrix to buffer the antimicrobial solution and to entrap *L. monocytogenes* into muscle fibers shielding the bacteria from LA and UV-C radiation [9,29].

LAB reduction by LA and UV-C was adjusted to a linear model in which the factors LA and UV-C were both significant ($p < 0.05$) and independent (no significative interaction ($p > 0.05$)). The low precision (R^2 = 0.5774) of the model for predicting the response in the design space was a consequence of the variability among the five replicas of the central point (Table 1). This variability may be attributed to the natural diversity of the LAB present in the meat samples, which may have different sensitivity to LA and UV-C [14,20]. The maximum level of LAB reduction matched with the highest LA concentration and UV-C dose used suggesting that both factors can be further increased to achieve a higher level of reduction as shown in the 3D Response Surface plot (Figure 1b).

However, it was not feasible to increase LA concentration because (i) regulations of USDA/FSIS and European Commission do not allow concentrations greater than 5%, and (ii) high LA concentrations produced unwanted color changes in meat. Regarding increasing the UV-C dose, our data suggested that an increase in UV-C dose would not achieve a larger reduction in the *Listeria monocytogenes* population. Though our model did not allow us to predict outside the design space, the level of *L. monocytogenes* reduction obtained at the $+\alpha$ experimental point applying 2.5% LA and 368 mJ/cm^2 (Table 1) was similar to the level of reduction achieved with 2.5% LA and 330 mJ/cm^2. In agreement with this observation, McLeod, et al., 2017 [29] reported that the application of 3 J/cm^2 did not increase the level of *Listeria monocytogenes* reduction in chicken breast beyond the reduction level obtained with 0.3 J/cm^2. The antilisterial effect of doses higher than 330 mJ/cm^2 combined with 2.5% LA needs to be further studied, as well as the effects on meat quality.

Meat color change, expressed as Chroma value, was only related to LA acid concentration, and was fitted to a linear model with one factor (Figure 1c). Chroma value detrimental change was mostly due to the decrease in redness value (a*) (Table S1, Supplementary File), a well-known effect of lactic acid in beef [19]. The fact that UV-C doses applied did not have a significant effect on fresh meat color was in agreement with previous studies [15].

Using the models obtained for each response (*L. monocytogenes* reduction, LAB reduction and Color), RSM predicted that 2.6% LA concentration and 330 mJ/cm^2 of UV-C dose were the conditions that combined satisfied the constraints imposed (highest *L. monocytogenes* and LAB reduction and chroma value equal or larger than 20). The *L. monocytogenes* reduction predicted was 1.55 log. This reduction level was higher than the reduction levels obtained with the highest LA concentration or the highest UV-C dose alone.

Treatment with 2.6% LA and UV-C of 330 mJ/cm^2 at the time of packaging prevented the surviving fraction of the inoculated population of *L. monocytogenes* from thriving, showing a tendency to decrease with time when stored at 4 °C. In the untreated samples, the counts of *L. monocytogenes* increased (Figure 2a). LA and UV-C caused cellular injury in the fraction of survivors preventing them from overcoming the additional stress imposed by low oxygen and temperature. There are no reports of the combined application though a similar trend was reported for *L. monocytogenes* over time in beef treated with LA [27,30]. Regarding the behavior of *L. monocytogenes* in control samples, previous studies reported both growth [31,32] and inhibition [27,30,33] during storage at 4 °C. Differences in *L. monocytogenes* behavior may be due to variations in experimental conditions such as moisture and pH of meat samples, oxygen permeability of the vacuum bags and the *L. monocytogenes* strains used. In this study, the strain used was isolated from a refrigerated environment after having suffered different types of stress which, according to Skandamis et al., 2008, may affect subsequent stress tolerance.

LAB followed a sigmoidal growth curve [22]. Treatment prolonged the lag phase by two weeks, probably because a fraction of the remaining living cells were injured by UV-C and would have a slower growth rate or would be unable to replicate under stress [14]. The final LAB count in control samples reached 8 Log while in the samples treated with LA/UVC reached 7 Log. Though these results were relevant, more studies are needed to understand the impact on the combined LA/UV-C application on meat shelf life.

At time zero, redness was the only color component that was affected by the combined LA/UV-C application. As mentioned above, the decrease in initial a* is mainly due to the application of LA [19,34]. After eight weeks of storage, treated samples had different values of L* and a* with respect to samples without treatment, however the chroma value was similar in both samples. The greater decrease in the value of L* and a* at week 8 in the treated samples may be due to the fact that both LA and UVC have the capacity to oxidize myoglobin and to cause lipid oxidation with the consequent loss of color in meat [26,35]. However, more studies need to be done to assess the color of treated meats over time.

In summary, the combined application of LA 2.6% and UV-C 330 mJ/cm^2 contributed to improving safety of vacuum packed beef, with a low impact on color. Although, more studies must be carried out regarding the effects on other bacteria present on meat and other physicochemical changes such as lipid and protein oxidation.

5. Conclusions

The selected Central Composite design and response surface methodology were effective tools to optimize and study the effects of LA and UV-C parameters on the *L. monocytogenes* and LAB reduction in vacuum packed beef. The combined application of LA and UV-C radiation under the tested conditions proved to be a useful strategy to reduce *L. monocytogenes* and LAB population in meat without significantly affecting meat color. The treatment had an effect over time by preventing *L. monocytogenes* growth and delaying LAB growth. The latter might have an impact on vacuum beef shelf life. The maximum reduction on *L. monocytogenes* obtained without significant changes in color was 1.55 log CFU/g. Considering that the usual amount of *L. monocytogenes* in fresh meat is low, this level of reduction is significant for meat safety purposes.

Supplementary Materials: The following are available online at https://www.mdpi.com/article/10.3390/foods10061217/s1, Table S1: Central composite experimental design matrix with L*, a* and b* values for beef samples.

Author Contributions: Conceptualization, G.B. and C.R.; methodology, G.B. and C.R.; formal analysis, G.B., S.R. and J.R.; investigation, G.B., S.R. and J.R.; resources, C.R.; data curation, C.R.; writing—original draft preparation, C.R. and G.B.; writing—review and editing, C.R. and G.B.; supervision, C.R.; project administration, C.R.; funding acquisition, C.R. All authors have read and agreed to the published version of the manuscript.

Funding: This research received no external funding.

Data Availability Statement: Not applicable.

Acknowledgments: The authors thank Santiago Luzardo from INIA (Tacuarembó, Uruguay) for support for color measures, Eng. Alvaro Irigoyen and Eng. Juan Sanchez (Facultad de Ingeniería–UdelaR) for support with UV-C lamp intensity measures.

Conflicts of Interest: The authors declare no conflict of interest.

References

1. Buchanan, R.L.; Gorris, L.G.M.; Hayman, M.M.; Jackson, T.C.; Whiting, R.C. A review of Listeria monocytogenes: An update on outbreaks, virulence, dose-response, ecology, and risk assessments. *Food Control* **2017**, *75*, 1–13. [CrossRef]
2. Labadie, J. Consequences of packaging on bacterial growth. Meat is an ecological niche. *Meat Sci.* **1999**, *52*, 299–305. [CrossRef]
3. New and Established Carcass Decontamination Procedures Commonly Used in the Beef-Processing Industry. *J. Food Prot.* **1997**, *60*, 1146–1151. [CrossRef] [PubMed]
4. Nychas, G.J.E.; Skandamis, P.N.; Tassou, C.C.; Koutsoumanis, K.P. Meat spoilage during distribution. *Meat Sci.* **2008**, *78*, 77–89. [CrossRef] [PubMed]
5. Hope, J.; Klein, G.; Koutsoumanis, K.; Mclauchlin, J.; Müller-graf, C. Scientific Opinion on the evaluation of the safety and efficacy of lactic acid for the removal of microbial surface contamination of beef carcasses, cuts and trimmings. *EFSA J.* **2011**, 9. [CrossRef]
6. Guerrero-Beltrán, J.A.; Barbosa-Cánovas, G.V. Review: Advantages and limitations on processing foods by UV light. *Food Sci. Technol. Int.* **2004**, *10*, 137–147. [CrossRef]
7. FDA. Irradiation in the production, processing and handling of food: Ultraviolet radiation for the processing and treatment of food. *Food Drugs* **2000**, *65*, 71056–71058.
8. Theron, M.M.; Lues, J.F.R. Organic acids and meat preservation: A review. *Food Rev. Int.* **2007**, *23*, 141–158. [CrossRef]
9. DeGeer, S.L.; Wang, L.; Hill, G.N.; Singh, M.; Bilgili, S.F.; Bratcher, C.L. Optimizing application parameters for lactic acid and sodium metasilicate against pathogens on fresh beef, pork and deli meats. *Meat Sci.* **2016**, *118*, 28–33. [CrossRef]
10. Bucur, F.I.; Grigore-Gurgu, L.; Crauwels, P.; Riedel, C.U.; Nicolau, A.I. Resistance of Listeria monocytogenes to Stress Conditions Encountered in Food and food processing environments. *Front. Microbiol.* **2018**, *9*, 1–18. [CrossRef]
11. Skandamis, P.N.; Yoon, Y.; Stopforth, J.D.; Kendall, P.A.; Sofos, J.N. Heat and acid tolerance of Listeria monocytogenes after exposure to single and multiple sublethal stresses. *Food Microbiol.* **2008**, *25*, 294–303. [CrossRef] [PubMed]
12. Bintsis, T.; Litopoulou-Tzanetaki, E. Existing and potential applications of ultraviolet light in the food industry—A critical review. *J. Sci. Food Agric.* **2010**, *80*, 637–645. [CrossRef]
13. Sommers, C.H.; Cooke, P.H.; Fan, X.; Sites, J.E. Ultraviolet light (254 nm) inactivation of Listeria monocytogenes on frankfurters that contain potassium lactate and sodium diacetate. *J. Food Sci.* **2009**, *74*, M114–M119. [CrossRef] [PubMed]
14. Gailunas, K.M.; Matak, K.E.; Boyer, R.R.; Alvarado, C.Z.; Williams, R.C.; Sumner, S.S. Use of UV light for the inactivation of Listeria monocytogenes and lactic acid bacteria species in recirculated chill brines. *J. Food Prot.* **2008**, *71*, 629–633. [CrossRef] [PubMed]
15. Kalchayanand, N.; Bosilevac, J.M.; King, D.A.; Wheeler, T.L. Evaluation of UVC radiation and a UVC-ozone combination as fresh beef interventions against Shiga toxin-producing Escherichia coli, salmonella, and listeria monocytogenes and their effects on beef quality. *J. Food Prot.* **2020**, *83*, 1520–1529. [CrossRef]
16. Tarek, A.R.; Rasco, B.A.; Sablani, S.S. Ultraviolet-C Light Inactivation Kinetics of E. coli on Bologna Beef Packaged in Plastic Films. *Food Bioprocess Technol.* **2015**, *8*, 1267–1280. [CrossRef]
17. Faustman, C.; Cassens, R.G. The Biochemical Basis for Discoloration in Fresh Meat: A Review. *J. Muscle Foods* **1990**, *1*, 217–243. [CrossRef]
18. Zerby, H.N.; Belk, K.E.; Sofos, J.N.; McDowell, L.R.; Smith, G.C. Case life of seven retail products from beef cattle supplemented with alpha-tocopheryl acetate. *J. Anim. Sci.* **1999**, *77*, 2458–2463. [CrossRef]
19. Rodríguez-Melcón, C.; Alonso-Calleja, C.; Capita, R. Lactic acid concentrations that reduce microbial load yet minimally impact colour and sensory characteristics of beef. *Meat Sci.* **2017**, *129*, 169–175. [CrossRef]
20. Signorini, M.L.; Ponce-Alquicira, E.; Guerrero-Legarreta, I. Effect of lactic acid and lactic acid bacteria on growth of spoilage microorganisms in vacuum-packaged beef. *J. Muscle Foods* **2006**, *17*, 277–290. [CrossRef]
21. García, A.L.; Brugnini, G.; Rodriguez, S.; Mir, A.; Carriquiry, J.; Rufo, C.; Briano, B. Vida útil de carne fresca de res envasada al vacío a 0 °C y +4 °C. *Prod. Agropecu. Desarro. Sosten.* **2015**, *4*, 27–45. [CrossRef]
22. Small, A.H.; Jenson, I.; Kiermeier, A.; Sumner, J. Vacuum-packed beef primals with extremely long shelf life have unusual microbiological counts. *J. Food Prot.* **2012**, *75*, 1524–1527. [CrossRef] [PubMed]

23. Pennacchia, C.; Ercolini, D.; Villani, F. Spoilage-related microbiota associated with chilled beef stored in air or vacuum pack. *Food Microbiol.* **2011**, *28*, 84–93. [CrossRef] [PubMed]
24. Zimet, P.; Mombrú, Á.W.; Faccio, R.; Brugnini, G.; Miraballes, I.; Rufo, C.; Pardo, H. Optimization and characterization of nisin-loaded alginate-chitosan nanoparticles with antimicrobial activity in lean beef. *LWT Food Sci. Technol.* **2018**, *91*, 107–116. [CrossRef]
25. Martinez-Silveira, A.; Villarreal, R.; Garmendia, G.; Rufo, C.; Vero, S. Process conditions for a rapid in situ transesterification for biodiesel production from oleaginous yeasts. *Electron. J. Biotechnol.* **2019**, *38*, 1–9. [CrossRef]
26. Rosario, D.K.A.; Mutz, Y.S.; Castro, V.S.; Bernardes, P.C.; Rajkovic, A.; Conte-Junior, C.A. Optimization of UV-C light and lactic acid combined treatment in decontamination of sliced Brazilian dry-cured loin: Salmonella Typhimurium inactivation and physicochemical quality. *Meat Sci.* **2021**, *172*, 108308. [CrossRef]
27. Brugnini, G.; Acquistapace, M.J.; Rodriguez, S.; Rufo, C. Efecto de la aplicación de ácido láctico sobre Listeria monocytogenes en carne envasada al vacío en Uruguay. *Innotec* **2018**, *15*, 7–14. [CrossRef]
28. MacDougall, D.B. Changes in the colour and opacity of meat. *Food Chem.* **1982**, *9*, 75–88. [CrossRef]
29. McLeod, A.; Liland, K.H.; Haugen, J.E.; Sørheim, O.; Myhrer, K.S.; Holck, A.L. Chicken fillets subjected to UV-C and pulsed UV light: Reduction of pathogenic and spoilage bacteria, and changes in sensory quality. *J. Food Saf.* **2018**, *38*, 2421. [CrossRef]
30. Ariyapitipun, T.; Mustapha, A.; Clarke, A.D. Survival of Listeria monocytogenes Scott A on Vacuum-Packaged Raw Beef Treated with polylactic Acid, Lactic Acid and Nisin. *J. Food Prot.* **2000**, *63*, 131–136. [CrossRef]
31. Tsigarida, E.; Skandamis, P.; Nychas, G.J.E. Behaviour of Listeria monocytogenes and autochthonous flora on meat stored under aerobic, vacuum and modified atmosphere packaging conditions with or without the presence of oregano essential oil at 5 °C. *J. Appl. Microbiol.* **2000**, *89*, 901–909. [CrossRef]
32. Dickson, J.S. Survival and Growth of Listeria Monoctyogenes on Beef Tissue Surfaces As Affected By Simulated Processing Conditions. *J. Food Saf.* **1989**, *10*, 165–174. [CrossRef]
33. Duffy, G.; Walsh, D.; Sheridan, J.J.; Logue, C.M.; Harrington, D.; Blair, I.S.; McDowell, D.A. Behaviour of Listeria monocytogenes in the presence of Listeria innocua during storage of minced beef under vacuum or in air at 0 °C and 10 °C. *Food Microbiol.* **2000**, *17*, 571–578. [CrossRef]
34. Salim, A.P.A.; Canto, A.C.; Costa-Lima, B.R.; Simoes, J.S.; Panzenhagen, P.H.; Costa, M.P.; Franco, R.M.; Silva, T.J.; Conte-Junior, C.A. Inhibitory effect of acid concentration, aging, and different packaging on Escherichia coli O157:H7 and on color stability of beef. *J. Food Process. Preserv.* **2017**, *42*, 1–8. [CrossRef]
35. Faustman, C.; Sun, Q.; Mancini, R.; Suman, S.P. Myoglobin and lipid oxidation interactions: Mechanistic bases and control. *Meat Sci.* **2010**, *86*, 86–94. [CrossRef] [PubMed]

Article

Biomapping of Microbial Indicators on Beef Subprimals Subjected to Spray or Dry Chilling over Prolonged Refrigerated Storage

Diego E. Casas, Rosine Manishimwe, Savannah J. Forgey, Keelyn E. Hanlon, Markus F. Miller, Mindy M. Brashears and Marcos X. Sanchez-Plata *

International Center for Food Industry Excellence, Department of Animal and Food Sciences, Texas Tech University, Lubbock, TX 79409, USA; diego.casas@ttu.edu (D.E.C.); r.manishimwe@ur.ac.rw (R.M.); savannah.forgey@ttu.edu (S.J.F.); keelyn.hanlon@ttu.edu (K.E.H.); mfmrraider@aol.com (M.F.M.); mindy.brashears@ttu.edu (M.M.B.)

* Correspondence: marcos.x.sanchez@ttu.edu; Tel.: +1-(806)-834-6503

Abstract: As the global meat market moves to never frozen alternatives, meat processors seek opportunities for increasing the shelf life of fresh meats by combinations of proper cold chain management, barrier technologies, and antimicrobial interventions. The objective of this study was to determine the impact of spray and dry chilling combined with hot water carcass treatments on the levels of microbial indicator organisms during the long-term refrigerated storage of beef cuts. Samples were taken using EZ-Reach™ sponge samplers with 25 mL buffered peptone water over a 100 cm^2 area of the striploin. Sample collection was conducted before the hot carcass wash, after wash, and after the 24 h carcass chilling. Chilled striploins were cut into four sections, individually vacuum packaged, and stored to be sampled at 0, 45, 70, and 135 days (n = 200) of refrigerated storage and distribution. Aerobic plate counts, enterobacteria, *Escherichia coli*, coliforms, and psychrotroph counts were evaluated for each sample. Not enough evidence ($p > 0.05$) was found indicating the hot water wash intervention reduced bacterial concentration on the carcass surface. *E. coli* was below detection limits (<0.25 CFU/cm^2) in most of the samples taken. No significant difference ($p > 0.05$) was found between coliform counts throughout the sampling dates. Feed type did not seem to influence the ($p > 0.25$) microbial load of the treatments. Even though no immediate effect was seen when comparing spray or dry chilling of the samples at day 0, as the product aged, a significantly lower ($p < 0.05$) concentration of aerobic and psychrotrophic organisms in dry-chilled samples could be observed when compared to their spray-chilled counterparts. Data collected can be used to select alternative chilling systems to maximize shelf life in vacuum packaged beef kept over prolonged storage periods.

Keywords: refrigerated meat shelf life; microbial indicators; vacuum packaging; carcass chilling; hot water intervention

Citation: Casas, D.E.; Manishimwe, R.; Forgey, S.J.; Hanlon, K.E.; Miller, M.F.; Brashears, M.M.; Sanchez-Plata, M.X. Biomapping of Microbial Indicators on Beef Subprimals Subjected to Spray or Dry Chilling over Prolonged Refrigerated Storage. *Foods* **2021**, *10*, 1403. https://doi.org/10.3390/foods10061403

Academic Editor: Mohammed Gagaoua

Received: 5 May 2021
Accepted: 11 June 2021
Published: 17 June 2021

1. Introduction

The world beef market is heavily influenced by consumer demands and choices; therefore, the beef industry must adapt to the consumers' needs and concerns and provide meat products that fulfill such needs. Certain consumer demands have created niche opportunities for a variety of meat product offerings. An important market niche for beef products is the "fresh meat" "never frozen" alternatives. This has led meat processors to seek schemes for increasing the shelf life of fresh meats by combinations of proper cold management, barrier technologies, and application of antimicrobial interventions (chemical or physical) [1,2]. The growing demand for fresh products has put pressure on the cold supply chain and quality control at all steps in the processing plant [3]. Such a trend has evolved rapidly, and now regulatory agencies have developed a series of labeling requirements for never frozen meat and poultry products. The U.S. Department of Agriculture Food Safety and Inspection Service (USDA-FSIS) has defined that any poultry product below −3.3 °C (26 °F) or red meat that has ever been frozen cannot be labeled as fresh, not frozen [4]. To address these market trends, beef processors need to explore novel processing schemes, product protection options, and process modifications that have been properly validated in commercial settings to extend product shelf life, especially when long transport regimes are necessary under refrigerated conditions due to significant distances between production and market locations.

Australia is one of the leading beef exporters in the world. As an important market player, the Australian beef industry has been continuously assessing new market opportunities and has been exploring fresh, never frozen beef alternatives for competitive markets worldwide. In 2018, Australia's bovine meat exports accounted for 43.2% of animal product exports with a market value of approximately $6.47 billion for the country's economy. Australia's biggest beef export market is Japan with a market share of 36.8% in 2018. With the recent interest of the United States to significantly increase its beef exports to Japan [5,6], Australia has sought opportunities to expand its presence in the European Union. This high-income market shows significant consumer interest in the fresh, never frozen beef products [7]. Unfortunately, the distance between the meat source and the EU market has created a challenge, due to the long-haul transportation needs and rigorous chilled conditions necessary for product arrival and suitability for fresh distribution. Consequently, extending the shelf life of chilled meat products has become of the utmost interest.

Meat shelf life extension has been achieved through the use of several antimicrobial interventions, chilling methods, and barrier technologies in the past [1,2,8,9]. At the same time, these interventions and barrier technologies mitigate the growth of indicator and pathogenic bacteria that are responsible for product deterioration. The hot water wash of carcasses has been observed to reduce 2.7–3.0 log CFU/cm^2 of *L. monocytogenes*, *Salmonella*, and APC counts [10]. Dry aging has been observed to reduce over 2 log CFU/cm^2 of generic *E. coli* and *E. coli* O157:H7 in beef carcasses and subprimals [11,12]. Air chilling has been shown to reduce total viable counts by 03–0.7 log CFU/cm^2 [13], and up to 2 log CFU/cm^2 of *E. coli* [14]. Moreover, indicator and pathogenic microorganisms have been reduced after the air conventional chilling and blast chilling of carcasses [15,16]. Comparatively, spray chilling has been observed to have no immediate effect in microbial populations [17]. Thus, in this study, we evaluate the use of hot carcass washing and different carcass chilling systems to assess Australian chilled beef's extended shelf life in export settings that require product viability for more than 130 days of refrigerated storage and distribution.

2. Materials and Methods

2.1. Sample Collection

Samples were taken at Teys Australia beef processing plant located in Beenleigh, QLD, Australia. A total of 200 carcasses were evaluated. Swab samples were taken using EZ-Reach™ Sponge Samplers hydrated with 25 mL buffered peptone water (BPW, World BioProducts, Mundelein, Illinois) by swabbing an over 100 cm^2 area on the striploin region of each carcass. Samples were taken before the hot carcass wash, after the hot carcass wash for washed samples, and 24 h after being subjected to one of the chilling methods described below (spray vs. dry chilling). Edible ink was used to mark the area where the sample was taken to avoid re-sampling of the same surface. The hot water carcass wash consisted of spraying 85 ± 2 °C water onto the surface of the carcass through eight nozzle sprayers, four per side of the carcass. Water temperature was recorded on the pipes feeding the water to the sprayers right before sample collection. The chilling methods evaluated consisted of 18–24 h storage in a refrigerated chamber subjected to continuous spraying of water at 0–2 °C in the room at 15 min intervals, following the processing plant's protocols. Dry chilling consisted of 18–24 h storage in a refrigerated room at 0 °C with constant airflow while the sprayers were completely turned off. After 24 h chilling, either under water spray conditions or dry refrigerated storage, striploins were taken and cut into 4 sections. Individual sections were vacuum packaged and assigned a date for further sampling at 0, 45, 70, and 135 days of refrigerated storage. Samples collected for day 0 were analyzed in an in-plant laboratory setup at the processing facility. Striploins were shipped via sea to the ICFIE Food Microbiology Laboratory at Texas Tech University (TTU) in Lubbock, Texas, USA for the long-term shelf life section of the study corresponding to storage at days 45, 70, and 135 under refrigerated conditions. Striploins were kept at 0–4 °C from carcass fabrication to meat reception at TTU. On day 40, striploins were received at TTU and the refrigerated temperature was raised to 7 °C, simulating abusive counter temperatures common in retail stores. On each sampling day, striploin packages were opened with sterile scalpels and an area of 100 cm^2 of the product was swabbed for sample collection.

2.2. Sample Processing

Swab samples collected were stomached for 30 s at 230 rpm. Serial dilutions for each swab sample were made with 9 mL BPW tubes. A volume of 1 mL was plated onto Petrifilm™ plates (3M, Saint Paul, Minnesota) in duplicate corresponding to *Enterobacteriaceae* (EB), *Escherichia coli* (EC), coliforms (CO), and aerobic plate counts (APC). In addition, aerobic plate count Petrifilm was also used to estimate psychrotroph counts (PSY) by incubating separate plates at 20 °C for 72 h [18,19]. *Enterobacteriaceae* Petrifilms were incubated for 24 h at 37 °C before counting. Coliforms were counted after 24 h incubation at 37 °C. *Escherichia coli* counts were recorded after 48 h incubation at 37 °C following the manufacturer's recommendations. APC plates were incubated for 48 h at 37 °C.

2.3. Experimental Design and Statistical Analysis

The hot water wash section of the study had a completely randomized design with a factorial arrangement of 2 factors, feed regime and carcass wash, at 2 levels each: grass vs. grain and washed vs. not washed, respectively. Three sampling points were evaluated, before wash, after the washing stage, and after a 24 h chilling period. For each repetition, 10 samples were taken per treatment (Table S1) at each sampling point. A total of 5 repetitions were conducted.

The section of the study regarding the extended shelf life of the striploins was arranged in a completely randomized design with a factorial arrangement of three factors, feed regime, hot water wash application, and chilling method, at two levels each: grass vs. grain, washed vs. not-washed, and dry vs. spray-chilled, respectively. For each repetition, 5 samples per treatment were taken at each sampling date (Table S2). A total of 5 repetitions were conducted resulting in 200 samples per sampling date. An ANOVA by sampling date was used to analyze the data when parametric assumptions were satisfied. The Kruskal–Wallis (nonparametric ANOVA) test was used to analyze the data when parametric assumptions were not met. When the ANOVA or Kruskal–Wallis was significant, pairwise comparisons were done using a pairwise T-test on significant ANOVAs or a Wilcoxon rank-sum test on significant Kruskal–Wallis tests [20]. Statistical significance was evaluated at a 0.05 probability level.

3. Results

3.1. Hot Water Wash

The main effect of the feed type had no statistical significance throughout any of the sampling dates of the study. There was no significant difference ($p > 0.25$) on the bacterial counts observed between grain and grass-fed carcasses in the study; therefore, the main effect of the feed type was removed to better visualize differences due to the washing and chilling types' main effects and their interaction. The hot water wash carcass intervention significantly reduced ($p < 0.05$) APC on the carcass surface (Figure 1). However, no washed treatments presented lower aerobic plate counts than the washed counterparts. After a 24 h chilling period, there was an increase in PSY counts and a stalled growth of APC. Psychrotrophic bacteria were not significantly reduced by the hot water wash intervention

(Figure 2) and had growth after a 24 h chilling period. EB, EC, and CO counts were below the detection limit (<0.25 CFU/cm^2) in most samples taken at each sampling point assessed.

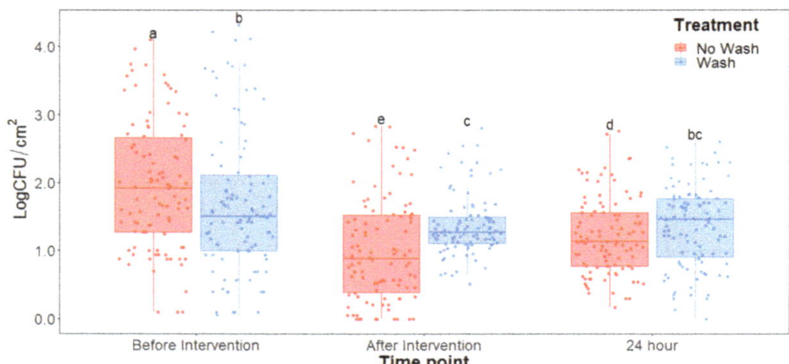

Figure 1. Aerobic plate counts of the beef carcass surface before and after the hot water wash intervention and 24 h chilling period. The horizontal line within the box plot represents the median. The box upper and lower limits represent the interquartile range, and the bars represent the 1.5xInterquartile Range. [a–e] Box plots with different letters are significantly different ($p < 0.05$).

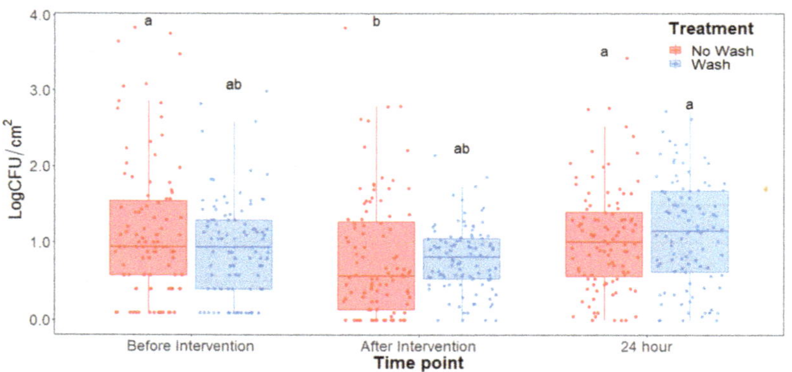

Figure 2. Psychrotroph counts of the beef carcass surface before and after the hot water wash intervention and 24 h chilling period. The horizontal line within the box plot represents the median. The box upper and lower limits represent the interquartile range, and the bars represent the 1.5xInterquartile Range. [a,b] Box plots with different letters are significantly different ($p < 0.05$).

3.2. Extended Shelf Life of Striploins

The statistical analysis indicates a significant effect of time for all the indicator microorganism loads assessed, as expected. Because of this, the statistical comparison between treatments was conducted within a per sampling date basis, rather than over the time of storage. The loads of each indicator microorganism evaluated was compared between treatments within each sampling date. On sampling day 0, no significant differences among treatments could be observed in any of the five microbial indicators quantified ($p > 0.05$) and the indicator bacteria were mostly below the detection limit (Table 1). Even though no immediate effect could be observed from spray and dry chilling at day 0, in the long term and throughout the additional sampling periods during refrigerated storage, significantly lower ($p < 0.05$) concentrations of APC, PSY, and EB can be observed in the dry chilling treatments (Figures 3–5) when compared to their spray-chilled counterparts.

Table 1. Summary table of microbial indicator microorganism counts in striploins before and after the intervention, chilling method, and evaluation at day 0, 45, 70, and 135 of refrigerated storage.

Microorganism	Treatment	Timepoint (LogCFU/cm^2 ± S.E. [1])					
		Before Wash	After Wash	Day 0	Day 45	Day 70	Day 135
Aerobic plate count	No wash Dry	1.96 ± 0.10	1.01 ± 0.04	1.23 ± 0.08	3.62 ± 0.29	3.19 ± 0.36	3.97 ± 0.38
	No wash Spray			1.23 ± 0.08	4.33 ± 0.20	5.1 ± 0.28	5.39 ± 0.31
	Wash Dry	1.72 ± 0.09	1.38 ± 0.08	1.31 ± 0.09	3.55 ± 0.25	4.51 ± 0.32	5.05 ± 0.35
	Wash Spray			1.42 ± 0.08	4.67 ± 0.24	5.41 ± 0.24	6.11 ± 0.24
Psychrotroph count	No wash Dry	1.16 ± 0.06	0.80 ± 0.04	1.02 ± 0.10	4.37 ± 0.27	4.09 ± 0.33	4.55 ± 0.38
	No wash Spray			1.10 ± 0.09	4.97 ± 0.20	5.54 ± 0.25	5.80 ± 0.26
	Wash Dry	0.93 ± 0.09	0.80 ± 0.08	1.17 ± 0.11	4.11 ± 0.24	4.96 ± 0.20	5.52 ± 0.30
	Wash Spray			1.11 ± 0.09	5.22 ± 0.24	6.00 ± 0.25	6.34 ± 0.23
Enterobacteriaceae count	No wash Dry	0.19 ± 0.05	0.04 ± 0.02	0.00 *	0.4 ± 0.12	1.23 ± 0.28	1.81 ± 0.38
	No wash Spray			0.01 ± 0.01	1.44 ± 0.22	2.18 ± 0.29	2.42 ± 0.35
	Wash Dry	0.02 ± 0.01	0.00 *	0.00 *	0.94 ± 0.20	1.39 ± 0.27	1.99 ± 0.35
	Wash Spray			0.00 *	1.71 ± 0.24	2.78 ± 0.33	3.28 ± 0.38
Coliform count	No wash Dry	0.02 ± 0.01	0.02 ± 0.01	0.12 ± 0.06	0.12 ± 0.08	0.22 ± 0.13	0.51 ± 0.24
	No wash Spray			0.19 ± 0.07	0.47 ± 0.07	0.47 ± 0.16	0.69 ± 0.25
	Wash Dry	0.00 *	0.00 *	0.04 ± 0.04	0.04 ± 0.04	0.24 ± 0.16	0.53 ± 0.23
	Wash Spray			0.17 ± 0.07	0.35 ± 0.07	0.66 ± 0.24	0.90 ± 0.29

[1] Standard Error. * Below detection limit (<0.25 CFU/cm^2).

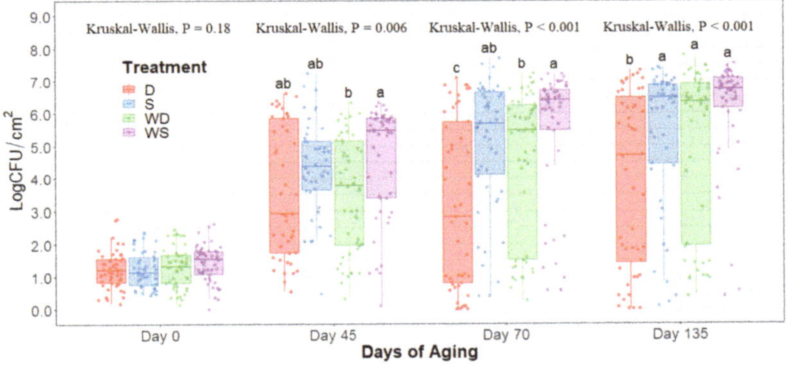

Figure 3. Aerobic plate counts of striploins at day 0, 45, 70, and 135 of refrigerated storage. The horizontal line within the box plot represents the median. The box upper and lower limits represent the interquartile range, and the bars represent the 1.5xInterquartile Range. D = No Wash Dry chill, S = No Wash Spray chill, WD = Wash Dry chill, WS = Wash Spray chill. [a–c] Box plots with different letters within each sampling date are significantly different ($p < 0.05$).

No significant differences on coliform counts between treatments at each sampling date could be found throughout the extended shelf life section of the study; however, significant growth over time was observed. *E. coli* counts on striploins were mostly below the detection limit (<0.25 CFU/cm^2) at the plant and throughout the extended shelf life. Thus, no significant growth of *E. coli* over time was observed. Furthermore, significant growth of EB was observed only after 45 days of wet aging, encountering significant differences between treatments during long-term storage.

Even though the hot water wash's main effect was not statistically significant throughout the extended shelf life study, a trend ($0.05 < p < 0.15$) of an increase in microbes quantified could be observed whenever the carcasses underwent the hot water wash intervention compared to their dry chilling counterparts. The highest microbial concentrations were consistently observed on the washed and spray-chilled striploins treatment and the lowest microbial loads were consistently observed in the no-washed dry-chilled striploins.

Although significant interaction between the main effects was not observed statistically, a trend in the interaction was observed ($0.05 < p < 0.15$).

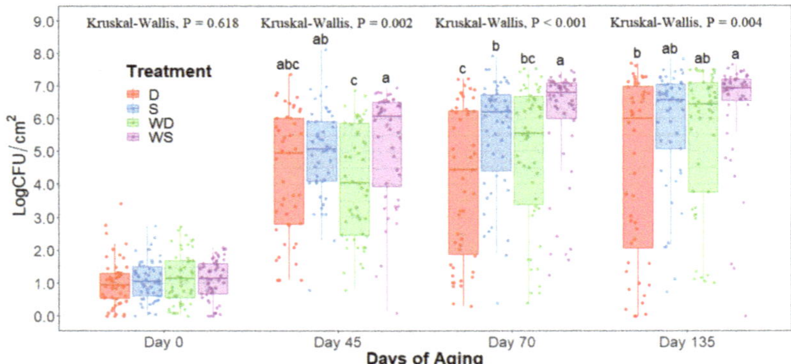

Figure 4. Psychrotroph counts of striploins at day 0, 45, 70, and 135 of refrigerated storage. The horizontal line within the box plot represents the median. The box upper and lower limits represent the interquartile range, and the bars represent the 1.5xInterquartile Range. D = No Wash Dry chill, S = No Wash Spray chill, WD = Wash Dry chill, WS = Wash Spray chill. [a–c] Box plots with different letters within each sampling date are significantly different ($p < 0.05$).

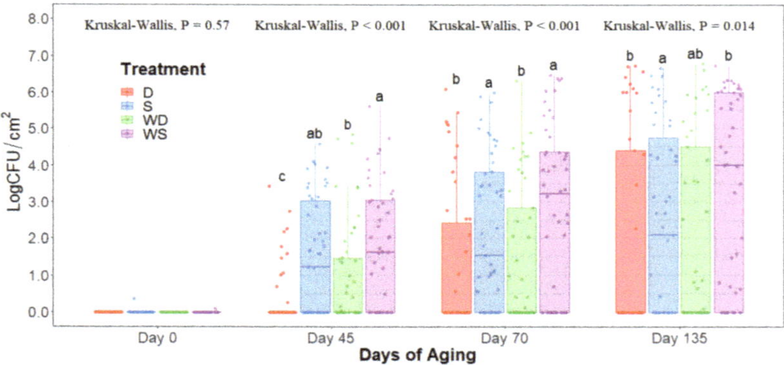

Figure 5. Enterobacteriaceae counts of striploins at day 0, 45, 70, and 135 of refrigerated storage. The horizontal line within the box plot represents the median. The box upper and lower limits represent the interquartile range, and the bars represent the 1.5xInterquartile Range. D = No Wash Dry chill, S = No Wash Spray chill, WD = Wash Dry chill, WS = Wash Spray chill. [a–c] Box plots with different letters within each sampling date are significantly different ($p < 0.05$).

EB counts were significantly different between treatments after long-term storage ($p < 0.05$). Dry chilling methods had their medians at 0 log CFU/cm^2, indicating a low concentration of EB even after 135 days of refrigerated storage. Moreover, the no-wash dry chilling treatment combination had the lowest concentration of EB across all times evaluated. The treatment's significant differences after prolonged refrigeration times become evident from day 45 of long-term refrigeration storage.

4. Discussion

Microbial indicator levels assessed in the hot carcass wash section of the study were substantially lower before the hot carcass wash intervention, demonstrating the efficacy of proper sanitary dressing procedures in the facility (Table 1). Since the initial concentration of microorganisms was so low, no significant reductions in indicator bacteria concentrations were observed in the early stages of sampling and no effects were observed after subjection to the different treatment combinations. Because of this, the concentration of *Enterobacteriaceae, E. coli*, and coliforms were below detection limits (<0.25 CFU/cm^2) after the hot water wash intervention in most samples collected at day 0. Under the conditions evaluated in this study, the hot water carcass intervention was not found to significantly reduce APC and PSY counts compared to no-wash treatments. This finding shows that despite significantly reducing a small number of bacteria on the surface of the carcass, washing the carcass may also redistribute the bacteria throughout the whole carcass surface and that can contribute to further differences during prolonged storage. This may pose a counterproductive result as bacteria will have a greater surface area of contact with the carcass and these may allow for more microbial attachment, growth, and development [21]. Furthermore, the washing of the carcass surfaces may increase available water for microbial growth which in the long term may allow a higher proliferation of bacteria in the striploins, a tendency observed in the long-term storage under refrigerating conditions [22]. Higher reductions may be achieved with alternative physical interventions such as steam vacuuming and trimming which do not use any chemicals for the reduction of bacteria [23,24].

When observing the hot carcass washing and type of carcass chilling, an immediate effect was not observed in any of the five bacteria quantified at day 0. Particularly EB, CO, and EC were all below detection limits within the in-plant sampling at day 0. By the time striploins had undergone shipment and distribution to export markets under refrigeration (day 45), a significant difference was observed between treatments in PSY, APC, and EB, where dry chilling treatments had lower bacterial counts overall. This trend was kept throughout the 135 days of refrigerated storage evaluated in this study.

APC evaluations show an overall count of mesophilic bacteria demonstrating a general picture of the total bacteria counts within the striploins. However, psychrotroph counts represent a more accurate bacterial load of meat, as meat is mostly stored under refrigerated conditions for a prolonged time. Previous research has demonstrated around a 0.5–1.0 log CFU increase in concentration on PSY counts compared to APC [18,25]. Most psychrotrophic enumeration methods require incubation at 7 °C for 10 days, or at 10 °C for 7 days, among others [19]. Furthermore, methods with incubation at 20 °C for 72 h have been used to enumerate carcass and meat subprimal psychrotrophic counts [18,26]. Due to variability in protocols for quantification of psychrotrophic bacteria counts, a trial comparing psychrotrophic counts using incubation at 7 °C for 10 days and 20 °C for 72 h was performed. Results led to a correlation of 0.96 and 0.98 at 45 and 70 days of aging, respectively (data not shown), thus validating the use of the protocol of incubation for psychrotrophic bacterial counts at 20 °C for 72 h. PSY counts were significantly lower on the dry chilling treatments, particularly in the no-washed dry-chilled treatment combination. Furthermore, package bloating and off-odors were less frequently found in striploins subjected to the dry-chilled treatment.

During long-term refrigerated storage of meat, Pseudomonas, Enterobacteria, and lactic acid bacteria become the main microorganisms that cause spoilage [27]. In the past, EB has also been used as an indicator of the risk of *Salmonella* spp. contamination. A higher concentration of EB may increase the risk of *Salmonella* spp. presence [28]. However, EB presence does not confirm *Salmonella* spp. presence. Similarly, pathogens of public health interest are within the EB family classification, such as *Shigella*, *E. coli*, and *Klebsiella*. In this context, all treatments were effective at mitigating EB presence at day 0 of storage; however, as the long-term storage continued, evident differences between treatments were observed, where dry chilling treatments more effectively mitigated EB proliferation in the striploins. This mitigation of EB growth throughout time may suggest that dry chilling not only prolongs shelf life but also further ensures food safety through lower rates of bacterial growth with the potential of increasing bacterial injury [8,29,30].

No difference between treatments was observed in CO and EC due to the low concentration encountered within sampling dates throughout the trial. EB and EC counts serve as a Gram-negative indicator of fecal contamination. Generic *E. coli* serves as an indicator of process control; the FSIS has published minimal sampling requirements for beef processors indicating that final carcasses must have negative results of *E. coli* to be considered acceptable; moreover, if more than four samples are between 1–100 CFU/cm^2 in a window of 13 consecutive samples or a sample is over 100 CFU/cm^2 a corrective action is warranted [31]. In this context, the beef processing plant is well within the acceptable limits of *E. coli* enumeration, having over 95% of the carcasses below the detection limit and all below 100 CFU/cm^2 after harvest and throughout the long-term storage of striploins under refrigerated conditions. This is an outstanding indicator of proper hygiene procedures and sanitary dressing procedures within the plant. Previous research effectively validated process controls within beef slaughter operations using EC as an indicator of process control alongside APC and EB counts [26,32].

Overall, dry chilling procedures prolonged the shelf life of striploins more effectively. Dry aging has been observed to reduce *E. coli* O157:H7 and generic *E. coli* concentrations in carcasses [11,12]. Moreover, air chilling and blast chilling of carcasses have both shown similar results in the reduction of indicator microorganisms in beef and pork carcasses [15,16]. Contrastingly, spray-chilled carcasses have been shown to have no immediate effect on the microbial load of carcasses [17,30], as observed in this study. However, in the long term, spray-washed treatments consistently had higher microbial loads throughout all the treatments and no-washed dry-chilled treatments had consistently significantly lower microbial concentrations, suggesting different slopes for growth curves of microorganisms under different treatments. Importantly, dry chilling procedures are known to reduce cold carcass weight due to the loss of moisture from the carcass surface during chilling, resulting in economic loss for the processing plant [33,34]. The optimization of interventions, chilling techniques, and barrier technologies depending on the end consumer and shelf life requirements of the meat can result in minimization of economic loss, a better microbial quality, and a safer meat product.

5. Conclusions

A hot water wash prior to carcass chilling did not significantly reduce microorganisms assessed under the conditions evaluated in this study. Dry chilling of carcasses can potentially increase the shelf life of meat products as it delays the growth of bacteria under the refrigerated conditions of storage during transport and distribution. Data collected can be used to select chilling systems to maximize shelf life, especially in long-term refrigerated storage conditions of never frozen beef products. The optimal shelf life of striploins can be achieved using dry chilling air systems, which will guarantee the required 130 days of shelf life for the export of fresh, never frozen beef from Australia to the EU. The use of spray chilling schemes increases available water for the growth of bacteria resulting in higher growth rates of bacteria during the long-term refrigerated storage and therefore a reduced shelf life. This extended quality preservation over an extended shelf life period allows more flexibility in beef exports, especially for major producers that are far from target consumer markets. Understanding the best parameters for beef carcass processing and storage will allow the beef industry to select optimized chilling schemes for long-term storage and increased consumer acceptability.

Supplementary Materials: The following are available online at https://www.mdpi.com/article/10.3390/foods10061403/s1. Table S1. Experimental design of no washed and hot water washed carcasses in a beef processing facility at each sampling point, before and after carcass wash, and 24-h carcass chilling, Table S2. Experimental design at each sampling date for the extended shelf-life evaluation of beef striploins. Table S1. Experimental design of no washed and hot water washed carcasses in a beef processing facility at each sampling point, before and after carcass wash, and 24-h carcass chilling. Table S2. Experimental design at each sampling date for the extended shelf life evaluation of beef striploins.

Author Contributions: Conceptualization, M.M.B., M.X.S.-P. and M.F.M.; methodology, D.E.C., R.M., S.J.F., K.E.H., M.M.B. and M.X.S.-P.; validation, D.E.C., M.F.M. and M.X.S.-P.; formal analysis, D.E.C. and M.X.S.-P.; investigation, D.E.C., R.M., S.J.F., K.E.H., M.M.B., M.F.M. and M.X.S.-P.; resources, M.M.B., M.F.M. and M.X.S.-P.; data curation, D.E.C.; writing—original draft preparation, D.E.C.; writing—review and editing, D.E.C., M.X.S.-P., M.F.M. and M.M.B.; visualization, D.E.C., and M.X.S.-P.; supervision, M.M.B., M.F.M. and M.X.S.-P.; project administration, D.E.C., K.E.H., M.M.B., M.F.M. and M.X.S.-P.; funding acquisition, M.M.B., M.F.M. and M.X.S.-P. All authors have read and agreed to the published version of the manuscript.

Funding: This research was funded through the joint efforts of the International Center for Food Industry Excellence at Texas Tech University and beef processor Teys Australia.

Institutional Review Board Statement: Not Applicable.

Informed Consent Statement: Not Applicable.

Data Availability Statement: All data from the research conducted are available on request from the corresponding author. The data are not publicly available due to the privacy of our industry collaborator that allowed for the project to be conducted.

Acknowledgments: We would like to acknowledge the hard work of the Texas Tech University Food Microbiology Laboratory personnel, who made the processing and storing of striploins possible. Furthermore, we would like to acknowledge the plant personnel who helped the project and made everything within the harvest and fabrication floor run smoothly, as well as the great hospitality throughout our stay in Brisbane, Australia.

Conflicts of Interest: The authors declare no conflict of interest.

References

1. Sun, X.D.; Holley, R.A. Antimicrobial and Antioxidative Strategies to Reduce Pathogens and Extend the Shelf Life of Fresh Red Meats. *Compr. Rev. Food Sci. Food Saf.* **2012**, *11*, 340–354. [CrossRef]
2. Zhou, G.H.; Xu, X.L.; Liu, Y. Preservation technologies for fresh meat—A review. *Meat Sci.* **2010**, *86*, 119–128. [CrossRef] [PubMed]
3. Anonymous. 'Fresh, Never Frozen' Trend Puts Pressure on the Cold Supply Chain | ProFood World. Available online: https://www.profoodworld.com/processing-equipment/heating-cooling-and-freezing/article/13279984/fresh-never-frozen-trend-puts-pressure-on-the-cold-supply-chain (accessed on 28 October 2020).
4. Food Safety Inspection Service Fresh, "Not Frozen" and Similar Terms when Labeling Meat and Poultry Products. Available online: https://www.fsis.usda.gov/wps/portal/fsis/topics/regulatory-compliance/labeling/claims-guidance/fresh-not-frozen-and-similar-terms (accessed on 28 October 2020).
5. Daniels, J. Japan Ends Longstanding Trade Restrictions on American Beef: USDA. Available online: https://www.cnbc.com/2019/05/17/japan-ends-longstanding-trade-restrictions-on-american-beef-usda.html (accessed on 28 October 2020).
6. Foreign Agriculture Service International Agricultural Trade Report United States Agricultural Exports to Japan Remain Promising. Available online: https://www.fas.usda.gov/data/united-states-agricultural-exports-japan-remain-promising (accessed on 28 October 2020).
7. Meat and Livestock Australia Market Snapshot | Beef and Sheepmeat European Union. Available online: https://www.mla.com.au/globalassets/mla-corporate/prices--markets/documents/os-markets/red-meat-market-snapshots/2019/mla-ms-european-union-beef-sheep-2019.pdf (accessed on 28 October 2020).
8. Algino, R.J.; Ingham, S.C.; Zhu, J. Survey of antimicrobial effects of beef carcass intervention treatments in very small state-inspected slaughter plants. *J. Food Sci.* **2007**, *72*, M173–M179. [CrossRef] [PubMed]
9. Hur, S.J.; Jin, S.K.; Park, J.H.; Jung, S.W.; Lyu, H.J. Effect of modified atmosphere packaging and vacuum packaging on quality characteristics of low grade beef during cold storage. *Asian Australas. J. Anim. Sci.* **2013**, *26*, 1781–1789. [CrossRef]
10. Castillo, A.; Lucia, L.M.; Goodson, K.J.; Savell, J.W.; Acuff, G.R. Use of hot water for beef carcass decontamination. *J. Food Prot.* **1998**, *61*, 19–25. [CrossRef]
11. Ingham, S.C.; Buege, D.R. 6-Day Dry-Aging as a Beef Slaughter Intervention Treatment. Available online: https://meathaccp.wisc.edu/assets/beef_carcass_dry-aging.pdf (accessed on 2 June 2021).
12. Dashdorj, D.; Tripathi, V.K.; Cho, S.; Kim, Y.; Hwang, I. Dry aging of beef; Review. *J. Anim. Sci. Technol.* **2016**, *58*. [CrossRef]
13. Nortje, G.L.; Naude, R.T. Microbiology of Beef Carcass Surfaces. *J. Food Protecion* **1981**, *44*, 355–358. [CrossRef]
14. Bacon, R.T.; Belk, K.E.; Sofos, J.N.; Clayton, R.P.; Reagan, J.O.; Smith, G.C. Microbial populations on animal hides and beef carcasses at different stages of slaughter in plants employing multiple-sequential interventions for decontamination. *J. Food Prot.* **2000**, *63*, 1080–1086. [CrossRef]
15. McEvoy, J.M.; Sheridan, J.J.; Blair, I.S.; McDowell, D.A. Microbial contamination on beef in relation to hygiene assessment based on criteria used in EU Decision 2001/471/EC. *Int. J. Food Microbiol.* **2004**, *92*, 217–225. [CrossRef]

16. Chang, V.P.; Mills, E.W.; Cutter, C.N. Reduction of bacteria on pork carcasses associated with chilling method. *J. Food Prot.* **2003**, *66*, 1019–1024. [CrossRef]
17. Kinsella, K.J.; Sheridan, J.J.; Rowe, T.A.; Butler, F.; Delgado, A.; Quispe-Ramirez, A.; Blair, I.S.; McDowell, D.A. Impact of a novel spray-chilling system on surface microflora, water activity and weight loss during beef carcass chilling. *Food Microbiol.* **2006**, *23*, 483–490. [CrossRef]
18. Vargas, D.A.; Miller, M.F.; Woerner, D.R.; Echeverry, A. Microbial Growth Study on Pork Loins as Influenced by the Application of Different Antimicrobials. *Foods* **2021**, *10*, 968. [CrossRef]
19. Jay, J.M. A Review of Aerobic and Psychrotrophic Plate Count Procedures for Fresh Meat and Poultry Products. *J. Food Prot.* **2002**, *65*, 1200–1206. [CrossRef]
20. Conover, W.J. *Practical Nonparametric Statistics*; Wiley Series: Hoboken, NJ, USA, 1998; ISBN 9780471160687.
21. Castillo, A.; Lucia, L.M.; Goodson, K.J.; Savell, J.W.; Acuff, G.R. Comparison of Water Wash, Trimming, and Combined Hot Water and Lactic Acid Treatments for Reducing Bacteria of Fecal Origin on Beef Carcasses. *J. Food Prot.* **1998**, *61*, 823–828. [CrossRef]
22. Fletcher, B.; Mullane, K.; Platts, P.; Todd, E.; Power, A.; Roberts, J.; Chapman, J.; Cozzolino, D.; Chandra, S. Advances in meat spoilage detection: A short focus on rapid methods and technologies. *CYTA J. Food* **2018**, *16*, 1037–1044. [CrossRef]
23. Hardin, M.D.; Acuff, G.R.; Lucia, L.M.; Oman, J.S.; Savell, J.W. Comparison of Methods for Decontamination from Beef Carcass Surfaces. *J. Food Prot.* **1995**, *58*, 368–374. [CrossRef] [PubMed]
24. Phebus, R.K.; Nutsch, A.L.; Schafer, D.E.; Wilson, J.R.C.; Riemann, M.J.; Leising, J.D.; Kastner, C.L.; Wolf, J.R.; Prasap, R.K. Comparison of Steam Pasteurization and Other Methods for Reduction of Pathogens on Surfaces of Freshly Slaughtered Beef. *J. Food Prot.* **1997**, *60*, 476–484. [CrossRef]
25. Goepfert, J.M. The Aerobic Plate Count, Coliform and Escherichia coli Content of Raw Ground Beef at the Retail Level. *J. Milk Food Technol.* **1976**, *39*, 175–178. [CrossRef]
26. Dormedy, E.S.; Brashears, M.M.; Cutter, C.N.; Burson, D.E. Validation of acid washes as critical control points in hazard analysis and critical control point systems. *J. Food Prot.* **2000**, *63*, 1676–1680. [CrossRef]
27. Sumner, J.; Vanderlinde, P.; Kaur, M.; Jenson, I. The Changing Shelf Life of Chilled, Vacuum-Packed Red Meat. In *Food Safety and Quality-Based Shelf Life of Perishable Foods*; Springer International Publishing: Berlin, Germany, 2021; pp. 145–156.
28. Ghafir, Y.; China, B.; Dierick, K.; De Zutter, L.; Daube, G. Hygiene indicator microorganisms for selected pathogens on beef, pork, and poultry meats in Belgium. *J. Food Prot.* **2008**, *71*, 35–45. [CrossRef] [PubMed]
29. Buege, D.; Ingham, S. Small Plant Intervention Treatments to Reduce Bacteria on Beef Carcasses at Slaughter. Available online: https://meathaccp.wisc.edu/validation/assets/SmallPlantAntimicrobialIntervention.pdf (accessed on 10 June 2020).
30. Calicioglu, M.; Kaspar, C.W.; Buege, D.R.; Luchansky, J.B. Effectiveness of spraying with tween 20 and lactic acid in decontaminating inoculated *Escherichia coli* O157:H7 and indigenous *Escherichia coli* biotype I on beef. *J. Food Prot.* **2002**, *65*, 26–32. [CrossRef]
31. Food Safety and Inspection Service Sampling Requirements to Demonstrate Process Control in Slaughter Operations. Available online: https://www.fsis.usda.gov/wps/portal/fsis/topics/regulatory (accessed on 4 January 2021).
32. Casas, D.E.; Vargas, D.A.; Randazzo, E.; Lynn, D.; Echeverry, A.; Brashears, M.M.; Sanchez-Plata, M.X.; Miller, M.F. In-Plant Validation of Novel On-Site Ozone Generation Technology (Bio-Safe) Compared to Lactic Acid Beef Carcasses and Trim Using Natural Microbiota and Salmonella and E. coli O157:H7 Surrogate Enumeration. *Foods* **2021**, *10*, 1002. [CrossRef]
33. Campañone, L.A.; Roche, L.A.; Salvadori, V.O.; Mascheroni, R.H. Monitoring of Weight Losses in Meat Products during Freezing and Frozen Storage. *Food Sci. Technol. Int.* **2002**, *8*, 229–238. [CrossRef]
34. Zhang, Y.; Mao, Y.; Li, K.; Luo, X.; Hopkins, D.L. Effect of Carcass Chilling on the Palatability Traits and Safety of Fresh Red Meat. *Compr. Rev. Food Sci. Food Saf.* **2019**, *18*, 1676–1704. [CrossRef]

Article

In-Plant Validation of Novel On-Site Ozone Generation Technology (Bio-Safe) Compared to Lactic Acid Beef Carcasses and Trim Using Natural Microbiota and Salmonella and *E. coli* O157:H7 Surrogate Enumeration

Diego E. Casas [1], David A. Vargas [1], Emile Randazzo [2], Dan Lynn [3], Alejandro Echeverry [1], Mindy M. Brashears [1], Marcos X. Sanchez-Plata [1,*] and Markus F. Miller [1]

1 International Center for Food Industry Excellence, Department of Animal and Food Sciences, Texas Tech University, Lubbock, TX 79409, USA; diego.casas@ttu.edu (D.E.C.); Andres.vargas@ttu.edu (D.A.V.); Alejandro.echeverry@ttu.edu (A.E.); mindy.brashears@ttu.edu (M.M.B.); mfmrraider@aol.com (M.F.M.)
2 Nebraska Beef Ltd., Omaha, NE 68107, USA; edazzo@nbeef.com
3 BioSecurity Technology, Omaha, NE 68107, USA; dan@biosecuritytechnology.com
* Correspondence: marcos.x.sanchez@ttu.edu; Tel.: +1-(979)-595-5208

Citation: Casas, D.E.; Vargas, D.A.; Randazzo, E.; Lynn, D.; Echeverry, A.; Brashears, M.M.; Sanchez-Plata, M.X.; Miller, M.F. In-Plant Validation of Novel On-Site Ozone Generation Technology (Bio-Safe) Compared to Lactic Acid Beef Carcasses and Trim Using Natural Microbiota and Salmonella and *E. coli* O157:H7 Surrogate Enumeration. *Foods* **2021**, *10*, 1002. https://doi.org/10.3390/foods10051002

Academic Editor: Declan J. Bolton

Received: 18 March 2021
Accepted: 1 May 2021
Published: 4 May 2021

Abstract: The purpose of this study was to evaluate the antimicrobial efficacy of an aqueous ozone (Bio-Safe) treatment and lactic acid solutions on natural microbiota and *E. coli* O157:H7 and Salmonella surrogates on beef carcasses and trim in a commercial beef processing plant. For every repetition, 40 carcass and 40 trim swabs (500 cm^2) were collected. Samples were taken using EZ-Reach™ swabs, and plated into aerobic plate count (APC), coliform, and *E. coli* Petrifilm™ for enumeration. In addition, a five-strain cocktail (MP-26) of *E. coli* surrogates was inoculated onto trim. For every trim surrogate repetition, 30 trim pieces were sampled after attachment and after ozone intervention. Samples were diluted and counts were determined using the TEMPO® system for *E. coli* enumeration. Ozone and lactic acid interventions significantly reduced ($p < 0.003$) bacterial counts in carcasses and trim samples. Moreover, lactic acid further reduced APC and coliforms in trim samples compared to ozone intervention ($p < 0.009$). In the surrogate trials, ozone significantly reduced ($p < 0.001$) surrogate concentration. Historical data from the plant revealed a reduction ($p < 0.001$) of presumptive *E. coli* O157:H7 in trim after a full year of ozone intervention implementation. The novel technology for ozone generation and application as an antimicrobial can become an alternative option that may also act synergistically with existing interventions, minimizing the risk of pathogens such as Salmonella and *E. coli* O157:H7.

Keywords: *Salmonella* spp.; *E. coli*; pathogen surrogates; ozone intervention; beef; beef trim

1. Introduction

Ever since the U.S. Department of Agriculture Food Safety and Inspection Service (FSIS) declared *E. coli* O157:H7 and Shiga toxin-producing *E. coli* (STEC) as adulterants in non-intact beef [1], the North American beef industry has continuously evaluated and implemented the use of antimicrobial interventions during beef harvest and processing. In addition to STECs, Salmonella presence on beef has also been identified as a significant threat to public health and an economic burden to the beef industry. Just recently, Salmonella has been linked to foodborne outbreaks and millions of pounds of ground beef have been recalled for risk of Salmonella presence in ground beef [2,3]. Despite the industry efforts to implement proper sanitary dressing procedures, best practices, and use of antimicrobial interventions, hides, and endogenous extra-intestinal sources of pathogens can contaminate beef carcasses [4]. Not one single intervention has been found to render a beef product completely safe. Thus, a multi-hurdle approach of a series of targeted antimicrobial interventions can more effectively reduce the risk of possible contamination through the slaughter process, consequently improving the microbial quality of carcasses [5]. A combination of physical and chemical interventions on beef carcasses and products may prove to be more effective than applying the same intervention at multiple stages of the slaughter and processing lines [6]. Therefore, exploring suitable and effective antimicrobial intervention alternatives may prove to be beneficial when finding synergies with already existing and implemented interventions that will further contribute to improving beef safety.

BioSecurity Technology has developed a novel ozone intervention known as BioSafe™ cleaning solution [7]. Aqueous ozone's oxidation-reduction potential grants it the capacity to be used as a disinfectant by causing cell lysis and damaging nucleic acids [8]. Although the antimicrobial properties of ozone are well documented [9], previous studies assessing ozone's potential as an intervention in beef carcasses have had contradictory results, where some have significantly reduced *E. coli* O157:H7 concentration whereas others have found no significant difference than water wash (28 °C) treatments [10,11]. Whether an intervention works in a laboratory environment or not, does not determine its feasibility or effectiveness in the beef processing plant environment, and therefore in-plant validation studies must be conducted in a particular commercial beef processing plant to assess its real effectiveness. Lactic acid is listed in FSIS Directive 7120.1 as a safe and suitable ingredient in the production of meat products. It may be used on beef subprimals at the amount of 2 to 5 percent solution not to exceed 55 °C (131 °F). The same Directive states that ozone is safe for use on all meat products per current industry standards. There are no labeling requirements on these single-ingredient items providing the use of the substance is consistent with the FDA's definition of a processing aid, and the application on meat meets all water retention requirements of 9 CFR 441.10.

Because foodborne pathogens should not be introduced into the beef processing environment under any circumstance, *E. coli* O157:H7 and Salmonella surrogates have been developed to validate antimicrobial interventions in commercial beef processing plants without compromising safety [12]. In this study, we hypothesize that the aqueous ozone intervention will significantly reduce indicator microorganisms naturally present in beef carcasses and trim in a commercial beef processing plant environment. Furthermore, we also evaluated if this intervention significantly reduces an *E. coli* O157:H7 and Salmonella surrogate-cocktail inoculated in beef trim in a commercial beef processing facility.

2. Materials and Methods

2.1. Intervention Parameters

Lactic acid operation parameters as applied in the plant used for this study included a spray treatment solution with a temperature of 110–130 °F (43–55 °C), at 2–5% lactic acid concentration with a spray pressure $\geq$15 psi. Bio-safe by BioSecurity Technology (Ozone) intervention operating parameters included ozone generators which utilize oxygen molecules from the air (O_2) and pass them through a corona field, splitting them into single atoms of oxygen (O_1). These atoms combine with an O_2 molecule to form a molecule of O_3 (Ozone). After the intervention and immediate reaction with organic matter, it turns back into oxygen, leaving no harmful byproducts or residuals according to manufacturer's description and proprietary technology developed. Oxidation-Reduction Potential (ORP) instrumentation is used to monitor and control the reactivity and effectiveness of the sanitizing power of ozonated water. The aqueous ozone treatment spray had incoming water maintained at 50–75 °F (10–24 °C), the concentration was 1.5–2.3 ppm and the ORP was measured by an in-line meter between 700 and 900 mV with a spray pressure of $\geq$20 psi. Ozone application consisted of a multiple hurdle carcass intervention system with three treatment cabinets using the following specifications: 52 spray nozzles delivering 24.6 gpm with 5 s treatment time, 62 spray nozzles delivering 34.6 gpm with 5 s contact time, and 36 spray nozzles delivering 13.6 gpm with 20 s contact time for each cabinet, respectively. The cumulative application used was 72.8 gpm with a total of 30 s contact time in carcasses. Moreover, the trim ozone intervention consisted on one treatment cabinet with 44 nozzles delivering 12.8 gpm with 18 s contact time.

2.2. Evaluation of Natural Microbiota on Carcass and Trim

For each repetition, in one production day, samples were randomly collected before and after treatment. A total of 20 carcasses were sampled before and after the final intervention. Of these carcasses, 10 were treated with lactic acid intervention and 10 with the ozone treatment intervention. Samples were taken before intervention at the harvest floor and after intervention at the hot box, for a total of 40 carcass swabs per repetition. The next day, trim was fabricated from the carcasses that were treated with the ozone intervention and lactic acid intervention, traced, and separated into different trim combos. Ten representative pieces of trim that came from the carcasses with the ozone intervention and 10 pieces of trim that came from carcasses with the lactic acid intervention were sampled before and after the trim intervention. The selected carcasses and trim were sampled on an area of 500 cm^2 using 25 mL buffered peptone water (BPW) EZ-Reach^{TM} swabs (World Bioproducts, Mundelein, IL, USA). Carcasses were sampled on the foreshank area, trim was sampled on several points until reaching approximately the target area of 500 cm^2. Samples were collected by Texas Tech University (TTU) trained personnel. Swab samples were immediately chilled and shipped overnight to the ICFIE-TTU Food Microbiology laboratory for microbiological analysis. Swab samples were homogenized in a stomacher (Model 400 circulator, Seward, West Sussex, UK) at 230 rpm for 1 min. Next, samples were serially diluted in 9 mL BPW (Millipore Sigma, Danvers, MA, USA) tubes and plated to determine total aerobic plate counts (APC), coliform counts, and *E. coli* counts using 3M^{TM} Petrifilm^{TM} (Saint Paul, MN, USA) plates. The counts of each sample were determined and converted to Log CFU/cm^2 for carcasses and Log CFU/sample for trim samples before statistical analysis. A total of six repetitions were conducted.

2.3. Salmonella and E. coli O157:H7 Surrogate Inoculation in Trim

2.3.1. Nonpathogenic Cocktail Preparation

Five non-pathogenic American Type Culture Collection (ATCC) Salmonella and *E. coli* O157:H7 surrogate strains were selected for this section of the study. These strains of non-virulent *E. coli* (BAA 1427, 1428, 1429, 1430, and 1431), when used as a cocktail, have been previously shown to mimic Salmonella and *E. coli* O157:H7 antimicrobial intervention behavior [13–16]. The use of surrogate strains to validate interventions in plant environments has been previously discussed and at times encouraged by FSIS USDA, which has allowed the use of such non-pathogenic surrogates with appropriate precautions [12]. The surrogate strains were independently propagated in a food grade biological safety level I (BSL-I) laboratory at TTU. Each ATCC strain was retrieved from a −80 °C freezer, separately transferred into 4 mL brain heart infusion (BHI; Becton, Dickinson and Company, Franklin Lakes, NJ, USA) tubes, and incubated at 37 °C for 18–24 h. Next, overnight enriched tubes were screened for *E. coli* O157:H7 and Salmonella presence using BAX® real-time *E. coli* O157:H7 Exact and Salmonella assays (Hygiena, Wilmington, DE). After found negative for both pathogen screenings, 500 μL of each enriched surrogate broth was transferred into 49.5 mL BPW tube and cleared to be used for the challenge study. Then, all five tubes were decanted onto a sprayer and mixed. The bottle sprayer was then used for trim target inoculation of 5–6 LogCFU/cm^2.

2.3.2. Trim Inoculation and Quantification

For each repetition, chuck and shank trim were randomly selected for inoculation. A total of 15 pieces of chuck and 15 pieces of trim were inoculated using the sprayer. Each piece of trim was sprayed with the *E. coli* O157:H7 and Salmonella surrogate cocktail and allowed for 30 min of cell attachment while at ambient temperature. After attachment time, an area of 100 cm^2 was sampled using a 25 mL BPW EZ-Reach^{TM} swab. Trim was next treated with the ozone treatment and immediately after intervention but before entering the production line, trim was sampled. All swabbed areas were marked with 100 cm^2 stamped area to ensure that the same area was not sampled repeatedly. Samples were collected by TTU trained personnel and shipped overnight to the TTU Food Microbiology laboratory for microbial enumeration. Swabs were homogenized in a stomacher at 230 RPM for 1 min. *E. coli* counts were determined using the TEMPO® system (Marcy-l'Étoile, France) following the manufacturer's instructions. TEMPO® cards were incubated at 35 °C for 22–28 h. *E. coli* counts were directly obtained from the TEMPO® Reader and converted to LogCFU/cm^2 before statistical analysis. A total of six repetitions were conducted.

2.4. Statistical Analysis

All data were analyzed using R (Version 4.0.3) Statistical analysis software to evaluate differences between lactic acid and the ozone intervention and testing for a significant reduction of microbial loads after each intervention in the natural microbiota setting was performed. A two-way ANOVA was done using intervention type (ozone and lactic acid), sampling point (before and after intervention), and their interaction as fixed effects. For the surrogate study, a two-way ANOVA was performed using trim type (chuck and shank), sampling point (before and after intervention), and their interaction as fixed effects. Post hoc analysis was done using a pairwise T-test with Bonferroni p-adjustment method for multiple comparisons. If parametric assumptions were not met, the Kruskal–Wallis test was used as a nonparametric alternative for the ANOVA, with post-hoc analysis using Wilcoxon rank-sum tests with a BH p-adjustment method for multiple comparisons. Significant differences were evaluated at the 0.05 alpha level. Historical data of *E. coli* O157:H7 presumptive positives from the commercial beef processing plant where the challenge study was conducted was shared with TTU researchers for information purposes. Chi-square comparison to identify the difference in prevalence before and after the ozone intervention application by year and on a per month basis was conducted.

3. Results

3.1. Natural Microbiota on Carcass

Both lactic acid and the ozone interventions significantly reduced ($p < 0.0001$) aerobic plate counts, coliform, and *E. coli* when applied to beef carcasses (Figure 1). Aerobic plate counts on carcasses were significantly reduced on average by 3.26 Log CFU/cm^2 and 3.83 LogCFU/cm^2 after ozone and lactic acid interventions, respectively. Coliform counts on carcasses were significantly reduced on average by 1.42 Log CFU/cm^2 and 1.37 Log CFU/cm^2 after ozone and lactic acid interventions, respectively. Likewise, *E. coli* counts on beef carcasses were significantly reduced by 1.29 LogCFU/cm^2 and 1.35 LogCFU/cm^2 after ozone and lactic acid intervention, respectively. Significant reduction of *E. coli* to undetectable levels was achieved after lactic acid and ozone interventions on beef carcasses. For each microorganism, there were no statistical differences in microbial populations between any of the two interventions.

3.2. Natural Microbiota on Trim

Coliforms and *E. coli* counts on the trim were substantially low when analyzed on a per cm^2 basis. When transformed to Log CFU/cm^2 for statistical analysis, most counts were below 1 CFU/cm^2, therefore resulting in negative Log CFU/cm^2 counts, making analysis and visualization more difficult. Thus, an analysis on a per sample (Log CFU/500 cm^2) basis was made to assess the effectiveness of the interventions. This conversion was achieved by multiplying the Log CFU/cm^2 by 500 cm^2 of area sampled, resulting in Log CFU/500 cm^2 which is equivalent to Log CFU/sample. On trim, both lactic acid and the ozone interventions significantly reduced ($p < 0.003$) aerobic plate counts, coliform, and *E. coli* when applied to trim (Figure 2). Moreover, lactic acid greatly reduced ($p < 0.009$) aerobic plate count and coliforms when compared to ozone. Aerobic plate counts on trim were significantly reduced on average by 0.74 Log CFU/sample and 2.08 Log CFU/sample after ozone and lactic acid interventions, respectively. Coliform counts on trim were significantly reduced on average by 0.93 Log CFU/sample and 2.13 Log CFU/sample after ozone and lactic acid interventions, respectively. Moreover, *E. coli* counts on beef trim were significantly reduced on average by 0.67 Log CFU/ sample and 1.08 Log CFU/sample after ozone and lactic acid interventions, respectively.

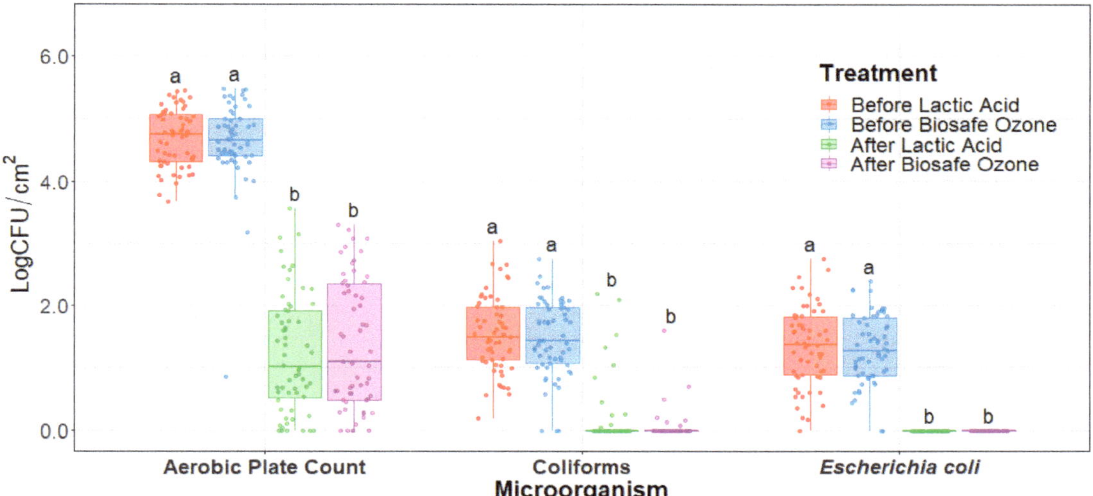

Figure 1. Carcass Aerobic plate count, coliform, and *Escherichia coli* counts (limit of detection < 0.05 CFU/cm^2) before and after the application of the interventions (LogCFU/cm^2). Horizontal line within the boxplot represents the median. The box upper and lower limit represents the interquartile range, and the bars represent 1.5xInterquartile Range. [a,b] Box plots with different letters within the same microorganism type represent statistical differences ($p < 0.05$).

Since trim natural microbiota encountered in coliforms and *E. coli* was substantially low, authors decided to inoculate *E. coli* O157:H7 and Salmonella surrogates on the trim and apply the ozone intervention to assess its efficacy. For both trim types, the ozone intervention significantly reduced ($p < 0.0001$) *E. coli* O157:H7 and Salmonella surrogate cocktail counts (Figure 3). Initial inoculation attachment was on average 5.67 Log CFU/cm^2 and 5.52 Log CFU/cm^2 for chuck and foreshank trim, respectively. *E. coli* cocktail attachment was well within target inoculation of 5–6 Log CFU/cm^2. On average, counts were reduced by 1.17 Log CFU/cm^2 after the ozone intervention. Reduction between trim types was similar ($p = 0.18$). Consequently, the intervention efficacy is expected to be the same when applied to different trim types.

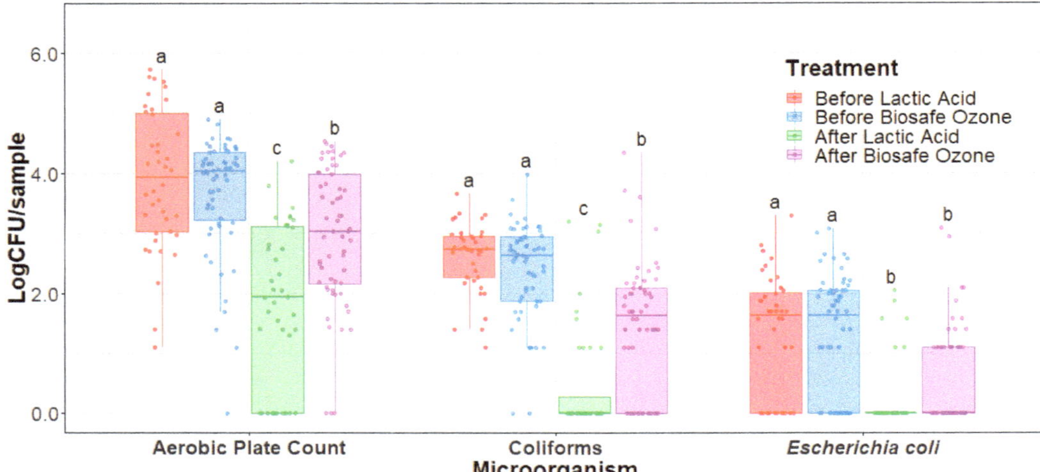

Figure 2. Trim aerobic plate count, coliforms, and *Escherichia coli* counts (limit of detection < 0.05 CFU/cm^2) before and after the application of the interventions (Log CFU/sample). Horizontal line within the boxplot represents the median. The box upper and lower limit represents the interquartile range, and the bars represent 1.5xInterquartile Range. [a,b] Box plots with different letters within the same microorganism type represent statistical differences ($p < 0.05$).

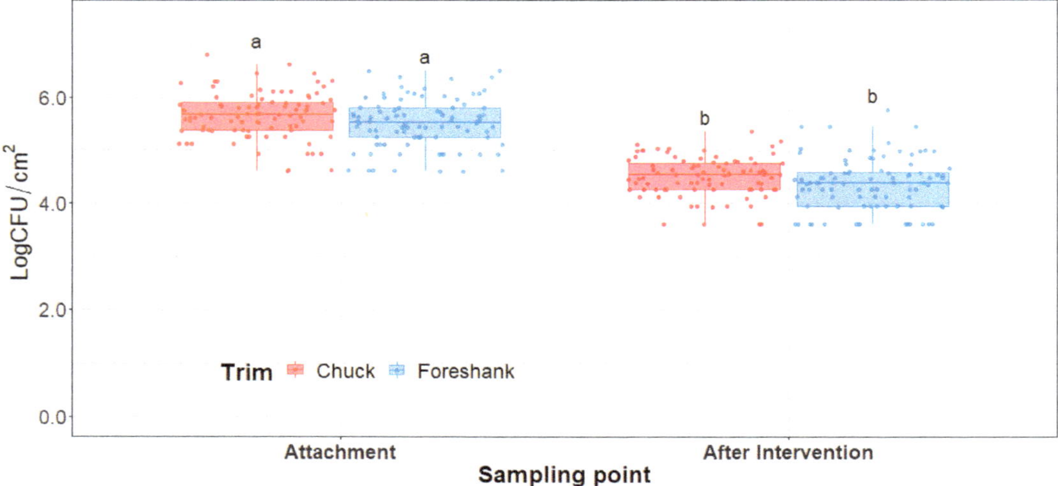

Figure 3. *Escherichia coli* surrogate attachment levels and after intervention counts (limit of detection < 4 CFU/cm^2) on LogCFU/cm^2 basis. Horizontal line within the boxplot represents the median. The box upper and lower limit represents the interquartile range, and the bars represent 1.5xInterquartile Range. [a,b] Box plots with different letters represent statistical differences ($p < 0.05$).

In the beef processing plant, the use of the ozone intervention was implemented on 11 October 2019. Chi-square analysis comparing the year prior (1.06%, 102/9,609) to implementation of Biosafe ozone intervention and the year after (0.26%, 25/9,439) implementation indicates statistical difference ($p < 0.0001$) in the percentage of presumptive positive rates of *E. coli* O157:H7 in trim per year. A month-by-month comparison can be observed in Figure 4. The year before implementation of the ozone intervention presented a 4.1 times greater incidence of presumptive *E. coli* O157:H7 than the year after implementation, indicating a potential 75.5% reduction of presumptive *E. coli* O157:H7 presence in trim.

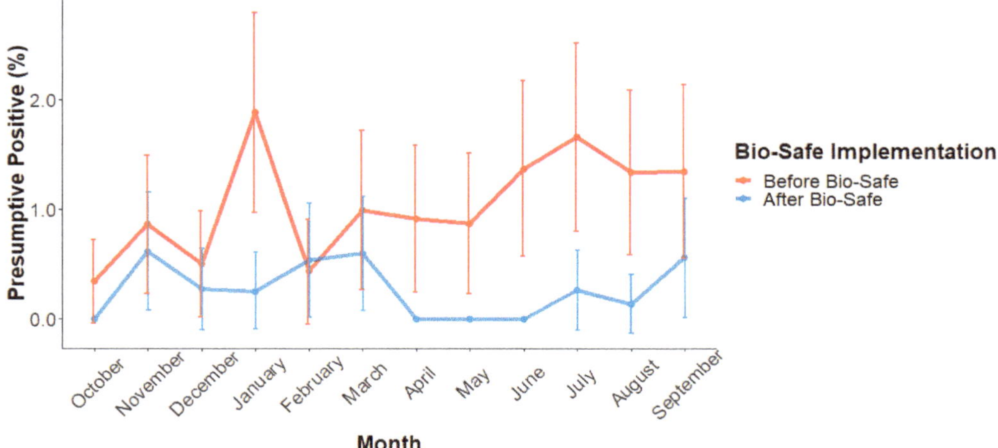

Figure 4. In-plant monthly Presumptive positive rate of *E. coli* O157:H7 in beef trim before and after implementation of the ozone intervention (N = 19,048). [I] Error bars represent 95% confidence intervals of the monthly incidence.

4. Discussion

The ozone intervention in carcasses significantly reduced indicator microorganisms studied in the commercial beef processing plant environment. This reduction was equivalent in magnitude to the reduction observed by using a final lactic acid carcass wash. The processing plant that allowed this study to be conducted, used 82 °C (180 °F) hot carcass wash prior to the lactic acid wash as their usual final harvest intervention before the carcasses entered the hot box. For this study, they left the hot water wash on and switched the lactic acid spray with the aqueous ozone treatment to evaluate the effect of ozone compared to that achieved with the use of lactic acid. Consequently, it can be observed that the multiple hurdle approach of using ozone after a hot water wash has equivalent reduction of APC, coliforms, and *E. coli* compared to using lactic acid after a hot water wash. Minimal sampling requirements to demonstrate process control in beef slaughter operations published by the FSIS require one generic *E. coli* sample for every 300 head of cattle harvested. A negative result is the acceptable outcome, but if in 13 subsequent generic *E. coli* tests there are more than three samples between 1 and 100 CFU/cm^2, the commercial processing plant fails the performance standards [17]. In this study, *E. coli* cell count was below the detection limit (<0.05 CFU/cm^2) after both final carcass interventions. Thus, the facility passed the performance standards and can demonstrate appropriate process control while using lactic acid or ozone interventions.

Ozone in an aqueous solution has been used in the past as a possible antimicrobial intervention in beef. Some studies have reported no significant reduction compared to a 28 °C water wash, whereas others have observed a significant reduction of 1.46 LogCFU/cm^2 of *E. coli* O157:H7 compared to 0.60 LogCFU/cm^2 reduction of water spray chill and a reduction of APC of 0.99 LogCFU/cm^2 [10,11]. In this study, a reduction of APC of 3.26 LogCFU/cm^2 was observed after hot water wash and ozone treatment. A multiple hurdle approach in the commercial plant environment is followed to more effectively eliminate pathogen presence in beef products [18,19]. Therefore, different interventions can act synergistically and more effectively to reduce the microbial load of beef in a commercial processing plant. Moreover, the recent development of an enhanced ozone technology and techniques to increase ozone half-life and reactivity in aqueous solution may increase the efficacy of ozone interventions in beef as observed in this study.

When comparing the ozone intervention against the lactic acid intervention in beef trim, we assessed the individual effect that the intervention has on trim. It is worth noting that the analysis in trim was done on a per-sample basis instead of a per-cm^2 basis due to substantially low coliform and *E. coli* presence in commercial samples. In this trim study, lactic acid further reduced APC and coliform counts compared to the aqueous ozone treatment. However, similar reductions were observed in generic *E. coli* when comparing both treatments. Lactic acid has been known to have a residual effect in the reduction of microbial load, where significant reductions in indicator microorganisms can be seen even after 12 days of treatment [20]. Contrastingly, ozone interventions have not yet been observed to have a residual effect in beef, since it is unstable and breaks down into oxygen shortly after generation and reaction with organic materials. Further research must be conducted to assess differences in shelf-life effects that ozone interventions may have in beef over extended storage times.

Generic *E. coli* has historically been used by processing plants to verify process control. The hazard analysis and critical control points system final rule of 1996 required generic *E. coli* testing [21]. *E. coli* presence is important to assess in beef because it is an indicator of fecal contamination as it is commonly found in the cattle gastrointestinal tract and hides. The gastrointestinal tract of cattle is also a possible reservoir of foodborne pathogens such as Salmonella and *E. coli* O157:H7 [17]. Therefore, if *E. coli* is found in beef, the risk of having Salmonella or pathogenic *E. coli* presence is likely to increase. In the trim sampled, over 90% of the trim had < 1 CFU/cm^2 of *E. coli*. Thus, to further validate the efficacy of the ozone treatment, the authors decided to conduct a Salmonella and *E. coli* O157:H7 surrogate inoculation study on the trim inside a commercial beef processing plant, to take into account the effects of commercial processing operations and actual equipment.

In the surrogate inoculation trial, ozone intervention significantly reduced the concentration of the *E. coli* cocktail. Foreshank and chuck trim were chosen as the "worst case scenario" for this section as, historically, these are the two types of trim that the commercial beef processing plant had more frequently found presumptive *E. coli* O157:H7 presence. These surrogates have been previously seen to mimic *E. coli* O157:H7 and Salmonella resistance to antimicrobial treatments when used as a cocktail in validation trials [13–16,22]. In some cases, reporting a slight increase in the magnitude of survival of the surrogate compared to Salmonella or *E. coli* O157:H7 for a relatively higher margin of safety. Thus, it can be inferred that the survival of the pathogens would be less than the one encountered with the surrogates. The surrogates are more on the conservative end of possible reduction since some of these strains might be slightly more resistant to an antimicrobial intervention than the actual pathogens [13,16]. In this context, the ozone intervention can significantly reduce *E. coli* O157:H7 and Salmonella average concentration by at least 1.17 LogCFU/cm^2, with further reductions potentially possible if subsequent sequential applications are considered and surface contact is enhanced. Furthermore, the antimicrobial intervention may cause sublethal injuries in cells that may hinder their ability to grow in selective media. Even though the samples were kept at refrigerating temperatures for approximately 24 h prior to processing in BPW while being shipped to the laboratory, bacteria may have not completely recovered from the intervention. However current sampling and quantification protocols used by the North American beef industry for *E. coli* follow quantification in selective media.

Historical data shared by the plant indicates a significant improvement since the implementation of the ozone intervention in the commercial facility. The year before ozone implementation, 102 lots of trim resulted in presumptive positive for *E. coli* O157:H7. After a year of ozone implementation, the plant observed a 75.5% reduction in positives, having only 25 presumptive positive lots. The improvement translates into a significant economic gain as substantially fewer lots of trim had to be disposed of or rerouted to fully cooked products at lower values. Ozone is known to have antimicrobial properties through direct oxidation of the cell wall resulting in cell lysis; however, it can also considerably damage DNA and produce reactions with oxygen radical by-products during its breaking down process [8]. Current methods for *E. coli* O157:H7 detection in beef, have screening procedures that use quantitative PCR for detection of a particular gene encoded in the DNA of the pathogen of interest [23]. In the multiple hurdle intervention setting, bacteria have been affected by a series of antimicrobial interventions, such as hot carcass washes, organic acid washes, carcass trimming, steam vacuuming, among others. By the time carcasses reach the chilling rooms, they have potentially undergone at least 2–4 antimicrobial interventions possibly reducing bacterial loads below detection limits, as it can be observed in coliform and *E. coli* counts in carcasses after interventions evaluated in this study. At that point, an ozone intervention may be able to further reduce bacterial concentration through cell lysis or other mechanisms; such as DNA damaging that has been reported [24,25] and ozone could have accessibility due to the synergistic effect on the bacterial membrane, that may be weakened from the prior antimicrobials used in the facility When cells undergo such damage, their proliferation becomes hindered under stressful conditions, such as refrigeration storage and distribution, enhancing beef safety in the value chain. Ozone's capacity for DNA degradation may be causing mutations in the bacterial genome rendering bacteria harmless and target genes of the real-time PCR screening procedures undetectable [24]. More research is needed to confirm cell damage and viability after the application of sequential ozone treatments, but these findings provide evidence that the aqueous ozone intervention evaluated in this study may play a significant role in controlling pathogen contamination in beef carcasses and trim.

5. Conclusions

The novel proprietary technology used to produce the high concentration, and stable reactivity of the aqueous ozone solution proved promising for the reduction of *E. coli* O157:H7 detection and indicator levels in beef. The findings encountered in this study indicate that the ozone intervention is not only effective but similar in performance to lactic acid in reducing bacterial load on carcasses and trim which will improve beef safety, therefore validating its use in the beef processing environment as an effective antimicrobial intervention. Bacterial surrogate studies become of utmost importance when trying to validate interventions in a commercial processing plant setting. They more accurately represent the specific effects that the antimicrobial intervention will have against pathogens they represent in a given environment, without compromising food safety. The evaluation of in-plant data for comparative purpose of intervention schemes gives additional support to the effectiveness of this technology, with ongoing control exerted over different seasons and processing months. Further research into multiple hurdle intervention interactions must be conducted to design the most effective ways of mitigating pathogen presence and ensure beef safety.

Author Contributions: Conceptualization, M.F.M., M.M.B., E.R., D.L., M.X.S.-P., and A.E.; methodology, M.X.S.-P., D.A.V., D.E.C., and M.F.M.; validation, D.E.C., M.X.S.-P., E.R., D.L., and D.A.V.; formal analysis, D.E.C. and D.A.V.; investigation, M.F.M., E.R., D.A.V., and D.E.C.; resources, E.R., D.L., M.M.B., M.F.M., and A.E.; data curation, D.E.C. and D.A.V.; writing—original draft preparation, D.E.C.; writing—review and editing, D.E.C., M.X.S.-P., A.E., M.F.M., and D.A.V.; supervision, D.L., A.E., M.M.B., and M.F.M.; project administration, D.E.C., D.A.V., and M.F.M.; funding acquisition, M.M.B., M.F.M., A.E., and E.R. All authors have read and agreed to the published version of the manuscript.

Funding: This research was funded through the collaboration of the International Center for Food Industry Excellence at Texas Tech University and our industry partner Nebraska Beef Ltd.

Data Availability Statement: Data available on request from the corresponding author. The data are not publicly available due to privacy from the beef processing partner that allowed the project to be conducted within their beef processing environment.

Acknowledgments: We would like to acknowledge the help of all ICFIE Food microbiology personnel that helped inside the laboratory in processing of samples.

Conflicts of Interest: The authors declare no conflict of interest.

References

1. Food Safety and Inspection Service. *Compliance Guidelines for Shiga Toxin Escherichia coli (STEC) Organisms Sampled and Tested Labeling Claims for Boneless Beef Manufacturing Trimmings ("Beef Trim")*; United States Department of Agriculture: Washington, DC, USA, 2014.
2. Center for Disease Control and Prevention. Outbreak of *Salmonella* Infections Linked to Ground Beef | Outbreak of *Salmonella* Infections Linked to Ground Beef | November 2019 | *Salmonella* | CDC. Available online: https://www.cdc.gov/salmonella/dublin-11-19/index.html (accessed on 21 October 2020).
3. Center for Disease Control and Prevention. Outbreak of *Salmonella* Infections Linked to Ground Beef | Outbreak of *Salmonella* Infections Linked to Ground Beef | October 2018 | *Salmonella* | CDC. Available online: https://www.cdc.gov/salmonella/newport-10-18/index.html (accessed on 21 October 2020).
4. Food Safety and Inspection Service Food Safety Research Priorities. Available online: https://www.fsis.usda.gov/wps/wcm/connect/fsis-content/internet/main/topics/science/food-safety-research-priorities/food-safety-research-priorities (accessed on 30 December 2020).
5. Bacon, R.T.; Belk, K.E.; Sofos, J.N.; Clayton, R.P.; Reagan, J.O.; Smith, G.C. Microbial populations on animal hides and beef carcasses at different stages of slaughter in plants employing multiple-sequential interventions for decontamination. *J. Food Prot.* **2000**, *63*, 1080–1086. [CrossRef] [PubMed]
6. Buege, D.; Ingham, S. Small Plant Intervention Treatments to Reduce Bacteria on Beef Carcasses at Slaughter. Available online: https://meathaccp.wisc.edu/validation/assets/SmallPlantAntimicrobialIntervention.pdf (accessed on 15 December 2020).
7. BioSecurity Technology Bio-Safe Cleaning Solution | BioSecurity Technology. Available online: https://biosecuritytechnology.com/biosafe-solution (accessed on 3 January 2021).
8. Environmental Protection Agency. *Wastewater Technology Fact Sheet Ozone Disinfection*; Office of Water: Washington, DC, USA, 1999.
9. Restaino, L.; Frampton, E.W.; Hemphill, J.B.; Palnikar, P. Efficacy of ozonated water against various food-related microorganisms. *Appl. Environ. Microbiol.* **1995**, *61*, 3471–3475. [CrossRef] [PubMed]
10. Kalchayanand, N.; Worlie, D.; Wheeler, T. A Novel Aqueous Ozone Treatment as a Spray Chill Intervention against *Escherichia coli* O157:H7 on Surfaces of Fresh Beef. *J. Food Prot.* **2019**, *82*, 1874–1878. [CrossRef] [PubMed]
11. Castillo, A.; Mckenzie, K.S.; Lucia, L.M.; Acuff, G.R. Ozone Treatment for Reduction of *Escherichia coli* O157:H7 and *Salmonella* Serotype Typhimurium on Beef Carcass Surfaces. *J. Food Prot.* **2003**, *66*, 775–779. [CrossRef] [PubMed]
12. Food Safety and Inspection Service. Use of Non-pathogenic *Escherichia coli* (*E. coli*) Cultures as Surrogate Indicator Organisms in Validation Studies. Available online: https://askfsis.custhelp.com/app/answers/detail/a_id/1392/~{}/use-of-non-pathogenic-escherichia-coli-%28e.-coli%29-cultures-as-surrogate (accessed on 2 January 2021).
13. Cabrera-Diaz, E.; Moseley, T.M.; Lucia, L.M.; Dickson, J.S.; Castillo, A.; Acuff, G.R. Fluorescent protein-marked *Escherichia coli* biotype I strains as surrogates for enteric pathogens in validation of beef carcass interventions. *J. Food Prot.* **2009**, *72*, 295–303. [CrossRef] [PubMed]
14. Marshall, K.M.; Niebuhr, S.E.; Acuff, G.R.; Lucia, L.M.; Dickson, J.S. Identification of *Escherichia coli* O157:H7 meat processing indicators for fresh meat through comparison of the effects of selected antimicrobial interventions. *J. Food Prot.* **2005**, *68*, 2580–2586. [CrossRef] [PubMed]
15. Niebuhr, S.E.; Laury, A.; Acuff, G.R.; Dickson, J.S. Evaluation of nonpathogenic surrogate bacteria as process validation indicators for *Salmonella* enterica for selected antimicrobial treatments, cold storage, and fermentation in meat. *J. Food Prot.* **2008**, *71*, 714–718. [CrossRef] [PubMed]
16. Keeling, C.; Niebuhr, S.E.; Acuff, G.R.; Dickson, J.S. Evaluation of *Escherichia coli* biotype I as a surrogate for *Escherichia coli* 0157:H7 for cooking, fermentation, freezing, and refrigerated storage in meat processes. *J. Food Prot.* **2009**, *72*, 728–732. [CrossRef] [PubMed]
17. Food Safety and Inspection Service. *Sampling Requirements to Demonstrate Process Control in Slaughter Operations*; Food Safety and Inspection Service: Washington, DC, USA, 2020.
18. Eastwood, L.C.; Arnold, A.N.; Miller, R.K.; Gehring, K.B.; Savell, J.W. Impact of Multiple Antimicrobial Interventions on Ground Beef Quality. *Meat Muscle Biol.* **2018**, *2*, 46. [CrossRef]
19. Kang, D.-H.; Koohmaraie, M.; Siragusa, G.R. Application of Multiple Antimicrobial Interventions for Microbial Decontamination of Commercial Beef Trim †. *J. Food Prot.* **2001**, *64*, 168–171. [CrossRef] [PubMed]
20. Dickson, J.S.; Acuff, G.R. Maintaining the Safety and Quality of Beef Carcass Meat. In *Ensuring Safety and Quality in the Production of Beef*; Dickson, J.S., Acuff, G.R., Eds.; Burliegh Dodds Science Publishing: Sawston, UK, 2017; Volume I, pp. 145–162.
21. Food Safety and Inspection Service. *Final Rule on Pathogen Reduction and Hazard Analysis and Critical Control Points (HACCP) Systems*; Food Safety and Inspection Service: Washington, DC, USA, 1996.

22. Thomas, C.L.; Stelzleni, A.M.; Rincon, A.G.; Kumar, S.; Rigdon, M.; McKee, R.W.; Thippareddi, H. Validation of antimicrobial interventions for reducing shiga toxin–producing *Escherichia coli* surrogate populations during goat slaughter and carcass chilling. *J. Food Prot.* **2019**, *82*, 364–370. [CrossRef] [PubMed]
23. Food Safety and Inspection Service. FSIS Procedure for the Use of *Escherichia coli* O157:H7 Screening Tests for Meat Products and Carcass and Environmental Sponges. In *Microbiology Laboratory Guidebook*; U.S. Department of Agriculture: Athens, GA, USA, 2014.
24. Cataldo, F. DNA degradation with ozone. *Int. J. Biol. Macromol.* **2006**, *38*, 248–254. [CrossRef] [PubMed]
25. Theruvathu, J.; Flyunt, R.; Aravindakumar, C.; Von Sonntag, C. Rate constants of ozone reactions with DNA, its constituents and related compounds. *J. Chem. Soc. Perkin Trans. 2* **2001**, 269–274. [CrossRef]

MDPI
St. Alban-Anlage 66
4052 Basel
Switzerland
Tel. +41 61 683 77 34
Fax +41 61 302 89 18
www.mdpi.com

Foods Editorial Office
E-mail: foods@mdpi.com
www.mdpi.com/journal/foods